Sustainable Textiles: Production, Processing, Manufacturing & Chemistry

Series Editor

Subramanian Senthilkannan Muthu, Head of Sustainability, SgT and API, Kowloon, Hong Kong

This series aims to address all issues related to sustainability through the lifecycles of textiles from manufacturing to consumer behavior through sustainable disposal. Potential topics include but are not limited to: Environmental Footprints of Textile manufacturing; Environmental Life Cycle Assessment of Textile production; Environmental impact models of Textiles and Clothing Supply Chain; Clothing Supply Chain Sustainability; Carbon, energy and water footprints of textile products and in the clothing manufacturing chain; Functional life and reusability of textile products; Biodegradable textile products and the assessment of biodegradability; Waste management in textile industry; Pollution abatement in textile sector; Recycled textile materials and the evaluation of recycling; Consumer behavior in Sustainable Textiles; Eco-design in Clothing & Apparels; Sustainable polymers & fibers in Textiles; Sustainable waste water treatments in Textile manufacturing; Sustainable Textile Chemicals in Textile manufacturing. Innovative fibres, processes, methods and technologies for Sustainable textiles; Development of sustainable, eco-friendly textile products and processes; Environmental standards for textile industry; Modelling of environmental impacts of textile products; Green Chemistry, clean technology and their applications to textiles and clothing sector; Eco-production of Apparels, Energy and Water Efficient textiles. Sustainable Smart textiles & polymers, Sustainable Nano fibers and Textiles; Sustainable Innovations in Textile Chemistry & Manufacturing; Circular Economy, Advances in Sustainable Textiles Manufacturing; Sustainable Luxury & Craftsmanship; Zero Waste Textiles.

More information about this series at https://link.springer.com/bookseries/16490

Ali Khadir · Subramanian Senthilkannan Muthu
Editors

Polymer Technology in Dye-containing Wastewater

Volume 2

 Springer

Editors
Ali Khadir
Western University
London, ON, Canada

Subramanian Senthilkannan Muthu
SgT Group and API
Hong Kong, Kowloon, Hong Kong

ISSN 2662-7108   ISSN 2662-7116 (electronic)
Sustainable Textiles: Production, Processing, Manufacturing & Chemistry
ISBN 978-981-19-0888-0   ISBN 978-981-19-0886-6 (eBook)
https://doi.org/10.1007/978-981-19-0886-6

This Springer imprint is published by the registered company Springer Nature Singapore Pte Ltd.
The registered company address is: 152 Beach Road, #21-01/04 Gateway East, Singapore 189721, Singapore

Contents

About the Editors

Ali Khadir is an environmental engineer and a member of the Young Researcher and Elite Club, Islamic Azad University of Shahre Rey Branch, Tehran, Iran. He has published several articles and book chapters in reputed international publishers, including Elsevier, Springer, Taylor & Francis and Wiley. His articles have been published in journals with IF of greater than 4, including *Journal of Environmental Chemical Engineering* and *International Journal of Biological Macromolecules*. He also has been the reviewer of journals and international conferences. His research interests center on emerging pollutants, dyes and pharmaceuticals in aquatic media, advanced water and wastewater remediation techniques and technology.

Dr. Subramanian Senthilkannan Muthu currently works for SgT Group as Head of Sustainability and is based out of Hong Kong. He earned his Ph.D. from The Hong Kong Polytechnic University and is a renowned expert in the areas of environmental sustainability in textiles and clothing supply chain, product life cycle assessment (LCA) and product carbon footprint assessment (PCF) in various industrial sectors. He has five years of industrial experience in textile manufacturing, research and development and textile testing and over a decade's of experience in life cycle assessment (LCA), carbon and ecological footprints assessment of various consumer products. He has published more than 100 research publications, written numerous book chapters and authored/edited over 100 books in the areas of carbon footprint, recycling, environmental assessment and environmental sustainability.

Poly(Vinyl Alcohol) (PVA)-Based Treatment Technologies in the Remediation of Dye-Containing Textile Wastewater

Farwa Mushtaq, Muhammad Anwaar Nazeer, Asim Mansha, Muhammad Zahid, Haq Nawaz Bhatti, Zulfiqar Ali Raza, Waleed Yaseen, Ammara Rafique, and Rubab Irshad

List of Abbreviations

CQDs	Carbon quantum dots
CMC	Carboxymethyl cellulose
CNC	Cellulose nanocrystals
CNTs	Carbon nanotubes
CR	Congo red
CS	Chitosan
ECH	Epichlorohydrin
FTIR	Fourier-transform infrared spectroscopy
GO	Graphene oxide
GA	Glutaraldehyde
MB	Methylene blue
MO	Methyl orange
MWCNTs	Multiwalled carbon nanotubes
PVA	Poly (vinyl alcohol)

F. Mushtaq · M. Zahid (✉) · H. N. Bhatti
Department of Chemistry, University of Agriculture Faisalabad, Faisalabad 38040, Pakistan
e-mail: rmzahid@uaf.edu.pk

F. Mushtaq · M. A. Nazeer
School of Engineering and Technology, National Textile University, Faisalabad, Pakistan

A. Mansha
Department of Chemistry, Government College University, Faisalabad, Pakistan

Z. A. Raza · A. Rafique · R. Irshad
Department of Applied Sciences, National Textile University, Faisalabad 37610, Pakistan

W. Yaseen
School of Chemistry and Chemical Engineering, Jiangsu University, Zhenjiang 212013, PR China

© The Author(s), under exclusive license to Springer Nature Singapore Pte Ltd. 2022
A. Khadir and S. S. Muthu (eds.), *Polymer Technology in Dye-containing Wastewater*,
Sustainable Textiles: Production, Processing, Manufacturing & Chemistry,
https://doi.org/10.1007/978-981-19-0886-6_1

PVDF   Poly (vinylidene fluoride)
VC   Vitamin C

1 Introduction

The textile industry is among the most significant polluters of water owing to the existence of various types of pollution streams generated by printing processes and textile dyeing. Finishing and dyeing wastes are highly colored in the textile industry. Dye's complicated chromophoric groups result in stability to heat/oxidizing agents and non-biodegradability. The generated wastewater contains large amounts of toxic dyes that require difficult treatment methods [1]. Dye effluents dispensed from paper-making, plastic, textile, dyestuff, and leather industries have mutagenic and toxic effects on humans and aquatic biota [2, 3]. It is critical to eliminate dyes from the effluent before it is discharged into natural water bodies. Traditional treatment methods, on the other hand, have not proved to be useful for colored effluents. Various techniques such as biological treatment, ozonation [4], coagulation [5], electrochemical oxidation [6], advanced oxidation processes [7], membrane filtration [8], adsorption in combination with photocatalytic degradation [9] have been established to get rid of several contaminants from water. Each of these techniques does have its own set of advantages and disadvantages. The adsorption method, for instance, is particularly appealing due to its low-cost, ease of design, simplicity, effectiveness, economic viability, insensitivity to toxins, trouble-free operation, environmental friendliness, and ease of recovery. Several attempts were made in the last years to evolve an effective and low-cost adsorbent [10, 11]. Powdery adsorbents including inorganic/organic hybrid catalysts, inorganic porous nanomaterials, functional porous polymeric materials are subjected to time-consuming sorbent separation, posing the danger of secondary pollution, and increasing costs. Although adsorbents with large-size configurations or magnetism can alleviate separation issues to a certain extent, most materials still are challenged by inadequate alkali/acid stability and unsatisfying adsorption capacities. Industrial-colored wastewater, on the other hand, is generally a complex combination containing a variety of dyes.

The great emphasis of scientific research is photocatalytic oxidation of contaminant dyes. Low operating expenses, the capacity to eliminate complicated chemicals, no need for extra materials, the use of free sunlight, and flexibility to complete the mission at ambient pressure and temperature are all advantages of this technology. According to the explanation above, effluent remedies include both adsorption and photodegradation of colors. Furthermore, simultaneous dye adsorption and degradation is a frequent method. Today's strict environmental rules for dye-containing discharges are prompting textile manufacturers to investigate the possibility of recovering the water using the latest technology. The membrane filtration method is also an efficient wastewater treatment alternate. To handle the complex-colored wastewater problem, improved catalyst-loaded materials are required. Several materials have been tested on the prospect of hazardous dye removal such as carbon nanotube [12], activated carbon [13], fly ash [14], zeolite [15], polymers [3], low-cost adsorbents [16], graphene oxide nanocomposites [17]. Most low-cost adsorbents, such as fly ashes, charcoal, biochar, activated carbon, and several inorganic adsorbent materials, are non-biodegradable which may cause particle contamination if they are not adequately removed from reaction media following the adsorption process [18]. Hence, it is needed to explore more cost-effective and efficient adsorbents. Clean, low-cost technologies using biodegradable materials might be effective tools for minimizing the environmental influence of textile effluents. As a result, polymeric adsorbents exhibiting large pore structures and surface area have recently been discovered to be the most promising materials for the adsorption, filtration, and degradation processes.

Adsorbents based on Polyvinyl alcohol (PVA) have significantly higher adsorption capacity than non-traditional cheap adsorbents. Many studies have been performed on the use of PVA-composites, gels, membranes for water treatment. This chapter is a brief evaluation of PVA-based materials work that has been published in the past decade for the removal of dyes outlining different methods (especially adsorption, degradation, and filtration) for dyes removal. PVA-based materials for water refinement are widely used for their outstanding characteristics, structure/properties, and oxygen-containing surface functional moieties; these active sites make different strong interactions with various kinds of dyes [19]. This chapter is aimed to highlight innovations related to polymer technology as a response to textile wastewater remediation.

2 Polyvinyl Alcohol (PVA): Structure and Properties

Polyvinyl alcohol (PVA) is a linear water-soluble thermoplastic synthetic-polymer exhibiting many distinguishing characteristics, including low-cost, outstanding thermal and chemical resistance, high crystallinity, good mechanical strength, non-toxicity, flexibility, non-carcinogenic, high intrinsic hydrophilicity, good biocompatibility, good electro spinnability, and good processability. It is made by hydrolyzing poly (vinyl acetate) [20–22]. This polymer is available in various hydrolysis degrees [23]. Figure 1 depicts the synthesis and chemical structure of PVA. Due to its unique physicochemical characteristics and flexibility in structure, PVA has a broad range of commercial applications in a variety of areas.

3 PVA-Based Materials

In fact, PVA has lately been ranked as one of the most significant polymers, owing to its high characteristics and broad variety of uses, as well as its ease of production, resistance to climatic conditions, visibility, affordability, and other

Fig. 1 Synthesis and structure of polyvinyl alcohol (PVA)

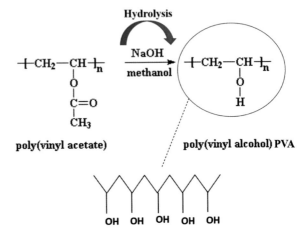

factors [24]. Unfortunately, PVA cannot be employed in wastewater treatment due to its water solubility. As a result, different crosslinking techniques (1) chemical crosslinking (by chemical agents/radiation crosslinking) or (2) physical crosslinking (by heat, physical interactions) must be used to transform it into a fully insoluble material exhibiting excellent mechanical characteristics. Crosslinked PVA has good mechanical strength and can be employed to make a range of materials such as hydrogels, fibers, films, composite particles, and membranes. Figure 2 describes various structural and morphological forms of materials derived from PVA. According to the most recent research, the structural and physicochemical properties of PVA are subjected to change when doped with different nano-fillers and other materials. Many writers have investigated the structure-property relationship, good reactivity, and compatibility of PVA with several other inorganic materials, biomaterials, polymers, carbonaceous and metallic materials [20, 25]. So far, various types of PVA-based structures such as nanofibers, membranes, gels, thin films, microspheres, and nanocomposites have been formulated by different techniques. Furthermore, in some studies, PVA is used as a capping agent fueling in nanocrystal formation [26]. It also acts as dispersing medium and avoids the agglomeration of particles. PVA has favorable characteristics, and its molecule contains many hydroxyl groups, giving it excellent hydrophilicity and interfacial compliance, allowing it to be mixed with a variety of materials to make different functional materials. It results in high-quality ultra-thin films (nano/micro) with outstanding tensile strength and elasticity, for a wide range of applications. PVA is more suitable as a membrane-material due to its excellent film forming and adhesive properties [23, 27].

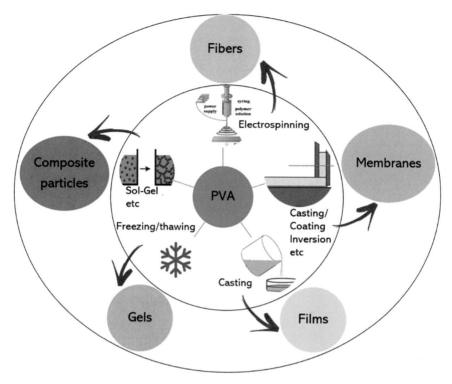

Fig. 2 Different categories of structural and morphological forms of PVA materials being used in water treatment

4 Interaction Mechanisms Between Dyes and PVA-Based Materials

A universal-adsorption process applicable to different dyes is required for broad-spectrum decontamination. In general, dyes adsorption requires hydrophilic adsorbent materials and depends heavily on several non-covalent contacts for example electrostatic interactions, hydrophobic interaction, hydrogen bonding, and Van der Waals forces with PVA-based functional materials of various morphological and structural forms including hydrogels, aerogels, membranes, composites, fibers and so on. The electrostatic interaction mechanism is typically only applicable to ionic (cationic or anionic) dyes, making it ineffective for removing both negatively and positively charged dyes together. Ionic and non-ionic (neutral) dyes having aromatic chemical structures benefit from the p-p interaction among dyes and PVA-based composite materials with graphene, GO, CNTs, MWCNTs, etc. [19, 20, 28, 29]. Figure 3 illustrates different non-covalent interactions between a variety of dyes and PVA-based materials. The blue rectangular area of the image contains various surface

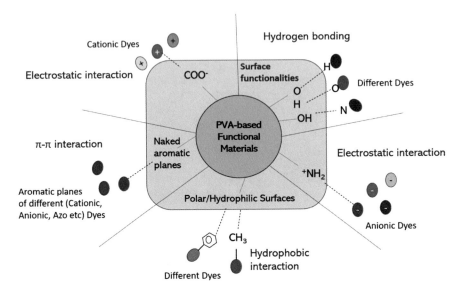

Fig. 3 Interaction mechanisms between textile dyes and PVA-derived materials

functionalities offered by different PVA materials for example PVA/chitosan blend would offer -OH and -NH$_2$ groups fo2r interaction with dyes.

In addition to the adsorbent's surface functionalities and type/structure/nature of the dye, the kind of non-covalent interaction between adsorbent and adsorbate (dyes) is greatly influenced by the reaction conditions also. The dye molecule's charge, as well as the surface charge of PVA-adsorbing material, may be affected by the pH of reaction solutions. Various research findings on the deletion of textile dyes using PVA materials have concentrated on the thorough assessment of factors such as pH, time, dye concentration, adsorbent/photocatalyst quantity, and so on [30–32]. Thus, for the removal of various types of dyes, different reaction conditions can be suitable which need to be optimized for practical implementations of these newly developed PVA-based materials [33].

5 Wastewater Treatment Technologies

In practice, the multicomponent system of textile wastewater containing various types of complex dyes is always simplified as a binary system of water/dye, because the goal of adsorbent-based treatment options is to recover clean water instead of achieving well-separated dyes. It is not viable because adsorption remediation only accomplishes contaminant transference from water to adsorbent, leaving the user with the choice of dumping the adsorption equilibrated adsorbent or generating

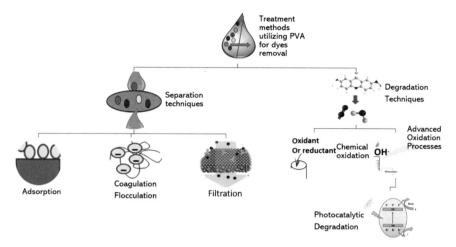

Fig. 4 Different treatment options and methodologies utilizing PVA-derived materials to remediate textile wastewater

extra complex-colored water for adsorbent reuse. Because of dye recovery, adsorbent reusability, economy, green process, selectively adsorbing materials, robust acid/alkali resistance, and increased adsorption capacity are desirable. Because of easy operation, low energy input, effective and clean semiconductor heterogeneous photocatalytic technique has been increasingly employed to degrade contaminants and water decomposition to produce hydrogen. It is more appealing to use filtering materials for removing recalcitrant dyes from water than other conventional standard approaches (such as biodegradation/adsorption/chemical methods) [34]. Filtration is a low-cost type of membrane technology, which is environmentally sound and energy-efficient. Membrane reactors featuring photocatalytic and bioreactors are also an alternative treatment technology for wastewater reuse applications [1]. Thus, PVA materials for water treatment applications can be executed in different methods. Figure 4 illustrates water treatment technologies utilizing PVA materials.

6 PVA-Based Materials for Dyes Removal Application

Multifunctional adsorbents featuring high efficiency, simple reusability, interactive components, and wide-ranging adsorption performance are required for the thorough treatment of various dye contaminants in water bodies. Pure PVA-based materials have many limitations in terms of poor mechanical properties and stability. Hybrid materials mitigate the drawbacks of native counterparts by combining the benefits of individual components. Adding component materials to PVA results in additional features and better applicability for dye removal. Combining various materials into composites may result in a variety of hybrid hydrogels, membranes, fibers, and films that can be used in a variety of water treatment settings satisfying various situations, e.g., degradation, filtration, and adsorption.

6.1 Composites Based on PVA

PVA is reported to be combined with a variety of materials including carbonaceous, metallic, and polymeric. These hybrids have the merits of single components and improved dye elimination capacity. A xylan/PVA/TiO$_2$ hybrid material was employed as a photocatalyst for the photocatalytic decolorization of Astrazon brilliant red-4G and ethyl-violet dyes under visible light illumination. The impacts of different operational factors were examined on the photodegradation process. More than 94% decolorization was accomplished. Positive holes (h$^+$) were discovered to be the most active radical component in the photodegradation process [35]. The preparation of glutaraldehyde crosslinked PVA/Vitamin-C-MWCNTs composite particles and their adsorption capability for cationic dye methylene blue is described. Because of the homogeneous dispersal of VC-MWCNTs in the polymer matrix, the thermal stabilities of PVA/VC-MWCNTs of 3 and 5 wt% were higher than native PVA. The PVA/VC-MWCNTs composite of 7 wt% could effectively remove approximately 97.3% of dye in 45 min from aqueous solution at room temperature, according to adsorption tests. However, the electrical conductivity test revealed that high conductivity was obtained at a concentration of 3 wt% VC-MWCNTs [20]. The sol-gel technique was used to manufacture three-dimensional networked PVA/sodium alginate/GO spherical porous composites for methylene blue adsorption in an aqueous solution. Batch studies were also used to examine the adsorption characteristics of MB. The findings revealed that the spherical composites possessed 3D porous architectures. The self-assembly process, hydrogen bonding, physical interweaving, and boric acid and Ca^{2+} crosslinking's were utilized to fuse and connect PVA, GO, and sodium alginate together. The adsorption capacities and removal percentages are displayed in the Table 1 [36].

Under various experimental conditions, the adsorption/desorption behavior of dyes including direct red-31, direct blue-67, direct orange-26, and ever direct orange-3GL on native, rice husk modified PVA, alginate, carboxymethyl cellulose, and immobilized biomasses was examined. Various physical and chemical procedures were used to modify the adsorbents. The finding indicated that pretreating dyes with HCl significantly improved their sorption capability as compared to native biomass. The 2-step rate equation was obeyed for dye desorption kinetics, according to this finding. FTIR research disclosed that the -carboxylic, -hydroxyl, and -amino groups are involved in the adsorption of dye molecules onto biomasses. According to studies, the modified rice husk has the best capability for dyes elimination from textile effluents [37].

The efficacy of the chitosan/PVA/talc composite to remove Methyl orange, Congo red, and metal ions was tested. Flocculation was the most common phenomenon in the elimination of Congo red and Methyl orange at 100 ppm concentration. Furthermore,

Table 1 A summary of different materials based on PVA for dyes elimination from water by different removal methods

PVA-based material		Preparation method	Pollutant dye	Removal method	Removal capacity (mg/g or % removal)	Other parameters and results	Refs.
Unmodified PVA		Gelation in hot water	Methylene blue	Adsorption	13.8 mg/g	LGI, FDI and Temkin adsorption, PSO, non-spontaneous and exothermic	[52]
PVA-based composites	Xylan/PVA/TiO$_2$ composite		Ethyl violet and Astrazon brilliant red-4G	Photocatalytic degradation under visible light	93.65 and 92.71%	Positive holes (h$^+$) were found to be the major species involved in dye degradation	[35]
	PVA/Na-alginate/GO	Sol-gel	Methylene blue	Adsorption	759.3 mg/g	PSO, FDI, exothermic, multilayer adsorption	[36]
	PVA-immobilized modified rice husk	Immobilization technique	Orange-3GL, direct orange-26, direct red-31, direct blue-67, and ever direct	Adsorption	–	–	[37]
	Chitosan/PVA/talc composite	Solution casting	Congo red, methyl orange	Flocculation	94% 85% at 50 mg/L	PSO	[38]
	PVA/SBA-15 composites (30%)	Impregnation method	Methylene blue	Adsorption	77 mg/g	PSO, LGI	[15]

(continued)

Table 1 (continued)

PVA-based material		Preparation method	Pollutant dye	Removal method	Removal capacity (mg/g or % removal)	Other parameters and results	Refs.
	PVA-capped ZnO nanostructure	Low-temperature chemical precipitation method	Reactive red 141	Photocatalytic degradation under solar light		Holes and OH radicals are the two major reactive species playing role in degradation of azo dye	[40]
PVA-based gels	PVA-GA hydrogel beads	Chemical crosslinking	Congo red	Adsorption	34 mg/g	Chemical crosslinked via glutaraldehyde; seven times reused multi-layered adsorption was revealed by Harkins Jura model	[47]
	Magnetic chitosan/PVA hydrogel beads	Freezing-thawing	Methyl orange	Adsorption	6,936 mg/g at pH 4.0	Chemical crosslinked by Fe_3O_4, PFO, LGI	[48]
	PVA/CNCs/graphene composite gel	Freeze-drying	Methyl blue	Adsorption	110.9 mg g^{-1}	Physical crosslinked, PSO, LGI	[45]
	ZnO loaded PVA/cellulose nanofibril aerogel	Freeze-thaw	Methylene blue	Combined adsorption and catalytic degradation utilizing sodium borohydride	24.53 mg/g	Chemical crosslinked by Glutaraldehyde, LGI suggested monolayer dye adsorption	[49]

(continued)

Table 1 (continued)

PVA-based material		Preparation method	Pollutant dye	Removal method	Removal capacity (mg/g or % removal)	Other parameters and results	Refs.
	PVA/AA–TiO$_2$ microgel	Radiation grafting technique	Metanil Yellow	Photocatalytic degradation under UV irradiation	~ 100% in 180 min	Chemically crosslinked by electron beam irradiation, 3 times reused after repeated use, the gel content reduced by 4% from the original value	[34]
PVA-based Fibers	PVA nanofibers PVA/Chitosan nanofibers PVA/Chitosan/SiO$_2$ nanofibers	Electrospinning	Direct red 80	Adsorption	25.6 148 322.5 mg/g		[44]
	Zeolitic imidazolate framework-8 (ZIF-8)-coated chitosan/PVA nanofiber	Electrospinning	Malachite green	Adsorption	1000 mg/g	PSO, LGI, high chemical stability	[51]
PVA-based membranes	NH$_2$-functionalized mesoporous PVA/SiO$_2$ hybrid coated membrane	"Single-step" hydrolysis and co-condensation followed by dip coating	Methyl blue	Adsorption filtration	41.88 mg/g	Filtrate flux = 544.26 Lm^{-2} h^{-1} bar^{-1}) oil droplets and water-soluble contaminants, dye after five reusing cycles the relative removal efficiencies still higher than 86.89%	[22]
	Chitosan/PVA/zeolite composite nanofibrous membrane	Electrospinning	Methyl orange	Adsorption	153 mg/g,	PSO, FDI	[53]

(continued)

Table 1 (continued)

PVA-based material		Preparation method	Pollutant dye	Removal method	Removal capacity (mg/g or % removal)	Other parameters and results	Refs.
	Nano-TiO$_2$ modified PVA/PVDF hollow fibrous composite membranes	Dip coating	Congo red, methyl orange, methylene blue	Filtration/desalination		Rejections 9472.57%,52.172.45%, and 9272.20%	[54]
PVA-based films	Ag-TiO$_2$-PVA thin film	A combination of sol-gel and spin coating	Methylene blue	Photocatalytic degradation under sunlight	91%	Higher crystallinity hydroxyl radicals involved	82]
	PVA/CQDs nanocomposites films	Solution casting	Methylene blue	Adsorption	$97 \pm 1\%$ after 40 min	Non-linear PFO, FDI reused five times	[55]

PSO pseudo second-order model, *PFO* pseudo first order, *LGI* Langmuir isotherm model, *FDI* Freundlich isotherm model kinetic studies revealed that a pseudo second-order equation was followed by these adsorbents [38].

The impregnation technique was used to make incredibly hydrophilic PVA/SBA-15 composites for methylene blue adsorption. Various proportions of PVA were impregnated into surfaces of mesoporous silica (SBA-15) to investigate the impact of PVA on the structural and textural characteristics of prepared composites. The acquired findings revealed that following PVA impregnation, the structure of SBA-15 swelled, indicating PVA dispersion within the pores. Higher PVA percentages cause the lattice parameters to drop, resulting in the development of PVA aggregation beyond the surface. The hydrophilicity of the resulting composite has improved because of the growing number of hydroxyl groups. Regarding adsorption capacity, the impacts of PVA content, contact time, adsorbent weight, and initial dye concentration were studied and addressed [39]. Utilizing either PVA or sodium dodecylsulfate (surfactant) as a capping agent, two kinds of ZnO nanostructures were effectively produced in large yield via a simple and low-temperature chemical precipitation technique and used for azo dye degradation. The surfactant had a significant Impact on the ZnO photocatalyst's shape, surface area, and photocatalytic activity. This provided some insights into the advancement of technical applications linked to industrial dye photodegradation. The method can be used to make additional II-VI type semiconductors on a big scale. The morphology of PVA-capped ZnO was found to be spherical. The photodegradation efficiency of the produced ZnO nanomaterial aided by surfactant was significantly greater than PVA-capped ZnO, with approximately 95% and 88% of the Reactive red 141 destroyed following irradiation in 240 min under UV and sunlight, respectively. This is owing to the higher specific surface area of the thin nanorod and plate-like morphological features [40].

6.2 PVA-Based Gel Materials

A gel is defined to be a non-fluid polymer or colloidal network expanded by a fluid all through its entire volume. Polymer gel is a kind of macromolecular network having a major portion of its structure made up of a solvent. Gels have a large surface area, ideal mechanical properties, customizable chemical structure, regeneration ability under moderate circumstances, feasibility, and effective porosities as few characteristics. Gels may be manufactured in bulk (macro), or they can take the micro/nano-particle forms. As a linear polymer, PVA has been used in separation procedures for a long time. Chemical crosslinking is necessary to build different gel materials. Furthermore, the presence of PVA in gels increases the susceptibility of crosslinking through intermolecular or/and intramolecular hydrogen bonding by H-groups [10, 19, 41].

PVA as a potential candidate in polymeric hydrogels is used as a crosslinker, rheological modifier, and matrix template owing to its simplicity of use and compatibility, however, its use as the hydrogel's main backbone is restricted due to brittle/soft resulting hydrogels with difficult recovery, poor mechanical properties, and fluctuating dimensions with changes in water content. The mechanical characteristics of hydrogels based on PVA can be primarily increased by the creation of an inorganic/organic network structure, which can efficiently disperse stress over a large area and boost the hydrogel's load-carrying capacity. Incorporating Fe_3O_4 and TiO_2 into hydrogels, for instance, results in magnetic and photocatalyzed hybrid hydrogels, respectively. Finally, most hybrid hydrogels exhibit high mechanical characteristics and can withstand external, violent stirring, which is critical for adsorbent reusing. Because of their intriguing mechanical qualities, non-toxicity, biocompatibility, and the inclusion of PVA into other natural or manufactured (e.g., poly (ethylene glycol) (PEG), polyacrylamides PAM), polymer frameworks are of great interest. The addition of a biopolymer increases not only the network stability, but also the hydrophilicity, biocompatibility, biodegradability, and rheological characteristics of PVA-based hydrogels [10, 18, 42]. In recent times, Zhang et al. [43] successfully developed yeast comprising PVA/CMC hybrid hydrogel capable of methylene blue dye degradation. CMC offers structural stability with Fe^{3+} or Al^{3+} and pH-sensitive higher swelling capacity.

6.2.1 Physical Gels

Hydrogen bonding, electrostatic interactions, ionic interactions, van der Waal forces, chain entanglements, and hydrophobicity, are all physical interacting factors contributing to the formation of physical gels. A binding force defined by an equilibrium between the van der Waal and electrostatic interactions can be used to physically interlink different charged polymeric groups/fibers/molecules. Physically crosslinked hydrogels are often sustainable and reusable, making them an environmentally beneficial option. By breaking the dynamic force among fibers, such reversible interactions may simply be reversed.

A smart hydrogel made of graphite-like carbon nitride (g-C_3N_4) and cellulose nanocrystals (CNC) reinforced sodium carboxymethylcellulose (CMC-Na)/PVA was recently made to exhibit outstanding tensile strength (~ 648 kPa), toughness (340 kJ m^{-3}), and elongation (1169%). The hydroge's equilibrium swelling capability is strong, resulting in improved methylene blue adsorption capacity (198.6 $^{mg\ g-1}$). The kinetic and isotherm models are listed in Table 1. The hydroge's high adsorption efficiency is attributed to $\pi-\pi$ stacking and hydrogen bonding [44]. Wu et al. [45], synthesized PVA/CNCs/graphene hybrid gel by suspension titration of t-butanol solution substitution and freeze-drying method. The effectiveness of gel in removing methyl blue from wastewater was tested. In the adsorption reaction phase, the optimal concentration of gel was determined to be 2.0 g L^{-1}, and maximal adsorption was achieved in 120 min. The greater the elimination action, the higher the starting pH value of the MB solution.

6.2.2 Chemical Gels

Chemical crosslinking results in robust and long-lasting fibers and is achieved with covalent bonding either through radical polymerization, enzymatic, irradiation, and chemical reactions. Chemical hydrogels are created by covalent crosslinking, electrostatic, or hydrophobic interactions with added chemical species. Chemical crosslinkers for example epichlorohydrin (ECH), citric acid, aldehydes, succinic anhydride, metal ions, and many others can covalently link PVA. Among aldehydes, glutaraldehyde (GA) has indeed been frequently employed for chemical crosslinking of PVA accomplished by an effective reaction between the -OH groups of PVA and the carbonyl (C=O) groups of GA in an acidic medium [20, 46]. Native PVA hydrogel beads crosslinked using glutaraldehyde were produced for Congo Red removal from a prototype industrial effluent by fixed-bed reactor and batch techniques. The highest adsorption capacity (34 mg g^{-1}) was reached using this newly developed adsorbent (at pH 6 and 45 °C). Multi-layered adsorption regulated by the Harkins Jura model was discovered in batch investigations. The adsorption kinetics, on the other hand, were regulated by intraparticle diffusion processes and followed a fractal-like pseudo second-order model. When low bed height and high dye influent rates were employed in fixed-bed investigations, steeper break-through curves were seen during removal operations. Thomas mathematical model best described this behavior. According to the research, internal and external mass diffusion became no longer rate-limiting throughout these studies [47].

Novel magnetic chitosan/PVA hydrogel beads were synthesized by a freeze-thawing process for methyl orange adsorption. The initial concentration of dye and pH played the greatest impact on the removal process. The dye adsorption onto the beads was quick. Furthermore, the low-cost magnetic adsorbent can be renewed and reused for color removal in basic media [48]. The freeze-thaw method was used to make glutaraldehyde crosslinked PVA/cellulose nanofibril aerogel nanocomposites, succeeded by in-situ production of ZnO particles using precursors. Methylene blue in an aqueous solution was decolorized using the aerogels through combined adsorption and catalytic degradation using sodium borohydride. After 10 cycles of treatments, the inclusion of ZnO particles in the highly durable and recyclable crosslinked PVA-based aerogel enhanced MB removal by twofold. PVA improved the mechanical properties and structural integrity of aerogels, acting as a reinforcing agent and binder. To speed up the degradation of organic dyes, reducing agents including benzoin radicals, diphenyl phosphine, and sodium borohydride (NaBH$_4$) were employed. In the existence of NaBH4, the generated ZnO-filled PVA-based aerogel was utilized as a bi-functional catalytic substrate for concurrent dye adsorption and discoloration [49].

Although the immobilization process may increase TiO$_2$ separation ability, it typically reduces the overall photocatalytic effect when compared to dispersed TiO$_2$ owing to a reduced surface/volume ratio and substantial depletion of photocatalytic active surface sites. As a result, a new synthesis technique to create TiO$_2$ photocatalysts that not only have strong photocatalytic performance but could also be reliably separated following photocatalytic events is required. The surface area of the photocatalyst is needed to be extended to optimize photocatalytic effectiveness. The surface/volume ratio of micro and nanospheres is quite high. They exhibit colloidal characteristics at the smaller end of their size distribution range. Micro or nanospher's interfacial characteristics are particularly essential and often indicate their activity. The following are the most essential physicochemical properties that may be regulated during microsphere production: size and dispersion of particles and molecular weight of the polymer. Microgels, particularly ones with large size, maybe easily isolated from the system via sedimentation or filtration, which is a benefit when compared to nanostructured materials, and has been used and forecasted as reliable catalysts [34].

6.3 Fibers Based on PVA

PVA-based fibers are dipped in the dye solution to get them adsorbed onto fibers. Mahmoodi et al. [50], used a mesoporous PVA/chitosan/SiO$_2$ hybrid nanofiber as a potential dye removing material. The optimum conditions of pH $= 2$, adsorbent dosage of 0.015 g, initial dye concentration of 515 mg/L, and maximum adsorption capacity of 322 mg/g were found in this report. Metal-organic frameworks have attracted researchers in the past decade because of their well-known features such as extreme porosity, high adsorption capacity, chemical, and physical stability, huge surface area, and other well-known qualities. The development of bio-nanocomposites based on MOF for water treatment applications has been more popular recently. In this regard, Mahmoodi et al. [51], coated ZIF-8@CS/PVA electrospun nanofiber, a zeolitic imidazolate framework-8 crystal was deposited onto chitosan/PVA blend nanofiber. The composite fiber followed Langmuir adsorption isotherm.

6.4 PVA-Based Membranes

PVA-based composite membranes can be used as adsorbing as well as a filtering material for dyes due to their surface functionalities and porous texture. These membranes can also be used in photocatalytic reactors if modified by a photocatalytic component, e.g., TiO$_2$ or Fe$_3$O$_4$. The electrospun nanofibrous composite membrane made of PVA, chitosan, and zeolite was used to adsorb methyl orange. Due to the addition of zeolite to blended chitosan/PVA, the membrane's Young's Modulus improved by $> 100\%$. Most of the dye was adsorbed onto the membrane in 6 min, according to UV–VIS measurements. With the increase in pH value, the adsorption performance dropped. After many cycles with methyl orange, the resultant nanofiber became less effective. Chitosa's binding sites are -hydroxyl and -amino groups, which aid in the interaction of organic dyes. Under harsh conditions, however, it is less stable. PVA acted as a crosslinker also that helped chitosan stay more stable. It strengthens chitosan by forming hydrogen bonding, resulting in additional binding sites for dye interaction. Thus, an appropriate filler material like zeolite can improve the nanofibrous membran's tensile strength. Zeolites also provide the opportunity for ion exchange during dyes interaction due to their aluminosilicate framework containing cations that may be exchanged. Thus, the membrane can serve as a powerful adsorbent because of its porous architecture and cation exchange capacity [53].

A dip coating technique was used to create novel PVA/poly (vinylidene fluoride) hollow fiber hybrid membranes amended by nano-TiO_2 for dye wastewater treatment and desalination. To upgrade the chemical/mechanical/thermal stability, the composite membrane was crosslinked employing glutaraldehyde. The dye/salt separation process of the composite membrane was significantly affected by pH, dye and salt concentration, and feed solution temperature. With Congo red, NaCl, and Na_2SO_4 as model chemicals, the fouling, separation, and thermal stability were also assessed. According to the separation results, the 1 g/L nano-TiO_2 amended membrane had the best accomplishment regarding CR, MB, and MO rejections, being 9472.57%, 9272.20%, and 52.172.45%, respectively. Furthermore, none of the PVA-composite membranes were salt rejecting (NaCl, Na_2SO_4). The nano-TiO_2 amended PVA-composite membrane has better antifouling, separation efficiency, and thermal properties than the unmodified PVA-composite membrane. The membrane can be used under high-temperature conditions and strong acid dye solution treatments [54].

A new method has been devised for eliminating water-soluble impurities and oil droplets from stabilized oil/water emulsions at the same time. The PVA/SiO_2 hybrid coatings were made by a single-step hydrolysis/co-condensation process, resulting in a mesoporous structure with a nano-sized pore width (9.31 nm) and an abundance of -NH_2 groups. PVDF membranes stuck strongly by hybrid coatings, possess superhydrophilicity, extremely high-water flux, underwater superoleophobicity, excellent antifouling activity, extremely effective oil-in-water emulsion segregating capability, and gravity-driven separation capacity. Furthermore, methyl blue and copper ions can be efficiently absorbed from the aqueous phase, with adsorption capacity values of 41.88 mg/g and 52.8 mg/g, respectively. The prepared membranes are a worthy material in practical implementation for water treatment, filling the gap between oil/water separation and contaminants adsorption, thanks to their ultra-low oil adhesive properties, significant lasting chemical durability, and outstanding reusability [22].

6.5 PVA-Based Films

Films can be used for the adsorption and degradation of dyes. The adsorbent films or photo-catalytically active films are usually immersed in dye contaminated solution and either adsorption and/or degradation involving radicals species takes place in reaction media. For a variety of technological applications, the synthesis of plasmonic nanocomposite coatings with adjustable photocatalytic and optical characteristics is critical. The photocatalytic, optical, and microstructural, characteristics of an Ag-TiO_2-PVA thin film nanocomposite have been modified. Ag-TiO_2-PVA thin films were formed on silica glass. With increased ion fluence, the morphological development of thin films caused by ion irradiation showed intriguing variations in the particle size distribution of nanostructures. Ion irradiation at $1 * 1013$ ions/cm^2 significantly improved the photocatalytic activity of thin film nanocomposite for methylene blue photodegradation [56]. New PVA/CQDs film nanocomposites for methylene blue elimination were prepared. These films were obtained by combining PVA and CQDs nanoparticles (prepared in zero-dimension using microwave heating). The significant association between PVA and CQDs nanoparticles through hydrogen bonding was verified by XRD and FTIR spectra. The quantities of CQDs implanted in the PVA matrix are accountable for the adjustable structural and optical characteristics of nanocomposite film. Nanocomposites including (PVA/CQDs 2 wt%) can remove a higher concentration of dye (30 mg/L) from the solution. Anti-ultraviolet, desalination and catalytic water decomposition are all anticipated benefits of PVA/CQDs films [55].

7 Conclusion

PVA has a great potential for the elimination of toxic textile dyes from water, due to its enormous compatibility to be combined with several materials including conventional adsorbents (activated carbons, fly ashes, clays, biomasses), biomaterials (biopolymers, biochar, etc.), polymers (synthetic and natural), carbonaceous materials (graphene, graphene oxide, graphitic carbon nitride, CNTs/MW-CNTs, carbon quantum dots), waste-derived materials (fly ashes, used biomasses, etc.), metallic materials (semiconductors, metal-oxides, metal-sulfides, etc.), and advanced materials (Metal-organic frameworks, layered double hydroxides, etc.). PVA-based hybrid materials show much-improved performance than their native counterparts. Moreover, these materials can be transformed into various forms, e.g., membranes, fibers, composites, films, and gels for different treatment technologies, e.g., filtration, adsorption, and degradation. However, discussions of all these materials' detailed applications for dyes removal are beyond the scope of this chapter. The chapter outlined the important material categories and removal methods employed for the elimination of dyes. The research is continued to develop, optimize, and implement these polymer-based technologies for dye decontamination.

References

1. Mansor ES, Abdallah H, Shaban A (2020) Fabrication of high selectivity blend membranes based on poly vinyl alcohol for crystal violet dye removal. J Environ Chem Eng 8(3):103706
2. Jun LY et al (2020) Modeling and optimization by particle swarm embedded neural network

for adsorption of methylene blue by jicama peroxidase immobilized on buckypaper/polyvinyl alcohol membrane. Environ Res 183:109158

3. Khan F et al (2021) Photocatalytic polymeric composites for wastewater treatment. In: Aquananotechnology. Elsevier, pp 457–480

4. Ulson SMDAG, Bonilla KAS, de Souza AAU (2010) Removal of COD and color from hydrolyzed textile azo dye by combined ozonation and biological treatment. J Hazardous Mater 179(1–3):35–42

5. Hanif MA et al (2021) Applications of coagulation-flocculation and nanotechnology in water treatment. In: Aquananotechnology. Elsevier, pp 523–548

6. Wu W, Huang Z-H, Lim T-T (2014) Recent development of mixed metal oxide anodes for electrochemical oxidation of organic pollutants in water. Appl Catal A 480:58–78

7. Tabasum A et al (2018) Fe_3O_4-GO composite as efficient heterogeneous photo-Fenton's catalyst to degrade pesticides. Mater Res Express 6(1):015608

8. Zahid M et al (2021) Role of polymeric nanocomposite membranes for the removal of textile dyes from wastewater. In: Aquananotechnology. Elsevier, pp 81–93

9. Mushtaq F et al (2020) $MnFe_2O_4$/coal fly ash nanocomposite: a novel sunlight-active magnetic photocatalyst for dye degradation. Int J Environ Sci Technol 17:4233–4248

10. Jing G et al (2013) Recent progress on study of hybrid hydrogels for water treatment. Colloids Surf A 416:86–94

11. Gong G et al (2015) Facile fabrication of magnetic carboxymethyl starch/poly (vinyl alcohol) composite gel for methylene blue removal. Int J Biol Macromol 81:205–211

12. Rajabi M, Mahanpoor K, Moradi O (2017) Removal of dye molecules from aqueous solution by carbon nanotubes and carbon nanotube functional groups: critical review. RSC Adv 7(74):47083–47090

13. Sandeman SR et al (2011) Adsorption of anionic and cationic dyes by activated carbons, PVA hydrogels, and PVA/AC composite. J Colloid Interface Sci 358(2):582–592

14. Mushtaq F et al (2019) Possible applications of coal fly ash in wastewater treatment. J Environ Manage 240:27–46

15. Sabarish R, Unnikrishnan G (2018) Polyvinyl alcohol/carboxymethyl cellulose/ZSM-5 zeolite biocomposite membranes for dye adsorption applications. Carbohyd Polym 199:129–140

16. Crini G (2006) Non-conventional low-cost adsorbents for dye removal: a review. Biores Technol 97(9):1061–1085

17. Zahid M et al (2019) Metal ferrites and their graphene-based nanocomposites: synthesis, characterization, and applications in wastewater treatment. In: Magnetic nanostructures. Springer, pp 181–212

18. Sarkar N, Sahoo G, Swain SK (2020) Graphene quantum dot decorated magnetic graphene oxide filled polyvinyl alcohol hybrid hydrogel for removal of dye pollutants. J Mole Liquids 302:112591

19. Arabkhani P, Asfaram A (2020) Development of a novel three-dimensional magnetic polymer aerogel as an efficient adsorbent for malachite green removal. J Hazard Mater 384:121394

20. Mallakpour S, Rashidimoghadam S (2019) Poly (vinyl alcohol)/vitamin C-multi walled carbon nanotubes composites and their applications for removal of methylene blue: advanced comparison between linear and nonlinear forms of adsorption isotherms and kinetics models. Polymer 160:115–125

21. Rongrong L et al (2011) The performance evaluation of hybrid anaerobic baffled reactor for treatment of PVA-containing desizing wastewater. Desalination 271(1–3):287–294

22. Liu H et al (2019) Amino-functionalized mesoporous PVA/SiO_2 hybrids coated membrane for simultaneous removal of oils and water-soluble contaminants from emulsion. Chem Eng J 374:1394–1402

23. Babu J, Murthy Z (2017) Treatment of textile dyes containing wastewaters with PES/PVA thin film composite nanofiltration membranes. Sep Purif Technol 183:66–72

24. El-Shamy AG (2020) An efficient removal of methylene blue dye by adsorption onto carbon dot@ zinc peroxide embedded poly vinyl alcohol ($PVA/CZnO_2$) nano-composite: a novel reusable adsorbent. Polymer 202:122565

25. Abou Taleb MF, Abd El-Mohdy H, Abd El-Rehim H (2009) Radiation preparation of PVA/CMC copolymers and their application in removal of dyes. J Hazard Mater 168(1):68–75

26. Aslam M, Kalyar MA, Raza ZA (2018) Polyvinyl alcohol: a review of research status and use of polyvinyl alcohol based nanocomposites. Polym Eng Sci 58(12):2119–2132

27. Yin H et al (2020) A novel Pd decorated polydopamine-SiO$_2$/PVA electrospun nanofiber membrane for highly efficient degradation of organic dyes and removal of organic chemicals and oils. J Clean Prod 275:122937

28. Xiao J et al (2017) Multifunctional graphene/poly (vinyl alcohol) aerogels: in situ hydrothermal preparation and applications in broad-spectrum adsorption for dyes and oils. Carbon 123:354–363

29. Dai J et al (2016) High structure stability and outstanding adsorption performance of graphene oxide aerogel supported by polyvinyl alcohol for waste water treatment. Mater Des 107:187–197

30. Anitha T (2016) Synthesis of nano-sized chitosan blended polyvinyl alcohol for the removal of Eosin yellow dye from aqueous solution. J Water Process Eng 13:127–136

31. Bhat SA et al (2020) Efficient removal of Congo red dye from aqueous solution by adsorbent films of polyvinyl alcohol/melamine-formaldehyde composite and bactericidal effects. J Clean Prod 255:120062

32. Jaseela P, Garvasis J, Joseph A (2019) Selective adsorption of methylene blue (MB) dye from aqueous mixture of MB and methyl orange (MO) using mesoporous titania (TiO$_2$)–poly vinyl alcohol (PVA) nanocomposite. J Mole Liquids 286:110908

33. Bayat A et al (2021) Electrospun chitosan/polyvinyl alcohol nanocomposite holding polyaniline/silica hybrid nanostructures: an efficient adsorbent of dye from aqueous solutions. J Mole Liquids 331:115734

34. Abd El-Rehim HA, Hegazy E-SA, Diaa DA (2012) Photo-catalytic degradation of Metanil Yellow dye using TiO$_2$ immobilized into polyvinyl alcohol/acrylic acid microgels prepared by ionizing radiation. Reactive Functional Polym 72(11):823–831

35. Liu Z et al (2019) Photocatalytic degradation of dyes over a xylan/PVA/TiO$_2$ composite under visible light irradiation. Carbohydrate Polym 223:115081

36. Ma Y-X et al (2020) Fabrication of 3d porous polyvinyl alcohol/sodium alginate/graphene oxide spherical composites for the adsorption of methylene blue. J Nanosci Nanotechnol 20(4):2205–2213

37. Bhatti HN et al (2020) Efficient removal of dyes using carboxymethyl cellulose/alginate/polyvinyl alcohol/rice husk composite: adsorption/desorption, kinetics and recycling studies. Int J Biol Macromol 150:861–870

38. Kalantari K, Afifi AM (2018) Novel chitosan/polyvinyl alcohol/talc composite for adsorption of heavy metals and dyes from aqueous solution. Sep Sci Technol 53(16):2527–2535

39. Abid Z et al (2019) Preparation of highly hydrophilic PVA/SBA-15 composite materials and their adsorption behavior toward cationic dye: effect of PVA content. J Mater Sci 54(10):7679–7691

40. Kakarndee S, Nanan S (2018) SDS capped and PVA capped ZnO nanostructures with high photocatalytic performance toward photodegradation of reactive red (RR141) azo dye. J Environ Chem Eng 6(1):74–94

41. Shoueir KR et al (2016) Macrogel and nanogel networks based on crosslinked poly (vinyl alcohol) for adsorption of methylene blue from aqua system. Environ Nanotechnol Monitoring Manage 5:62–73

42. Yue Y et al (2016) Cellulose nanofibers reinforced sodium alginate-polyvinyl alcohol hydrogels: core-shell structure formation and property characterization. Carbohyd Polym 147:155–164

43. Zhang M et al (2020) A novel Poly (vinyl alcohol)/carboxymethyl cellulose/yeast double degradable hydrogel with yeast foaming and double degradable property. Ecotoxicol Environ Safety 187:109765

44. Wang H et al (2020) Eco-friendly polymer nanocomposite hydrogel enhanced by cellulose nanocrystal and graphitic-like carbon nitride nanosheet. Chem Eng J 386:124021

45. Wu Y et al (2019) Study on the preparation and adsorption property of polyvinyl alcohol/cellulose nanocrystal/graphene composite aerogels (PCGAs). J Renew Mater 7(11):1181–1195
46. Zhang YS, Khademhosseini A (2017) Advances in engineering hydrogels. Science 356(6337)
47. Jain P et al (2021) Color removal from model dye effluent using PVA-GA hydrogel beads. J Environ Manage 281:111797
48. Wang W et al (2018) Facile preparation of magnetic chitosan/poly (vinyl alcohol) hydrogel beads with excellent adsorption ability via freezing-thawing method. Colloids Surf A 553:672–680
49. Azmi A et al (2021) Zinc oxide-filled polyvinyl alcohol–cellulose nanofibril aerogel nanocomposites for catalytic decomposition of an organic dye in aqueous solution. Cellulose 28(4):2241–2253
50. Mahmoodi NM, Mokhtari-Shourijeh Z, Abdi J (2019) Preparation of mesoporous polyvinyl alcohol/chitosan/silica composite nanofiber and dye removal from wastewater. Environ Prog Sustain Energy 38(s1):S100–S109
51. Mahmoodi NM et al (2020) Synthesis of pearl necklace-like ZIF-8@ chitosan/PVA nanofiber with synergistic effect for recycling aqueous dye removal. Carbohydrate Polym 227:115364
52. Umoren S, Etim U, Israel A (2013) Adsorption of methylene blue from industrial effluent using poly (vinyl alcohol). J Mater Environ Sci 4(1):75–86
53. Habiba U et al (2018) Adsorption study of methyl orange by chitosan/polyvinyl alcohol/zeolite electrospun composite nanofibrous membrane. Carbohyd Polym 191:79–85
54. Li X et al (2014) Desalination of dye solution utilizing PVA/PVDF hollow fiber composite membrane modified with TiO_2 nanoparticles. J Membr Sci 471:118–129
55. El-Shamy AG, Zayied H (2020) New polyvinyl alcohol/carbon quantum dots (PVA/CQDs) nanocomposite films: structural, optical and catalysis properties. Synthetic Metals 259:116218
56. Singh J, Sahu K, Mohapatra S (2019) Ion beam engineering of morphological, structural, optical and photocatalytic properties of $Ag-TiO_2$-PVA nanocomposite thin film. Ceram Int 45(6):7976–7983

Polymeric Membranes Nanocomposites as Effective Strategy for Dye Removal

Rabia Nazir, Yaseen Ayub, and Muhammad Ibrar

Abbreviations

AMTD	2-Amino-5-mercpto-1,-3,-4-thiadiazole
CA	Cellulose acetate
CNT	Carbon nanotubes
CPM	Clay-polymeric hybrid membranes
CX	Carbon-based xerogels
GO	Graphene oxide
IPN	Interpenetrating networks
MF	Microfiltration
MMT	Montmorillonite
MWCO	Molecular weight cut off
NF	Nanofiltration
NPs	Nanoparticles
PAN	Polyaniline
PDMS	Polydimethylsiloxanes
PEG	Polyethylene glycol
PEI	Polyethyleneimine
PES	Polyethersulfone

R. Nazir (✉)
Pakistan Council of Scientific and Industrial Research Laboratories Complex, Ferozepur Road, Lahore, Pakistan
e-mail: rabiapcsir@yahoo.com

Y. Ayub · M. Ibrar
Department of Chemistry, Forman Christian College, Zahoor Elahi Rd, Gulberg III, Lahore 54600, Punjab, Pakistan

R. Nazir · Y. Ayub · M. Ibrar
Department of Chemistry, Lahore Garrison University, Sector C, DHA Phase 6, Lahore, Pakistan

© The Author(s), under exclusive license to Springer Nature Singapore Pte Ltd. 2022
A. Khadir and S. S. Muthu (eds.), *Polymer Technology in Dye-containing Wastewater*, Sustainable Textiles: Production, Processing, Manufacturing & Chemistry, https://doi.org/10.1007/978-981-19-0886-6_2

PMMA   Polymethyl methacrylate
PPSU   Polyphenylsulfone
Psf   Polysulfone
PT   Polyester textile
PVA   Polyvinyl alcohol
PVDF   Polyvinylidene fluoride
RO   Reverse osmosis
SDS   Sodium dodecyl sulfate
SEM   Scanning electron microscope
TPC   Triphthaloyldechloride
UF   Ultra-filtration
UV   Ultraviolet
ZIF67   Zeolitic like framework-67

1 Introduction

Conventional water technologies have long been supplanted and superseded by polymer technology. The techniques have been refurbished by advent of nanotechnology that facilitated supply of quality water through use of safe and modern technologies.

Dyes are an important class of organic compounds, having functional groups responsible for color, (Fig. 1a) that can be divided into various classes as per description of color index (Fig. 1b). These dyes especially the azo ones ($-N=N-$) which accounts for 65–70% of total dyes production, are regarded as one of the potential contaminants that can deteriorate the water quality even at small quantities and owing to their biodegradable nature pose serious threats to the environment as well as aquatic system [73]. The world dye consumption data as presented in Chemical Economics Handbook (2021) determines it to be population based with Asia as highest (Fig. 1c).

Dyes are consumed in various industries like textile, leather, cosmetics, pharmaceutical etc. but the largest consumption is in textile sector [67]. The textile industry waste water is deemed as a complex system consisting of dyes, salts, heavy metals, oils and greases, solvents, detergents etc. based on the processing protocols. Furthermore, high COD (chemical oxygen demand) and pH values offer much challenges in water processing [107]. There are number of methods that are used for treatment of dye containing effluent which include coagulation, flocculation, adsorption, electrolysis, membranes, oxidation etc. [102]. Among these, the membrane technologies are considered as one of the promising techniques that have potential of incorporating advance technologies and removal of large amounts of water while maintaining the desired efficacy and quality of water. This chapter hence focuses on latest developments pertaining to membrane technology while focusing on dye removal.

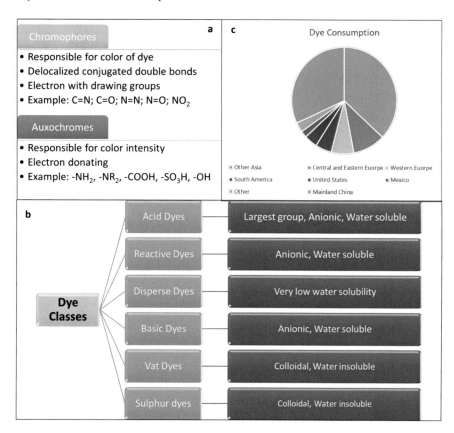

Fig. 1 Dyes' **a** functional groups, **b** classification, **c** consumption pattern in the world

2 Polymeric Membranes

One of the noble challenges to assist the novel society is to defend adequate resources of water of desirable quality for several designated purpose. To compensate this challenge for water treatment polymeric membrane played a significant role in treatment of drinking brackish water and waste water for their reuse [110]. A membrane can be stated as an intermixing of two phases behaving as a selective barrier that allows some molecules, ions and particles to pass through while curtailing others [31]. Henceforth, enabling removal of different inorganic and organic molecules including low molecular weight solutes from solution [94]. Polymeric membranes are usually categorized on basis of applications: (1) Ultra-filtration (UF) (2) Microfiltration (MF) (3) Nanofiltration (NF) and (4) Reverse osmosis (RO) which also becomes the criteria of classification based on pore-size and applied pressure (Fig. 2).

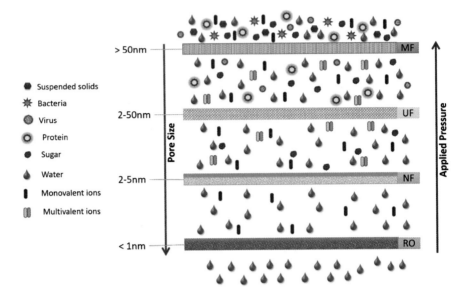

Fig. 2 Different polymeric membranes with **a** their structure, pore size, applied pressure and **b** selective filtration

These synthetic membranes are available in diverse nature and forms [26, 27] and offer certain advantages in essence that they are easy to handle, easy to scale-up, less costly, low energy consumption and flexible in use and applications but they also suffer from some critical drawbacks like chemical attack and fouling. Henceforth, most of the studies have been focusing on thorough blending and modifications of these polymeric materials to improvise and revamp the their properties with special focus on salt rejection, water flux, antifouling and chemical resistant properties [13, 30]. This offers added advantages of improved mechanical strength, higher stability, hydraulics performance, high stability, tunable properties with focus on pore structure, surface functionalization and skeleton chemistry [72].

Polyimide ultrafiltration membrane fabricated from 2,4,6-trimethyl-1,3-phenylenediamine, 4,4′-diaminodiphenylmethane and 1,2,4,5-benzenetetracarboxylic dianhydride molecular weight cut off (MWCO) cut off of 9320 Da resulting in passing of both monovalent and divalent salts (NaCl and Na_2SO_4) and rejection of dyes (Direct red 23 > Congo red > Coomassie brilliant blue > Evans blue) in range of 93–98%. The operating temperature range of 20–90 °C was deemed suitable for textile industry but these membranes show de-structuring at pH > 13 [107]. On other hand, the surface functionalization of polymeric membrane (Cardo poly (arylene ether ketone) equipped with hydrophilic carboxylic acid groups) depicted improved uptake for all the dyes (99.8–98%) with dye rejection order: Congo red dye > Direct red 23 > Coomassie brilliant blue > Evans blue at 0.4 MPa. This ultrafiltration membrane possesses dye permeation flux of 100.9 L $m^{-2\,h-1}$ and resulted in complete penetration of monovalent ions with MW cutoff of

9260 Da but resulted in 10% rejection of divalent salts. The membrane is operatable till 95 °C and pH range of 1–10 with added advantage of possessing anti-fouling behavior as well as anti-dye adsorption properties with a high flux recovery ratio [51, 53]. Utilization of pre-synthesized GN polymeric membrane (MWCO ~ 10,000) was only able to reject 78.95% of Direct 15 dye at operating pH range of 2–11, pressure 276–1379 kPa and temperature of 50 °C. The cake resistance of 4.66 × 10^{-10} m^{-1} was observed when 20 ppm dye concentration was used [1].

Ultrafiltration polymeric membrane prepared from blending of polyphenylsulfone (PPSU) with increasing concentration of polyethersulfone (PES) helps in improving porosity, enhancing hydrophilicity and decreasing the contact angle enabling better uptake of acid black 210 (99% at 3 bar) at 35 ppm dye concentration with PES concentration of 4%. The rejection rate was not impacted by changes in pH and temperature [4]. Similar studies on blended PPSU-PES also confirmed the influence of blending ratio on contact angle and consequently dye elimination efficiency. The increase in blending ratio of PES (i.e. 0, 1, 2, 3, 4%) was also observed to increase pore size, pore density and roughness. The roughness was also found to be related to antifouling and with 4% PES, roughness value increased to about 93% with significant reduction in contact angle. Room temperature studies of the prepared membrane PPSU-PES (16:4 wt./wt%) showed 96.62% dye (Drupel Black NT) separation at 50 ppm [28].

Modification of polymeric membranes with surfactant was found important factor in controlling dye rejection rate. Cellulose acetate modified PVA (CA-PVA)-based polymeric membranes with improved mechanical strength, enhanced antifouling and surface wettability were produced by employing different ratios of PVA. With different doping ratios of PVA, the resulting membranes showed different morphological changes that impact the rejection rate as well as flux. The membrane produced at blending ratio of 6% having molecular weight of 800 Daltion showed good durability for long-term applications resulting in almost 100% rejection of crystal violet dye at flux of 17 LHM. The rejection rate was observed to be highly pH dependent which increases as the pH increased from 3 to 11 [56].

Polyvinylidene fluoride (PVDF) membranes modified by Brijj-58 depicted improved hydrophilic character as compared to the one modified by using same weight percentage of PEG-400. The water contact angle decreases in the order Pristine membrane > PEG 400 additive (2%) > PEG 400 additive (4%) > Brij-58 additive (2%) > Brij-58 (4%) which can be attributed to structural differences between the two surfactants i.e. molecular weight and hydrophobic tailing. Generally, dye rejection faces reduction with decrease in membrane hydrophilicity but in current case the opposite scenario was observed which can be due to development of enhanced porosity induced with increase in surfactants ratio which also effects the molecular weight cutoff factor of these modified membranes [68].

3 Metal and Metal Oxide-Based Polymeric Membranes

Amalgamation of Metal/metal oxide into the polymeric materials is another area that is exploited by researchers to explore new horizons in water treatment applications. These composite membranes are much appreciated for their enhanced durability, better selectivity, tunable properties, improved antifouling properties, structural flexibility and photoluminescence behavior. Inspite of their added advantages these membranes pose some difficulty owing to issues with homogeneous dispersion of metal/metal oxide NPs into the polymer matrix which can be further aggravated by aggregation of NPs. This can be attributed to multiple factors like liquid–solid surface interactions, NPs concentration and solution's pH and ionic strength. Hence, in the fabrication of polymeric membranes the synthetic protocols, nature of functional groups present in the backbone of polymer and metal/metal oxide nature are deemed very important [65, 69]. Two synthetic protocols are usually preferred for synthesis of metal/metal oxide polymeric composite membranes: (i) addition of NPs to casting solution resulting in Phase inversion and (ii) immersion of polymeric membrane into NPs aqueous suspension [15]. There is a huge list of NPs that have been incorporated into the polymeric matrix (Table 1), few of which are highlighted in this chapter.

3.1 Silver NP

Silver ions are long recognized as effective antimicrobial agent, even at very low dosages, making them suitable material for handling fouling of membranes from microbial growth [88]. Most of the reported literature, hence, is targeted on antimicrobial properties of the Ag-doped membranes while few of them address dye rejection potential of these membranes.

Nanofiltration membrane decorated with Ag NPs was prepared by polyvinylidene fluoride (PVDF), polyethylene glycol (PEG), zeolitic like framework-67 (ZIF67), ethylenediamine and polyester textile (PT) support. The membrane i.e. (PT/Ag/PVDF-PEG/ZIF-67) selectively removes rose bengal dye (96%) in presence of interfering dyes (amido black and methylene blue) with selectivity factor of 12.7 and 14.4, respectively [60]. Another hybrid membrane is prepared by incorporation of Ag-AgBr into triethanolamine modified PAN (PAN-ETA) with help of chitosan-TiO_2. The resulting polymeric membrane (PAN-ETA/CS-TiO_2/Ag-AgBr) was able to reject methyl orange (88%) > Congo red (95%) > methylene blue (97%) > rose bangal (~ 100%). When the membrane is irradiated in visible light, it resulted in high photocatalytic degradation of dyes (90–97%) which was attributed to formation of electron-hole pair in Ag/Ag-Br. These electrons and holes will result in radicle formation i.e. $\cdot O_2^-$, $\cdot Br^-$ and $\cdot OH$ which are responsible for degradation of dye molecules [101].

Table 1 Metal and MO-based polymeric membranes and their application in removal

Polymer	Metal/metal oxide NPs	Synthetic approach	Dye	Removal capacity	References
PVDF	Ag	–	Rose Bengal	96%	[60]
Psf	Pd (2%)	Phase inversion	Crystal violet	99%	[33]
PVDF	TiO$_2$ (1.5%); anatase	Phase inversion	Methylene blue	99%	[66]
PVDF	TiO$_2$ (0–1%)	–	Brilliant green; indigo carmine	81; 89%	[95]
PES	TiO$_2$ (13%) mixture of anatase and rutile	Phase inversion	Methyl orange	90%	[40]
PES	TiO$_2$ with N and Pd (7%)	Phase inversion	Eosin yellow	92%	[47]
Psf	PANI modified TiO$_2$	Phase inversion	Reactive black 5; reactive orange 16	81.5; 96.5%	[78]
PES	ZnO (0.035%)	Phase inversion	Methyl blue	82.3%	[14]
Polyethylene	ZnO	Thermal induced phase separation and chemical bath deposition method	Blue indigo dye	99%	[97]
PSf/PVA	ZnO (0.5%)	–	Congo red	53.5%	[46]
Psf/PVA	SiO$_2$	–	Congo red	Negligible	[46]
Polyamide (polydopamine modified)	ZnO–TiO$_2$ (1:1–2:1)	One step and two step	Methylene blue	30–51%	[12]
-do-	TiO$_2$ (0.01–0.05%)		Methylene blue	30–35%	
-do-	ZnO–TiO2 (1:1–2:1)		Methylene blue	31–53%	
PVDF	ZnO–La (37%)	Electrospinning	Methylene blue Rhodamine B	96.3% 3.4%	[76]
PES	ZnO (17%)	Phase inversion	Methyl orange	100%	[62]
PVDF	Lead-zinc oxide	Non-solvent induced phase separation	Reactive black 5	98%	[24]

(continued)

Table 1 (continued)

Polymer	Metal/metal oxide NPs	Synthetic approach	Dye	Removal capacity	References
PVDF	ZnO	Phase inversion	Methylene blue	> 90%	[25]
Polypyrrole	ZnO	Immersion	Methylene blue	96.1%	[113]
Chitosan	ZnO/CuO	–	Fast green	91.2%	[6]
Chitosan	ZnO	–	Fat green	71.4%	
Cellulose acetate-polyurethane	ZnO	Solution dispersion blending method	Reactive red 11; reactive orange 84	–	[91]
Polyethyleneimine (PEI)	SiO_2 (0.1%)	–	Crystal violet	99%	[45]
Chitosan/PVA	SiO_2 (1.0%)	Electrospinning	Direct red 23	98%	[41]
PES-polyvinyl pyrrolidone	SiO_2 coated magnetite	Phase inversion	Rhodamine B, direct black 38, reactive green 19	90%	[98]
PES	$Fe_3O_4@SiO_2\text{-}NH_2$	Phase inversion	Methyl red	97%	[43]
PVA	SiO_2	One-step hydrolysis and co-condensation	Methyl blue	41.8 mg/g	[52]
PVA	Pd decorated polydopamine-SiO_2	Electrospinning	Methylene blue Congo red	99% 99%	[109]
Vinyl-modified mesoporous poly(acrylic acid)	SiO_2	Electrospinning	Malachite green	220.5 mg/g	[105]

3.2 TiO$_2$ NPs

TiO$_2$ NPs is known for its photocatalytic activity and have been used in various water treatment applications. TiO$_2$ exists in three forms (Rutile, Anatase and Brookit) and most stable form is rutile [15]. Dye removal is mediated by synergistic effects of adsorption and photodegradation and hence resulting in improved rejection.

Polyvinylidene fluoride (PVDF) membranes decorated with anatase TiO$_2$ showed incremental increase in porosity of membrane with enhancement in TiO$_2$/PVDF ratio [66, 95] which consequently decreases the rejection capability of doped membranes (21% from 45%) [66]. Further increase in TiO$_2$/PVDF (> 0.5) led to collapsing of structure and consequently decrease in porosity [95]. Different treatments of membrane like ethanol conditioning or addition of sodium dodecyl sulfate (SDS) into the methylene blue solution significantly improve dye rejection rate. In case of

SDS, rejection rate jumps to 99% from 21% which is attributed to SDS deposition and gel layer formation onto membrane surface resulting of clogging of pores [66]. In case of ethanol, structural changes of membrane are not evidences but enhancement in contact angle was observed owing to wetting of both external as well as internal pore surfaces which increases dye rejection [95].

UV-cleaning of membranes showed significant rejection recovery (91%) and degradation of methylene blue by cleavage of C=N bond [66]. Increase in membrane porosity with increase in TiO_2 loading (0–13 wt%) was also observed for TiO_2–PES film while further rise results to pores blockage. Contrary to previous finding increase in TiO_2 content led to high photocatalytic degradation of dye by the composite membrane under UV irradiation [40]. Visible light irradiation for 180 min was found successful to degrade 92% eosin yellow dye when PES matrix was co-doped with 7% N, Pd and TiO_2. Moreover, same relation between porosity, wt% of doped content and dye degradation hold good in this case as well [47].

Modification of TiO_2 by various means is also known to impact photodegradation of dyes. Polyaniline (PANI) modified TiO_2 NPs, when incorporated in polysulfone (Psf) resulted in hollow fiber membranes which dye rejection rate directly depends on PANI-TiO_2 concentration. The NH^+ of the polymeric membrane interacts electro-statistically with anionic dyes i.e. Reactive Black 5 and Reactive Orange 16 leading to rejection of 81.5% and 96.5%, respectively [78].

Selectivity of the polymeric membranes toward specific dye can also be enhanced by modification of TiO_2 NPs with molecular imprinting polymer. The resulting membrane TiO_2/Psf, promotes selective uptake and consequently degradation of methylene blue (90%) owing to electrostatic attraction between dye molecule and polymeric surface [59]. Self-cleaning properties were also inculcated in membranes by modification of TiO_2 with two different Polydimethylsiloxanes (PDMS) i.e. Vinyl-terminated (AHV) and hydroxyl-terminated (AHH). AHV gives hydrophobic character to the membrane while AHH resulted in hydrophilic surface. The AHV-PDMS/TiO_2 membrane depicted almost similar kind of dye photodegradation activity as was noticed by TiO_2 NPs suggesting the presence of exposed TiO_2 regions on the polymeric surface promoting direct interaction of dye molecules with TiO_2. On the other hand, continuous polymeric films result in complete layering on TiO_2 which consequently doesn't allow direct contact of dye molecules with NPs and hence no photocatalytic degradation of dyes (rhodamine B). Such surfaces which are not completely covered by polymer layer help in degradation of dyes which contributes to self-cleaning property of these membranes which facilitates their reuse-ability with almost same efficacy for number of cycles [64].

3.3 ZnO NPs

ZnO NPs' capability to perform under sunlight radiation, high photocatalytic activity, enhanced stability and less toxicity makes these materials excellent candidate for polymeric composite membranes [91].

Impregnation of ZnO into PES membrane at lower concentration results in homogeneous distributed membranes having microvoids. Increase in ZnO loading (> 17%) causes increase in agglomeration of ZnO and reduction in pores which consequently damage the membrane structure [62]. Similar studies performed by varying both ZnO and PES weight percentages showed that enhancement in ZnO resulted in formation of hydrophilic membranes while on other hand increase in polymer concentration induces hydrophobic character to the membranes. Furthermore, rise in PES ratio also led to reduction in macrovoids which together with ZnO induced charge effects and hydrophilicity enables rejection of dyes with molecular weight smaller than 400 Da [14].

Surface modification of polyethylene with ZnO layer was affected by activation of PE either with KMnO$_4$ or plasma by varying immersion time, Zn and KMnO$_4$ concentration. The thickness of leafy structure obtained by immersion of polymer into Zn(NO$_3$)$_2$ solution gets more dense with increase in immersion time which indicates higher concentration of Zn deposition on the PE layer. Furthermore, reduction in contact angle and improved hydrophilicity was also found connected with Zn deposition. Significant variations in pure water flux were recorded with change in immersion time as well as Zn(NO$_3$)$_2$ and KMnO$_4$ concentration which attains maximum value at 40 min, 0.5 M and 5 M, respectively resulting in almost 99% dye rejection which stays considerably constant even after 5 cycles [97].

Doping of rare earth/ZnO nanostructures into PVDF matrix impacts pore size in order La > Er > Sm. Higher color removal was achieved by increasing the rare earth concentration while the maximum was achieved at La/ZnO (37%) concentration leading to 96 and 93% removal of Methylene blue and rhodamine B after 360 min irradiation. The membrane still retains its capability to 98% even after 10 cycles [76]. Transition metal (Pb) co-doping with ZnO also effects the dye rejection capacity of polymeric membranes. Sol-gel prepared ZnOPb, when immersed in PVDF at varying ratios resulted in formation of membrane with finger like macrovoids which increase as the loading ratio rises. Contrary to the earlier results, increase in ZnOPb upto 10% weight didn't impact porosity and hydrophilic character of membranes but after that increase in dopant amount reduces the hydrophilicity of membrane, while no significant change in porosity was observed. There is appreciable increase in dye rejection with increase in doping of NPs as compared to neat membrane which follows the order: PVDF (68.5%) < PVDF/ZnOPb-5% (86.6%) < PVDF/ZnOPb-10% (95.5%) < PVDF/ZnOPb-20% (98.9%) [24].

3.4   SiO$_2$ NPs

SiO$_2$ nanoparticles are another metal oxide that is fancied by the researchers involved in water treatment technologies. These NPs have strong surface energy and are capable of forming membranes that offer high selectivity, permeability, hydrophilicity, chemical stability with appreciable flux and antifouling potential [17].

Interfacial polymerization was PEI with triphthaloyldechloride (TPC) resulted in positively charged surfaces, further incorporation of SiO_2 at different weight concentrations resulted in hydrophilic polymeric membranes. The contact angle reduces with increase in SiO_2 doping which consequently improves their fouling resistance and dye rejection rate. Maximum dye (crystal violet) uptake i.e. 100 and 99% was observed for 0.1% loading of NPs in water and 2-propanol, respectively [45]. Amino functionalized PVA/SiO_2 gel layering onto PVDF induces hydrophilic character resulting in easy wettability of the coated membranes as compared to the uncoated ones (PVA/PVDF). These coated membranes interact with Methylene blue at acidic pH as a result of electrostatic interaction with membranes' protonated primary amine groups ($-NH_3 +$) at acidic pH resulting in uptake capacity of 41.88 mg/g and high removal efficiencies (99.76%) even after 5 cycles [52]. The modification of poly(acrylic acid)/SiO_2 by triethoxyvinylsilane to enable higher uptake of dye i.e. 220 mg/g owing to formation of nanofibrous vinyl modified polymeric membrane which was found absent in unmodified membrane (Fig. 3a, b). Three mechanisms were proposed for probable interaction of polymeric membrane and dye molecule (malachite green) as presented in Fig. 3c, (1) Electrostatic interaction between silica (negatively charged $-Si-O^-$ and $-COO^-$) and malachite green (positively charged $-N^+$), (2) physical adsorption and (3) Conjugation effect between membrane (vinyl or carbonyl group) and dye's delocalized π bond.

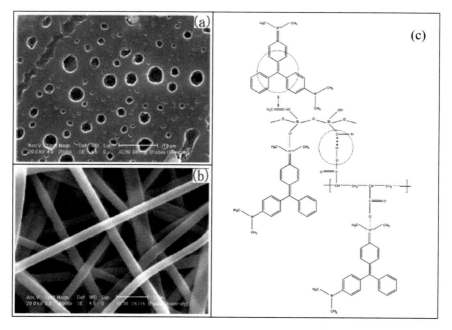

Fig. 3 SEM images of **a** unmodified and **b** vinyl modified poly(acrylic acid)/SiO_2 and **c** mechanism of interaction with modified one with malachite green [105]

Modifications of membranes through metallic impurity doping were also tried in an effort to alter the membranes properties and consequently dye rejection rate. Co-deposition of dopamine and Pd NPs onto PVA/SiO$_2$ resulted in formation of Pd decorated polydopamine-SiO$_2$/PVA membrane which showed degradation efficacy of 99% toward methyl blue and Congo red dyes in presence of NaBH$_4$ which was proposed to donate electron to Pd that were onwards transmitted to dye molecule to affect its degradation. The developed membranes show superhydrophilicity with contact angle of 0° enabling their good recyclability upto 5 cycles while maintaining same degradation potential i.e. 99% [109].

4 Clay-Based Polymeric Membranes

Clays have gained considerable attention owing to their high surface area, porous and layered structure and nanoscale sizes. Performance of the polymeric membranes can be enhanced by their combination with clay materials resulting in advanced and tailored properties of the resulting composite membranes like pore size, roughness and hydrophilicity [16, 85] which are useful for removal of dyes. There are number of inorganic clays (Table 2) that are used for the purpose after surface modifications to enhance their compatibility with organic polymer [39]. These clay-polymeric hybrid membranes (CPM) can be classified into three types: (i) Exfoliated nanocomposites, (ii) intercalated nanocomposites, (iii) micro-composites [7].

The ratio of clay nanoparticles to that of polymer, polymer nature and presence of other additives have considerable impact on the properties of membranes like pore structure, fouling potential, water permeability [8, 85]. In addition to these, polymerization approach used also affects the membrane properties as well as dye removal. Increasing the clay to monomer ratio initially results in increasing the adsorption capacity but further increase results in aggregation of clay particles into the polymeric matrix making it difficult for dye molecules to reach the ionic sites and hence reduction in dye removal [49].

Chitosan/PVA composite prepared by taking varying ratios (30:70; 50:50; 70:30) which when physically blended with 5% MMT resulted in polymeric membranes with different swelling ratio and dye adsorption. Swelling ratio increase with increase in PVA in MMT/chitosan/PVA membrane but in general addition of clay into the chitosan/PVA was observed to decrease the swelling ratio owing to intercalation of clay with polymeric structure which makes membranes more rigid. In case of dye adsorption addition of MMT to the chitosan/PVA membranes slightly reduces the uptake capacity of the hybrid membrane. Freundlich model best describes the dye adsorption by the membrane [8].

Intercalation of organically modified clay MMT with Polyacrylonitrile (PAN) resulted in nanofibrous polymeric membranes which morphology, diameter and dye degradation potential are highly dependent on MMT ratio added to the polymer. There is incremental decrease in the average diameter of pure PAN (350–375 nm) with increasing ratio of MMT i.e. 325 nm (1%), 240 nm (3%) and 138 nm (5%) was

Table 2 Clay-based polymeric membranes and their targeted pollutants

Polymer	Clay	Clay: monomer concentration	Clay modifier	Synthetic approach	Dopant	Dye	Adsorption potential/% removal	Refs.
PMMA	MMT (Kunipia F)	–	–	Simple mixing	–	Methyl violet	–	[49]
PMMA	MMT (Kunipia F)	0.5	–	Emulsion polymerization	–	Methyl violet	95%	
PMMA	MMT (PK-802)	0.6	–	Emulsion polymerization	–	Methyl violet	95%	
Polystyrene/poly ethylene-vinyl acetate	Bentonite	33%	Aniline monomer	Gamma irradiation	–	Toluidine blue, amido black, remazol red	53.30 mg/g 63.12 mg/g 84.81 mg/g	[5]
Polypropylene	Green clay	–	–	–	–	Methylene blue	0.0090 mg/g	[89]
Karaya gum	MMT	10%	–	Microwave assisted free radical polymerization	–	Methylene blue, toluidine blue, crystal violet azure B	155.85, 149.64, 137.77, 128.78 mg/g	[82]

(continued)

Table 2 (continued)

Polymer	Clay	Clay: monomer concentration	Clay modifier	Synthetic approach	Dopant	Dye	Adsorption potential/% removal	Refs.
PVA	MMT	5%	–	Physical blending	Chitosan	Reactive red	81.34 mg/g	[8]
PAN	MMT	5%	Cetyltrimethyl ammonium bromide	Electrospinning and spin coating	TiO$_2$	Methylene blue	97.2%	[100]
Polyvinylidene fluoride	Halloysite nanotubes		Dopamine	Phase inversion	–	Direct red 28; direct yellow 4; direct blue 14	86.5; 85; 93.7	[112]
Polyvinylpyrrolidone	-do-	20%	3-aminopropyltriethoxysilane	Interfacial polymerization	–	Setazol red reactive dye; reactive orange dye	99.7; 99.7%	[70]
Polyetherimide	-do-	–	3-(2-Aminoethylamino)propyl]trimethoxysilane	–	–	Methylene blue; rhodamine B	20.4; 19.6 mg/g	[39]

recorded with formation of beaded structure owing to aggregation (Fig. 4a–d). This decrease in diameter is associated with increase in surface area and porosity which consequently enhanced the dye removal with maximum uptake of 97% at 5% loading ratio. Introduction of TiO$_2$ multiple layers (3–6) by spin coating onto the PAN/MMT

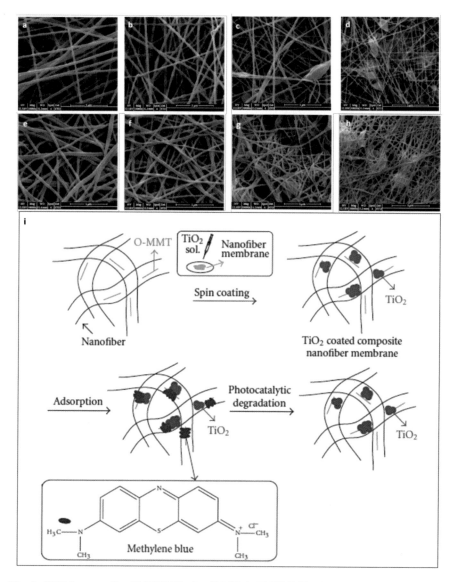

Fig. 4 SEM images of **a** PAN/MMT—0wt% MT; **b** PAN/MMT—1wt% MT; **c** PAN/MMT—3wt% MT; **d** PAN/MMT—5wt% MT; **e** PAN/MMT—0wt% MT and TiO$_2$; **f** PAN/MMT—1wt% MT and TiO$_2$; **g** PAN/MMT—3wt% MT and TiO$_2$; **h** PAN/MMT—5wt% MT and TiO$_2$ and **i** PAN/MMT/TiO$_2$ synthesis and degradation mechanism for methylene blue [100]

membranes further reduce the average diameters of membranes introducing steric hindrance (Fig. 4e–h) that resulted in delayed adsorption but no significant impact on dye adsorption was recorded (98.66% with 3 layers). On other hand presence of TiO_2 significantly enhances the dye degradation when membranes having adsorbed dyes were UV irradiated owing to formation of oxidative species which makes membranes recyclable (Fig. 4i) [100].

Halloysite, a volcanic clay material with high silica and alumina content, is also a material of choice for preparing CPM with and without modifications. Depending upon the application various modifiers are used to improve their surface properties [34]. Dopamine treatment of halloysite nanotubes considerably affected the Polyvinylidene fluoride (PVDF)-based polymeric membranes' hydrophilicity, surface roughness, porosity, contact and antifouling properties. The dye rejection was recorded to be 86.5, 85 and 93.7% for Direct Red 28, Direct Yellow 4 and Direct Blue 14, respectively. Studies showed that the membrane prepared by halloysite without dopamine modification showed comparable dyes uptake [112]. Amine modification of the halloysite with [3-(2-Aminoethylamino)propyl]trimethoxysilane resulted in polyetherimide membrane with uniform distribution of the clay. Hydrophilicity character of the hybrid membrane increases with rise in clay content leading to dye uptake capacity of 20.4 and 19.6 mg/g for methylene blue and rhodamine B, respectively [39].

Supporting thin film membrane of Polyvinylpyrrolidone into spongy polysulfone followed by intercalation with halloysite nanotubes (10, 20 and 30 wt%) resulted in membranes with thickness of 108, 122 and 228 nm, respectively and negatively charged surfaces. The dye rejection rate was dependent on temperature while pH shown to have insignificant effect. 20% Doping of nanotubes resulted in slightly enhanced uptake of dyes (Setazol red reactive dye and Reactive orange dye) as compared to undoped membrane; below and above this concentration addition to nanotubes negatively impacts the removal rate [70].

5 Hydrogels, Xerogels and Aerogels-Based Membranes

5.1 Hydrogels-Based Polymeric Membranes

Typically, hydrogel is three-dimensional matrix constituted by linear/branched hydrophilic polymers which are cross-linked either chemical or physical bonding. These are characterized by their ability to absorb water with swelling ratio greater than 100 and keeping stable network even in their swollen state [35]. Hydrogels are hydrophilic in nature which are insoluble in water, have large water content and poor mechanical strength [48]. These can be synthesized both physically and chemically and in variety of forms, including slabs, nanoparticles, microparticles, films and coatings [83]. On industrial scale following four methods are employed for the synthesis of Hydrogels: (i) ultraviolet radiation [57], (ii) cross-linking through chemical [71] and (iii) through γ radiation [3], (iv) and cross-linking agent such as glutaraldehyde [93] and (v) through freezing–thawing of successive cycles [75]. The last method is preferred because of its biocompatible nature and convenience, while the chemical bonds (hydrogen bonds) in molecules formed by freezing thawing of polymeric aqueous solutions, act as crosslinks [37].

These hydrogels are deemed highly feasible materials for dye removal from water as they possess both cationic and anionic groups [22]. In this regard natural polymer-based hydrogels have sought attention of researchers as a potential material for effective removal of both cationic and anionic dyes. Gum Sharma et al. [90], carrageenan [22], hydroxyapatite [96], chitosan [87], wheat flour [58] are few of the materials that have been exploited as source of biopolymer.

Guar gum-based biodegradable hydrogel (i.e. Guar gum-cl-poly(AA)) was modified with aniline followed by its acid doping with HCl to result in undoped and doped Guar gum-cl-poly(AA-ipn-aniline) with conducting interpenetrating networks (IPN), respectively. The studies showed increase in surface roughness in order: Guar gum-cl-poly(AA) < Undoped Guar gum-cl-poly(AA-ipn-aniline) < Doped Guar gum-cl-poly(AA-ipn-aniline) which also impacted its uptake capacity for methylene blue dye. The semi IPN (Guar gum-cl-poly(AA)) resulted in 93% dye removal while the undoped and doped depicted reduction in sorption capacity to 88 and 85%, respectively at 70 °C and pH 10 [90]. A similar gum-based hydrogel prepared by grafting of poly(acrylic acid-co-N-vinylimidazole) on katira gum was found effective for removal of both cationic dyes (i.e. Methylene blue and methyl violet) and anionic dyes (i.e. Tartrazine, Carmoisine-A). The uptake is governed by pH resulting in maximum removal of cationic dyes at 7 and that of anionic at pH 2–3 owing to ionic interaction. In case of anionic dyes-SO_3Na and imidazolic nitrogen of hydrogel interact while for cationic dyes $Y-NR_2^+$ interacts with the $-COOH$ group of hydrogels as shown in Fig. 5. The removal of dyes by katira gum hydrogels follows the order MB (331.5 mg/g) > Methyl violet (286.0 mg/g) > Carmoisine-A (273.5 mg/g) > Tartrazine (201.5 mg/g) [42]. Contrary to that pH of dye (crystal violet cationic dye) solution had almost no impact on the adsorption capacity but depicted high dependence on *Kappa*-carrageenan (KC) ratio in the prepared polymer resulting in an increase with increasing ratio, highest being recorded PVA:KC (0.6:1.4) i.e. 78.2 mg/g [54]. Considerable impact on sorption capacity of crystal violet dye (198 mg/g) was also recorded by doping of nano silver chloride into KC-based hydrogel i.e. [22].

Chitosan, another natural cationic polymer, is extensively used for the hydrogels' synthesis [2, 61]. These chitosan-based hydrogels were found effective for removal against wide range of dyes for both cationic and anionic dyes (Table 3). Adsorption potential is determined by degree of cross-linking, more the cross-linking, more rigid the structure is which drastically reduces the adsorption potential as well as swelling ratio [21] (Table 3).

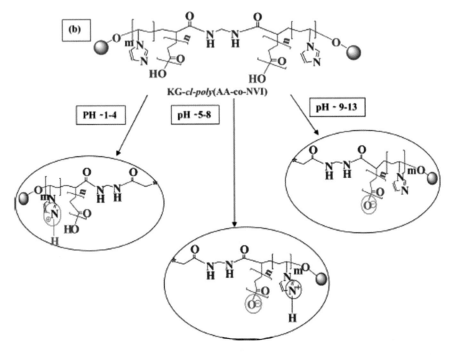

Fig. 5 Interaction of cationic and anionic dyes with katira gum-cl-poly(AA-co-NVI) hydrogel at different pHs [42]

5.2 Xerogels-Based Polymers

Xerogels, solid gel, is a porous structure produced from hydrogels through their slow drying by interchanging a liquid phase (solvent crystals) with a gaseous phase resulting in dense polymer with porous solid architecture [38, 63, 80]. Owing to their high surface area these materials have been extensively used for dye removal [38].

Carbon-based xerogels (CX) are very prominent because of the presence of conducting structures and enhancement in surface area and porosity resulting in better uptake properties [114]. In addition to that an added advantage of these is reusability making them cost effective [80]. The usual protocol of their synthesis is by organic reactants' sol-gel polymerization and polycondensation reactions. The resulting xerogel porosity and surface area are impacted to great deal by pH of initial precursor solutions. CXs were prepared from resorcinol, furfural and hexamethylenetetramine at 4 different pH ranging 6.96–9.93 followed by the carbonization at 900 °C. Xerogels hence obtained showed maximum BET area at pH 8.07 below and above that reduction was observed. Similar order was noticed for the adsorption of dyes which followed the order: Methylene blue (250 mg/g), Acid blue 40 (245 mg/g), Reactive black 5 (~ 25 mg/g). The materials prepared had the advantage of 70% regeneration [114]. Resorcinol to catalyst ratio as well as heat treatment

Table 3 Hydrogels and dye removal applications

Polymer name	Dopant	Synthetic approach	Contaminant dye	Removal capacity mg/g	References
Chitosan/poly(vinyl alcohol)	Fe_3O_4	Instantaneous gelation method	Congo red	470.1	[115]
Diammonium tartrate modified cross-linked chitosan	–	–	-do-	1597	[111]
Urea diammonium tartrate modified cross-linked chitosan	–	–	-do-	1447	
Hyper cross-linked Chitosan/poly(vinylbenzyl chloride-p-divinylbenzene)	–	Davankov procedure	Indigo Carmine Rhodamine 6G Sunset Yellow	~ 120 ~ 70 ~ 65	[87]
Cross-linked hydrogel with glutaraldehyde	–	–	Food blue 2 Food red 17	155 134	[32]
Hydrogel with activated carbon	–	–	Food blue 2 Food red 17	112 92.9	
Cross-linked hydrogel	–	–	Cobalt-tetrasulfonated phthalocyanine dye	50	[44]
Chitosan/magnetite-based composite hydrogel	Fe_3O_4		MB	23	[11]
Chitosan-poly (N-2-aminoethyl acrylamide) covalently cross-linked with glutaraldehyde			Congo red Direct blue 1	78.63×10^{-4} $79.46 \times 10^{-4}*$	[79]
Chitosan grafted poly (acrylamide-itaconic acid)			Crystal violet	170	[50]
GO–chitosan composite hydrogel	–	Tube inversion	MB Rhodamine B Methylene Orange Congo red	300 80 75 92	[99]

(continued)

Table 3 (continued)

Polymer name	Dopant	Synthetic approach	Contaminant dye	Removal capacity mg/g	References
GO–chitosan composite hydrogel	–	Hummers method	MB Eosin Y	300 320	[19]
GO–chitosan composite hydrogel	–	Hummers method	MB	402.6	[86]
Graphene oxide impregnated chitosan–PVA hydrogel			Congo red		
GO/polyethylenimine	–	Hummers method	MB Rhodamine B	334 131	[36]
GO-poly(ethylene glycol) diacrylate	–	Thiol-ene photopolymerization	MB	714	[51]
Hydrogel composite with starch, polyacrylamide, poly (2-acrylamido-2-methylpropane sulfonic acid)—GO	–	–	MB	156	[81]
GO/Chitosan–PVA hydrogel	–	–	Congo red	12.38	[23]
CS-Polyacrylicacid	TiO$_2$		MB	99%	[55]

* mmol/g; *MB* methylene blue

temperature also found to influence the BET surface area and consequently the dye uptake capacity. The former impacts the surface area more while the later has greater influence on the pore volume [29].

Activation and modification of carbon are also deemed as factors that play important role in the dye sorption capacity of the xerogels. The xerogel of resorcinol and formaldehyde modified by different ratios of KOH performed superior to those modified either with steam or H_3PO_4. The dye removal was maximized with KOH ratio (4:1) resulting in uptake capacity of 455 mg/g and 499 mg/g for Chromotrope 2R and Orange II, respectively [84]. Amine activation of xerogel obtained from hybrid material i.e. silica gel and diatomaceous earth promotes not only the scavenging activity of Eriochrome black T but also promotes its selective uptake. The adsorption takes place at acidic pH which protonates $-NH_2$ of xerogel to $-NH_3^+$ which interacts strongly with the sulphonic group of the dye molecule. On other hand other dyes (Malachite green, Chrysoidine Y and Methyl orange) having tertiary or primary amine get protonated at lower pH resulting in repelling of the dye molecules, hence no interaction between dye and xerogel and promoting of selective uptake of Eriochrome black [92]. Another hybrid material anilinepropylsilica xerogel with mesopores achieved from aniline, 3-chloropropyltrimethoxysilane and etraethylorthosilicate was found effective in pH range of 5.0 and 7.0 but resulted in poor adsorption of dye (22.6 mg/g) [77].

2-Amino-5-mercpto-1,-3,-4-thiadiazole (AMTD)-based Cu(II)-AMTD coordination polymeric xerogel having highly porous structure was obtained at 80 °C after 4 h drying (Fig. 6). The gel was checked against multiple dyes (i.e. Congo red, Acid fuchsin, naphthol green, methyl orange, thymol blue, azure I, crystal violet, methylene blue, brilliant green, and safranine T) and was effective toward sulphonic dyes only i.e. Thymol blue, Acid fuchsin and Methyl orange. The possible reason was attributed to possible interactions (hydrogen bonding and electrostatic interactions) between xerogel amino functional groups and dyes' sulfonic groups. The order of removal is as follows: Congo red [19.60 mg/g) > Acid fuchsin (16.12 mg/g) > naphthol green (15.88 mg/g) > methyl orange (14.86 mg/g) > thymol blue (10.50 mg/g)

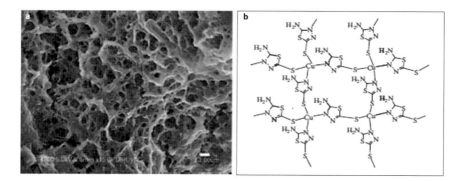

Fig. 6 Cu(II)-AMTD xerogel **a** SEM image and **b** proposed 2D structure [20]

> azure I (6.58 mg/g) > crystal violet (6.19 mg/g) > methylene blue (5.69 mg/g) > brilliant green (3.69 mg/g) > safranine T (2.60 mg/g)] [20].

5.3 Aerogel-Based Polymeric Membranes

Aerogels are sol-gels in which the internal part is occupied with gas (air). They are extensively branched, high porosity and high connectivity materials with density in range of 1–1000 kg/m^3. Almost 90–99% volume of these is occupied of air and very little solid [9, 63]. As in case of xerogels, carbon-based aerogels are most promising owing to their tunable properties and morphologies. They occur in different types i.e. CNT aerogels, graphene aerogels, nano-diamond-based aerogels, and composite of these [9].

Absorption capacity of carbon aerogels is dependent on their textural properties as well as on the surface functionalities which can be hampered by employing different carbonization temperature as well as surface modifications. Increase in carbonization temperature of carbon aerogel upto 500 °C caused decrease in adsorption capacity of aerogel irrespective of dye used. After that further increase in carbonization temperature lead to increase in dye uptake for all the four dyes (i.e. Methylene blue, rhodamine B, crystal violet, Acridine orange). The surface oxidation of these carbon aerogels by different agents enhances their oxygen level which follows the order: HNO_3 > air > H_2O_2. Higher absorption rates were noticed for HNO_3 and rest of the two had little impact on uptake capacity of gel [108]. Modification of the graphene oxide-based aerogels with cellulose acetate nanofibers resulted in selective adsorption of cationic dyes which recorded exponential increase in adsorption capacity i.e. 825 mg/g (Neutral red), 750 mg/g (methylene blue), 700 mg/g (methylene green), 650 mg/g (rhodamine B), 600 mg/g (crystal violet) while for anionic dyes (indigo carmine) no adsorption was observed. The phenomenon was attributed to negative charge on aerogel which interacts strongly with the cationic dyes [103]. A green approach was also tried for the synthesis of carbon aerogels using starch as raw materials. The resulting aerogel has both meso and micropores which facilitate the uptake of cationic dyes i.e. crystal violet (1515 mg/g), methyl violet (1423 mg/g) and methylene blue (1181 mg/g) [18].

An aerogel designed by using hydrothermal route i.e. Graphene/PVA aerogel was found effective in removing cationic, anionic and non-ionic dyes (i.e. Neutral red, rhodamine B, crystal violet, methylene blue, methylene green, indigo carmine, methyl orange, eosin Y, amido black 10B, oil 3 red O, acid fuchsin) owing to synergistic effects. In binary dye system, the adsorption capacity for dye reached > 900 mg/g. The gels were found solvent resistant and hence can be easily cleaned up resulting in 5 adsorption–desorption cycles with dye removal capacity staying around 75 and 67 mg/g for Neutral Red and Indigo carmine [104]. Effective separation of the adsorbents was facilitated by using Fe_3O_4 doping. Magnetic bacterial cellulose nanofiber/graphene oxide PVA aerogel prepared by freeze drying route showed dye (Malachite green) sorption percentage of 270 mg/g. The adsorption was facilitated by 3D interconnected porous network which disappears after dye adsorption and several interactions between the adsorbent and dye molecule as explained in Fig. 7. The adsorbent had the advantage of maintaining the adsorption capacity around 62.7% up to seven cycles but desorption solvents impact to little extant [10]. Effective separation of adsorbent was also facilitated by its foamy nature. The aerogel (mesoporous SnO_2 aerogel polystyrene composite) formed by use of emulsion polymerization in conjunction with epoxide-assisted gelation facilitated the formation of dense aerogel with mesopores occupied by the Sn. Degradation of methylene blue dye was facilitated by electron-hole pair and interfacial charge transfer promoted by Sn resulting increased degradation rate [74].

A GO-based aerogel polymer achieved by varying ratios of GO, chitosan and lignosulfonate led to formation of composite aerogel consisting of interconnected 3D porous structure. The uptake capacity of aerogels was observed to increase with increase in chitosan ratio as well as increase in pH. The surface of aerogel was equipped with carboxyl, hydroxyl and sulfonic acid groups which at lower pH get protonated and hence repel the cationic dye while with increase in pH the surface becomes negatively charged leading to electrostatic interaction between dye and aerogel and hence better adsorption [106].

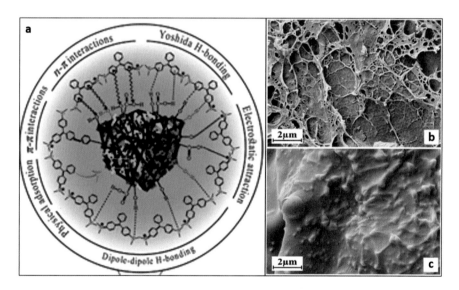

Fig. 7 **a** Probable interactions between dye and aerogel and SEM image of aerogel, **b** before and **c** after dye adsorption [10]

6 Conclusion

Polymeric membranes have undergone series of advancements over the year in essence to improve their properties, dye rejection, degradation, antifouling potential selectivity and cost-effectiveness. In this regard polymeric hybrid materials have offered a way forward with its tunable properties that enable the removal of dyes and in some cases its degradation as well that enhance their re-use. Incorporation of metal/metal oxides, graphene oxide and use of natural polymers are one of the few promising areas that have much more potential and depth to open new horizons of research. The researches done have limitation in aspect that much of the work presented is related to efficacy of developed membranes in simulated samples; as the situation is much more rigorous and different in actual applied setups the effectiveness of these membranes remains debatable.

References

1. Ahmad A, Puasa S, Abiding S (2006) Crossflow ultrafiltration for removing direct-15 dye from wastewater of textile industry. ASEAN J Sci Technol Dev 23(3):207–216
2. Ahmadi F, Oveisi Z, Samani SM, Amoozgar Z (2015) Chitosan based hydrogels: characteristics and pharmaceutical applications. Res Pharm Sci 10(1):1–16
3. Ajji Z (2005) Preparation of poly(vinyl alcohol) hydrogels containing citric or succinic acid using gamma radiation. Radiat Phys Chem 74(1):36–41
4. Al-Ani DM, Al-Ani FH, Alsalhy QF, Ibrahim SS (2021) Preparation and characterization of ultrafiltration membranes from PPSU-PES polymer blend for dye removal. Chem Eng Commun 208(1):41–59
5. Alshangiti DM, Ghobashy MM, Alkhursani SA, Shokr FS, Al-Gahtany SA, Madani MM (2019) Semi-permeable membrane fabricated from organoclay/PS/EVA irradiated by γ-rays for water purification from dyes. J Mater Res Technol 8(6):6134–6145
6. Alzahrani E (2018) Chitosan membrane embedded with ZnO/CuO nanocomposites for the photodegradation of fast green dye under artificial and solar irradiation. Anal Chem Insights 13:1177390118763361
7. Amari A, Alzahrani FM, Mohammedsaleh Katubi K, Alsaiari NS, Tahoon MA, Rebah FB (2021) Clay-polymer nanocomposites: Preparations and utilization for pollutants removal. Materials 14(6):1365
8. Amri N, Radji S, Ghemati D, Djamel A (2019) Studies on equilibrium swelling, dye adsorption, and dynamic shear rheology of polymer systems based on chitosan-poly (vinyl alcohol) and montmorillonite. Chem Eng Commun 206(6):716–730
9. Ansari MO, Kumar R, Pervez Ansari S, Abdel-Wahab Hassan MS, Alshahrie A, Barakat MAE-F (2019) Nanocarbon aerogel composites. Nanocarbon and its composites. In: Khan A, Jawaid M, Inamuddin, Asiri AM (eds) Woodhead Publishing, pp 1–26
10. Arabkhani P, Asfaram A (2020) Development of a novel three-dimensional magnetic polymer aerogel as an efficient adsorbent for malachite green removal. J Hazard Mater 384:121394
11. Ashok Kumar S, Srinivasan G, Govindaradjane S (2019) Development of a new blended polyethersulfone membrane for dye removal from synthetic wastewater. Environ Nanotechnol Monit Manage **12**:100238
12. Bahamonde Soria R, Zhu J, Gonza I, Van Der Bruggen B, Luis P (2020) Effect of (TiO$_2$:ZnO) ratio on the anti-fouling properties of bio-inspired nanofiltration membranes. Sep Purif Technol 251:117280

13. Baker RW (2012) Membrane technology and applications. Wiley
14. Balta S, Sotto A, Luis P, Benea L, Van Der Bruggen B, Kim J (2012) A new outlook on membrane enhancement with nanoparticles: the alternative of ZnO. J Membr Sci 389:155–161
15. Bet-Moushoul E, Mansourpanah Y, Farhadi K, Tabatabaei M (2016) TiO2 nanocomposite based polymeric membranes: a review on performance improvement for various applications in chemical engineering processes. Chem Eng J 283:29–46
16. Buruga K, Song H, Shang J, Bolan N, Jagannathan TK and Kim K-H (2019) A review on functional polymer-clay based nanocomposite membranes for treatment of water. J Hazard Mater 379:120584
17. Căprărescu S, Modrogan C, Purcar V, Dăncilă AM and Orbuleț OD (2021) Study of polyvinyl alcohol-SiO$_2$ nanoparticles polymeric membrane in wastewater treatment containing zinc ions. Polymers 13(11):1875
18. Chang X, Chen D, Jiao X (2010) Starch-derived carbon aerogels with high-performance for sorption of cationic dyes. Polymer 51(16):3801–3807
19. Chen Y, Chen L, Bai H, Li L (2013) Graphene oxide–chitosan composite hydrogels as broad-spectrum adsorbents for water purification. J Mater Chem A 1(6):1992–2001
20. Cheng Y, Feng Q, Ren X, Yin M, Zhou Y, Xue Z (2015) Adsorption and removal of sulfonic dyes from aqueous solution onto a coordination polymeric xerogel with amino groups. Colloids Surf A Physicochem Eng Asp 485:125–135
21. Crini G, Torri G, Lichtfouse E, Kyzas GZ, Wilson LD, Morin-Crini N (2019) Dye removal by biosorption using cross-linked chitosan-based hydrogels. Environ Chem Lett 17(4):1645–1666
22. Dargahi M, Ghasemzadeh H, Bakhtiary A (2018) Highly efficient absorption of cationic dyes by nano composite hydrogels based on κ-carrageenan and nano silver chloride. Carbohydr Polym 181:587–595
23. Das L, Das P, Bhowal A, Bhattacharjee C (2020) Synthesis of hybrid hydrogel nano-polymer composite using graphene oxide, chitosan and PVA and its application in waste water treatment. Environ Technol Innov 18:100664
24. Dassi RB, Chamam B, Méricq J, Faur C, El Mir L, Trabelsi I, Heran M (2020) Novel polyvinylidene fluoride/lead-doped zinc oxide adsorptive membranes for enhancement of the removal of reactive textile dye. Int J Sci Environ Technol:1–12
25. Dossin Zanrosso C, Piazza D, Lansarin MA (2020) Solution mixing preparation of PVDF/ZnO polymeric composite films engineered for heterogeneous photocatalysis. J Appl Polym Sci 137(9):48417
26. Farrell S, Sirkar K (1997) A reservoir-type controlled release device using aqueous-organic partitioning and a porous membrane. J Membr Sci 130(1–2):265–274
27. Farrell S, Sirkar KK (1999) A mathematical model of an aqueous-organic partition-based controlled release system using microporous membranes. J Control Release 61(3):345–360
28. Ghadhban MY, Majdi HS, Rashid KT, Alsalhy QF, Lakshmi DS, Salih IK, Figoli A (2020) Removal of dye from a leather tanning factory by flat-sheet blend ultrafiltration (UF) membrane. Membranes 10(3):47
29. Girgis BS, Attia AA, Fathy NA (2011) Potential of nano-carbon xerogels in the remediation of dye-contaminated water discharges. Desalination 265(1):169–176
30. Giwa A, Ahmed M and Hasan SW (2019) Polymers for membrane filtration in water purification.polymeric materials for clean water. Springer International Publishing, Cham, pp 167–190
31. Gohil J, Choudhury R (2019) Introduction to nanostructured and nano-enhanced polymeric membranes: preparation, function, and application for water purification:25–57
32. Gonçalves JO, Santos JP, Rios EC, Crispim MM, Dotto GL, LaA P (2017) Development of chitosan based hybrid hydrogels for dyes removal from aqueous binary system. J Mol Liq 225:265–270
33. Goswami R, Gogoi M, Borah HJ, Ingole PG, Hazarika S (2018) Biogenic synthesized Pd-nanoparticle incorporated antifouling polymeric membrane for removal of crystal violet dye. J Environ Chem Eng 6(5):6139–6146

34. Grylewicz A, Mozia S (2021) Polymeric mixed-matrix membranes modified with halloysite nanotubes for water and wastewater treatment: a review. Sep Purif Technol 256:117827
35. Guilherme MR, Aouada FA, Fajardo AR, Martins AF, Paulino AT, Davi MF, Rubira AF, Muniz EC (2015) Superabsorbent hydrogels based on polysaccharides for application in agriculture as soil conditioner and nutrient carrier: a review. Eur Polym J 72:365–385
36. Guo H, Jiao T, Zhang Q, Guo W, Peng Q, Yan X (2015) Preparation of graphene oxide-based hydrogels as efficient dye adsorbents for wastewater treatment. Nanoscale Res Lett 10(1):272
37. Hago E-E, Li X (2013) Interpenetrating polymer network hydrogels based on gelatin and PVA by biocompatible approaches: synthesis and characterization. Adv Mater Sci Eng 2013
38. Haye E, Job N, Wang Y, Penninckx S, Stergiopoulos V, Tumanov N, Cardinal M, Busby Y, Colomer J-F, Su B-L (2020) ZnO/carbon xerogel photocatalysts by low-pressure plasma treatment, the role of the carbon substrate and its plasma functionalization. J Colloid Interface Sci 570:312–321
39. Hebbar RS, Isloor AM, Inamuddin AMS, Ismail AF, Asiri AM (2018) Fabrication of polyetherimide nanocomposite membrane with amine functionalised halloysite nanotubes for effective removal of cationic dye effluents. J Taiwan Inst Chem Eng 93:42–53
40. Hir ZAM, Moradihamedani P, Abdullah AH, Mohamed MA (2017) Immobilization of TiO_2 into polyethersulfone matrix as hybrid film photocatalyst for effective degradation of methyl orange dye. Mater Sci Semicond Process 57:157–165
41. Hosseini SA, Vossoughi M, Mahmoodi NM, Sadrzadeh M (2018) Efficient dye removal from aqueous solution by high-performance electrospun nanofibrous membranes through incorporation of SiO_2 nanoparticles. J Clean Prod 183:1197–1206
42. Jana S, Ray J, Mondal B, Pradhan S, Tripathy T (2018) pH responsive adsorption/desorption studies of organic dyes from their aqueous solutions by katira gum-cl-poly (acrylic acid-co-N-vinyl imidazole) hydrogel. Colloids Surf A Physicochem Eng Asp 553:472–486
43. Kamari S, Shahbazi A (2020) Biocompatible Fe_3O_4@SiO_2-NH_2 nanocomposite as a green nanofiller embedded in PES–nanofiltration membrane matrix for salts, heavy metal ion and dye removal: long–term operation and reusability tests. Chemosphere 243:125282
44. Karimi AR, Rostaminezhad B, Khodadadi A (2018) Effective removal of a cobalt-tetrasulfonated phthalocyanine dye from an aqueous solution with a novel modified chitosan-based superabsorbent hydrogel. J Appl Polym Sci 135(16):46167
45. Kebria MRS, Jahanshahi M, Rahimpour A (2015) SiO_2 modified polyethyleneimine-based nanofiltration membranes for dye removal from aqueous and organic solutions. Desalination 367:255–264
46. Khumalo NP, Vilakati GD, Mhlanga SD, Kuvarega AT, Mamba BB, Li J, Dlamini DS (2019) Dual-functional ultrafiltration nano-enabled PSf/PVA membrane for the removal of Congo red dye. J Water Process Eng 31:100878
47. Kuvarega AT, Khumalo N, Dlamini D, Mamba BB (2018) Polysulfone/N, Pd co-doped TiO_2 composite membranes for photocatalytic dye degradation. Sep Purif Technol 191:122–133
48. Lee SJ, Kim SS, Lee YM (2000) Interpenetrating polymer network hydrogels based on poly (ethylene glycol) macromer and chitosan. Carbohydr Polym 41(2):197–205
49. Lin R-Y, Chen B-S, Chen G-L, Wu J-Y, Chiu H-C, Suen S-Y (2009) Preparation of porous PMMA/Na^+–montmorillonite cation-exchange membranes for cationic dye adsorption. J Membr Sci 326(1):117–129
50. Liu B, Zheng H, Wang Y, Chen X, Zhao C, An Y, Tang X (2018) A novel carboxyl-rich chitosan-based polymer and its application for clay flocculation and cationic dye removal. Sci Total Environ 640–641:107–115
51. Liu C, Mao H, Zheng J, Zhang S (2017a) Tight ultrafiltration membrane: Preparation and characterization of thermally resistant carboxylated cardo poly (arylene ether ketone) s (PAEK-COOH) tight ultrafiltration membrane for dye removal. J Membr Sci 530:1–10
52. Liu H, Yu H, Yuan X, Ding W, Li Y, Wang J (2019) Amino-functionalized mesoporous PVA/SiO_2 hybrids coated membrane for simultaneous removal of oils and water-soluble contaminants from emulsion. Chem Eng J 374:1394–1402

53. Liu J, Zhu K, Jiao T, Xing R, Hong W, Zhang L, Zhang Q, Peng Q (2017b) Preparation of graphene oxide-polymer composite hydrogels via thiol-ene photopolymerization as efficient dye adsorbents for wastewater treatment. Colloids Surf A Physicochem Eng Asp 529:668–676

54. Mahdavinia GR, Massoudi A, Baghban A, Shokri E (2014) Study of adsorption of cationic dye on magnetic kappa-carrageenan/PVA nanocomposite hydrogels. J Environ Chem Eng 2(3):1578–1587

55. Mahmoud GA, Sayed A, Thabit M, Safwat G (2020) Chitosan biopolymer based nanocomposite hydrogels for removal of methylene blue dye. SN Appl Sci 2(5):968

56. Mansor ES, Abdallah H, Shaban A (2020) Fabrication of high selectivity blend membranes based on poly vinyl alcohol for crystal violet dye removal. J Environ Chem Eng 8(3):103706

57. Martens P, Anseth K (2000) Characterization of hydrogels formed from acrylate modified poly (vinyl alcohol) macromers. Polymer 41(21):7715–7722

58. Md. Munjur H, Hasan MN, Awual MR, Islam MM, Shenashen MA, Iqbal J (2020) Biodegradable natural carbohydrate polymeric sustainable adsorbents for efficient toxic dye removal from wastewater. J Mol Liq 319:114356

59. Melvin Ng HK, Leo CP, Abdullah AZ (2017) Selective removal of dyes by molecular imprinted TiO₂ nanoparticles in polysulfone ultrafiltration membrane. J Environ Chem Eng 5(4):3991–3998

60. Mofradi M, Karimi H, Ghaedi M (2020) Hydrophilic polymeric membrane supported on silver nanoparticle surface decorated polyester textile: toward enhancement of water flux and dye removal. Chin J Chem Eng 28(3):901–912

61. Mohammadzadeh Pakdel P, Peighambardoust SJ (2018) Review on recent progress in chitosan-based hydrogels for wastewater treatment application. Carbohydr Polym 201:264–279

62. Mohd Hir ZA, Abdullah AH, Zainal Z, Lim HN (2017) Photoactive hybrid film photocatalyst of polyethersulfone-ZnO for the degradation of methyl orange dye: kinetic study and operational parameters. Catalysts 7(11):313

63. Nayak AK, Das B (2018) Introduction to polymeric gels. Polymeric gels. In: Pal K, Banerjee I (eds) Woodhead Publishing, pp 3–27

64. Neves JC, Mohallem NDS, Viana MM (2020) Polydimethylsiloxanes-modified TiO₂ coatings: the role of structural, morphological and optical characteristics in a self-cleaning surface. Ceram Int 46(8, Part B):11606–11616

65. Ng LY, Mohammad AW, Leo CP, Hilal N (2013) Polymeric membranes incorporated with metal/metal oxide nanoparticles: a comprehensive review. Desalination 308:15–33

66. Ngang HP, Ooi BS, Ahmad AL, Lai SO (2012) Preparation of PVDF–TiO₂ mixed-matrix membrane and its evaluation on dye adsorption and UV-cleaning properties. Chem Eng J 197:359–367

67. Nidheesh P, Zhou M, Oturan MA (2018) An overview on the removal of synthetic dyes from water by electrochemical advanced oxidation processes. Chemosphere 197:210–227

68. Nikooe N, Saljoughi E (2017) Preparation and characterization of novel PVDF nanofiltration membranes with hydrophilic property for filtration of dye aqueous solution. Appl Surf Sci 413:41–49

69. Opoku F, Kiarii EM, Govender PP and Mamo MA (2017) Metal oxide polymer nanocomposites in water treatments. Descr Inorganic Chem Res Metal Compounds:173–199

70. Ormanci-Acar T, Celebi F, Keskin B, Mutlu-Salmanlı O, Agtas M, Turken T, Tufani A, Imer DY, Ince GO, Demir TU (2018) Fabrication and characterization of temperature and pH resistant thin film nanocomposite membranes embedded with halloysite nanotubes for dye rejection. Desalination 429:20–32

71. Ossipov DA, Hilborn J (2006) Poly (vinyl alcohol)-based hydrogels formed by "click chemistry." Macromolecules 39(5):1709–1718

72. Pan B, Zhang X, Jiang Z, Li Z, Zhang Q, Chen J (2019) Polymer and polymer-based nanocomposite adsorbents for water treatment. Polym Mater Clean Water:93–119

73. Pang YL, Abdullah AZ (2013) Current status of textile industry wastewater management and research progress in Malaysia: a review. Clean-Soil Air Water 41(8):751–764

74. Parale VG, Kim T, Phadtare VD, Han W, Lee K-Y, Jung H-N-R, Choi H, Kim Y, Yadav HM, Park H-H (2019) SnO$_2$ aerogel deposited onto polymer-derived carbon foam for environmental remediation. J Mol Liq 287:110990

75. Park SE, Nho YC, Lim YM, Kim HI (2004) Preparation of pH-sensitive poly (vinyl alcohol-g-methacrylic acid) and poly (vinyl alcohol-g-acrylic acid) hydrogels by gamma ray irradiation and their insulin release behavior. J Appl Polym Sci 91(1):636–643

76. Pascariu P, Cojocaru C, Samoila P, Olaru N, Bele A, Airinei A (2021) Novel electrospun membranes based on PVDF fibers embedding lanthanide doped ZnO for adsorption and photocatalytic degradation of dye organic pollutants. Mater Res Bull 141:111376

77. Pavan FA, Dias SLP, Lima EC, Benvenutti EV (2008) Removal of Congo red from aqueous solution by anilinepropylsilica xerogel. Dyes Pigm 76(1):64–69

78. Pereira VR, Isloor AM, Zulhairun A, Subramaniam M, Lau W, Ismail A (2016) Preparation of polysulfone-based PANI–TiO$_2$ nanocomposite hollow fiber membranes for industrial dye rejection applications. RSC Adv 6(102):99764–99773

79. Perju M, Dragan E (2010) Removal of azo dyes from aqueous solutions using chitosan based composite hydrogels. Ion Exchange Lett 3:7–11

80. Pillai A, Kandasubramanian B (2020) Carbon xerogels for effluent treatment. J Chem Eng Data 65(5):2255–2270

81. Pourjavadi A, Nazari M, Kabiri B, Hosseini SH, Bennett C (2016) Preparation of porous graphene oxide/hydrogel nanocomposites and their ability for efficient adsorption of methylene blue. RSC Adv 6(13):10430–10437

82. Preetha BK, Vishalakshi B (2020) Microwave assisted synthesis of karaya gum based montmorillonite nanocomposite: characterisation, swelling and dye adsorption studies. Int J Biol Macromol 154:739–750

83. Ray M, Pal K, Anis A, Banthia A (2010) Development and characterization of chitosan-based polymeric hydrogel membranes. Des Monomers Polym 13(3):193–206

84. Ribeiro RS, Fathy NA, Attia AA, Silva AMT, Faria JL, Gomes HT (2012) Activated carbon xerogels for the removal of the anionic azo dyes orange II and chromotrope 2R by adsorption and catalytic wet peroxide oxidation. Chem Eng J 195–196:112–121

85. Rodrigues R, Morihama ACD, Barbosa IM, Leocã GN (2018) Clay nanoparticles composite membranes prepared with three different polymers: Performance evaluation. J Memb Separ Tech 7:1–11

86. Sabzevari M, Cree DE, Wilson LD (2018) Graphene oxide-chitosan composite material for treatment of a model dye effluent. ACS Omega 3(10):13045–13054

87. Salzano De Luna M, Castaldo R, Altobelli R, Gioiella L, Filippone G, Gentile G, Ambrogi V (2017) Chitosan hydrogels embedding hyper-crosslinked polymer particles as reusable broad-spectrum adsorbents for dye removal. Carbohydr Polym 177:347–354

88. Sawada I, Fachrul R, Ito T, Ohmukai Y, Maruyama T, Matsuyama H (2012) Development of a hydrophilic polymer membrane containing silver nanoparticles with both organic antifouling and antibacterial properties. J Membr Sci 387–388:1–6

89. Şen F, Demirbaş Ö, Çalımlı MH, Aygün A, Alma MH, Nas MS (2018) The dye removal from aqueous solution using polymer composite films. Appl Water Sci 8(7):1–9

90. Sharma R, Kaith BS, Kalia S, Pathania D, Kumar A, Sharma N, Street RM, Schauer C (2015) Biodegradable and conducting hydrogels based on Guar gum polysaccharide for antibacterial and dye removal applications. J Environ Manage 162:37–45

91. Sheikh M, Pazirofteh M, Dehghani M, Asghari M, Rezakazemi M, Valderrama C, Cortina J-L (2020) Application of ZnO nanostructures in ceramic and polymeric membranes for water and wastewater technologies: a review. Chem Eng J 391:123475

92. Sriram G, Bhat MP, Kigga M, Uthappa U, Jung H-Y, Kumeria T, Kurkuri MD (2019) Amine activated diatom xerogel hybrid material for efficient removal of hazardous dye. Mater Chem Phys 235:121738

93. Stevens KR, Einerson NJ, Burmania JA, Kao WJ (2002) In vivo biocompatibility of gelatin-based hydrogels and interpenetrating networks. J Biomater Sci Polym Ed 13(12):1353–1366

94. Stewart KK, Craig LC (1970) Thin-film dialysis studies with highly acetylated cellophane membranes. Anal Chem 42(11):1257–1260

95. Tahiri Alaoui O, Nguyen QT, Mbareck C, Rhlalou T (2009) Elaboration and study of poly(vinylidene fluoride)–anatase TiO$_2$ composite membranes in photocatalytic degradation of dyes. Appl Catal A 358(1):13–20

96. Varaprasad K, Nunez D, Yallapu MM, Jayaramudu T, Elgueta E, Oyarzun P (2018) Nano-hydroxyapatite polymeric hydrogels for dye removal. RSC Adv 8(32):18118–18127

97. Vatanpour V, Nekouhi GN, Esmaeili M (2020) Preparation, characterization and performance evaluation of ZnO deposited polyethylene ultrafiltration membranes for dye and protein separation. J Taiwan Inst Chem Eng 114:153–167

98. Vatanpour V, Shahsavarifar S, Khorshidi S, Masteri-Farahani M (2019) A novel antifouling ultrafiltration membranes prepared from percarboxylic acid functionalized SiO$_2$ bound Fe$_3$O$_4$ nanoparticle (SCMNP-COOOH)/polyethersulfone nanocomposite for BSA separation and dye removal. J Chem Technol Biotechnol 94(4):1341–1353

99. Vo TS, Vo TTBC, Suk JW, Kim K (2020) Recycling performance of graphene oxide-chitosan hybrid hydrogels for removal of cationic and anionic dyes. Nano Convergence 7(1):1–11

100. Wang Q, Gao D, Gao C, Wei Q, Cai Y, Xu J, Liu X, Xu Y (2012) Removal of a cationic dye by adsorption/photodegradation using electrospun PAN/O-MMT composite nanofibrous membranes coated with. Int J Photoenergy 2012

101. Wang Y, Dai L, Qu K, Qin L, Zhuang L, Yang H, Xu Z (2021) Novel Ag-AgBr decorated composite membrane for dye rejection and photodegradation under visible light. Front Chem Sci Eng 15(4):892–901

102. Wawrzkiewicz M, Hubicki Z (2015) Anion exchange resins as effective sorbents for removal of acid, reactive, and direct dyes from textile wastewaters. Ion Exchange-Stud Appl:37–72

103. Xiao J, Lv W, Song Y, Zheng Q (2018) Graphene/nanofiber aerogels: performance regulation towards multiple applications in dye adsorption and oil/water separation. Chem Eng J 338:202–210

104. Xiao J, Zhang J, Lv W, Song Y, Zheng Q (2017) Multifunctional graphene/poly(vinyl alcohol) aerogels: In situ hydrothermal preparation and applications in broad-spectrum adsorption for dyes and oils. Carbon 123:354–363

105. Xu R, Jia M, Zhang Y, Li F (2012) Sorption of malachite green on vinyl-modified meso-porous poly(acrylic acid)/SiO$_2$ composite nanofiber membranes. Micropor Mesopor Mat 149(1):111–118

106. Yan M, Huang W, Li Z (2019) Chitosan cross-linked graphene oxide/lignosulfonate composite aerogel for enhanced adsorption of methylene blue in water. Int J Biol Macromol 136:927–935

107. Yang C, Xu W, Nan Y, Wang Y, Hu Y, Gao C, Chen X (2020) Fabrication and characterization of a high performance polyimide ultrafiltration membrane for dye removal. J Colloid Interface Sci 562:589–597

108. Yang W, Wu D, Fu R (2008) Effect of surface chemistry on the adsorption of basic dyes on carbon aerogels. Colloids Surf A Physicochem Eng Asp 312(2):118–124

109. Yin H, Zhao J, Li Y, Huang L, Zhang H and Chen L (2020) A novel Pd decorated polydopamine-SiO$_2$/PVA electrospun nanofiber membrane for highly efficient degradation of organic dyes and removal of organic chemicals and oils. J Clean Prod 275:122937

110. Yin J, Deng B (2015) Polymer-matrix nanocomposite membranes for water treatment. J Membr Sci 479:256–275

111. Zahir A, Aslam Z, Kamal MS, Ahmad W, Abbas A, Shawabkeh RA (2017) Development of novel cross-linked chitosan for the removal of anionic Congo red dye. J Mol Liq 244:211–218

112. Zeng G, Ye Z, He Y, Yang X, Ma J, Shi H, Feng Z (2017) Application of dopamine-modified halloysite nanotubes/PVDF blend membranes for direct dyes removal from wastewater. Chem Eng J 323:572–583

113. Zhang L, He Y, Luo P, Ma L, Fan Y, Zhang S, Shi H, Li S, Nie Y (2020) A heterostructured PPy/ZnO layer assembled on a PAN nanofibrous membrane with robust visible-light-induced self-cleaning properties for highly efficient water purification with fast separation flux. J Mater Chem A 8(8):4483–4493

114. Zhou G, Tian H, Sun H, Wang S, Buckley CE (2011) Synthesis of carbon xerogels at varying sol–gel pHs, dye adsorption and chemical regeneration. Chem Eng J 171(3):1399–1405
115. Zhu H-Y, Fu Y-Q, Jiang R, Yao J, Xiao L, Zeng G-M (2012) Novel magnetic chitosan/poly(vinyl alcohol) hydrogel beads: preparation, characterization and application for adsorption of dye from aqueous solution. Bioresour Technol 105:24–30

Polymer Membrane in Textile Wastewater

Kanchna Bhatrola, Sameer Kumar Maurya, N. C. Kothiyal, and Vaneet Kumar

Abbreviations

CA	Cellulose acetate
CMC	Critical micelle concentration
COD	Chemical Oxygen Demand
DMF	Dimethylformamide
IP	Interfacial polymerization
MEUF	Micellar-enhanced ultrafiltration
NF	Nanofiltration
PA	Polyamide
PC	Polycarbonate
PE	Polyethylene
PET	Polyethylene naphthalate
PP	Polypropylene
PS	Polystyrene
PTFE	Polytetrafluoroethylene
RO	Reverse osmosis
TDS	Total dissolved solids
TFC	Thin-film composite
THF	Tetrahydrofuran
TMC	Trimesoyl chloride

K. Bhatrola (✉) · S. K. Maurya · N. C. Kothiyal
Department of Chemistry, Dr. B. R. Ambedkar National Institute of Technology Jalandhar, GT Road, Amritsar Bypass, Jalandhar, Punjab 144011, India
e-mail: bhatrolakanchan14@gmail.com

V. Kumar
Department of Applied Sciences, CTIEMT, CT Group of Institutions Jalandhar, Jalandhar, Punjab, India

UF   Ultrafiltration
VOCs   Volatile organic compounds

1 Introduction

Water is an essential and vital feature of our life because it covers around 71% of the earth's surface. In additional total amount of water contain 97.5% is saline water and approx. 2.5% is freshwater, hardly 0.007% drinking water is found [1]. The industries of wastewater such as textile industries produce different kinds of pollutants (i.e., inorganic and organic). In this industrialization era, the ejection of waste from textile industries can cause the serious problems in future [2]. Historically, textile industries used vast amounts of water and released huge- amounts of wastewater, that contained high salts and dyes which is an environmental concern. Dyes are non-biodegradable and toxic in nature causing major damage to human health and aquatic ecosystems. On the basis of TWIC 2016 database (Bangladesh PaCT (Partnership for Cleaner Textiles), "Feasibility Study for Setting Up CETP in the Konabari Cluster", TWIC Report, 2016) the textile effluents contain various types of chemicals as depicted in Fig. 1. Basically, the crisis of fresh water demand caused by textile industrial activities, along with that pollution and supplementary factors has become one of the soberest global troubles threatening human existence and improvement [3]. As a result of their complicated, stable chemical structures, hazardous nature, non-biodegradable, and high molecular weight of most of the effluents, the chemicals discharged cause major environmental issues, as shown in Table 1. Textile wastewater management has gotten a lot of attention from the previous year [4].

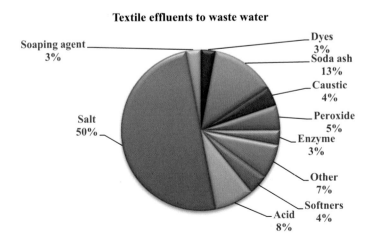

Fig. 1 Percentage division of effluents in textile wastewater

Table 1 Illustrated critical components of wastewater in textile process [5]

S. No.	Aspect/component	Range
1	pH	2–10
2	BOD (mg/L)	200–300
3	Color (mg/L)	> 300
4	COD (mg/L)	50–5000
5	Organic nitrogen (mg/L)	18–39
6	TSS (mg/L)	50–500
7	Total chromium (mg/L)	0.2–0.5
8	Total phosphorus (mg/L)	0.3–15
9	Temperature (°C)	30–80

Furthermore, colors are viewed as a potentially harmful earth poison, with even a small concentration (less than one part per million) having a significant impact on water bodies' trendy authenticity, straightforwardness, and gas dissolvability [6].

Textile manufacturers have thus transformed the conventional wastefulness pollution management mechanism into a profitable activity by recycling waste effluent [7]. This activity allows reusable chemicals and water to be recovered from a variety of processing sources in textile industry [4].

Various efforts have been taken over time to introduce textile waste water treatment solutions, such as coagulation-flocculation, biological treatment systems, and traditional filtering systems, among others. Existing technologies are also there to meet current reuse standards and discharge effluents. One of the textile waste water treatment technologies that have come in scenario of foremost advanced technology is term as membrane technology. The technology of membrane has developed considerably nowadays. On the basis of its nature it provides a better treatment of textile wastewater. Because of the lower energy usage, lower startup costs, and smaller equipment footprint, membrane technology has great promise for treating textile effluent [8].

The technology of membrane has been confirmed to be most enthusiastic choice in wastewater (textile) treatment processes in most modern. Although membrane technology is not a recent creation, untrustworthy behavior and there's still space for change when it comes to textile wastewater, in terms of requirements, efficiency, energy, technical skills, and quality of permeate requirements. After that, membrane components and membrane modules are established to enhance membrane reduction which is a significant task for these processes. Water treatment facilities are continuously looking for new ways to combine two or more membrane processes, such as adsorption and/or coagulation, in a hybrid fashion [9].

Membranes, in general, act as a barrier between two phases, restricting component mobility through it in a selective fashion [10]. The period of eighteenth century was known for existence of membranes. Since then, numerous advancements made for more favorable membranes in order to attain their several applications [11]. Based on their properties the membranes can be categorized as anisotropic and isotropic.

Isotropic membranes are homogeneous in physical structure and their chemical composition. They are microporous, which also means they have high permeation fluxes, or nonporous (dense), which means its applicability is greatly limited due to having lesser permeation fluxes.

In microfiltration membranes, an isotropic membrane is commonly applied. Similarly, the anisotropic membranes are uneven across the whole membrane surface and it composed of many layers with different chemical compositions and structures. Selected thin layer is covered with a thicker, highly spongy layer in these membranes. Which are predominantly useful in RO (reverse osmosis) systems [12]. According to their material nature, membranes are commonly classified as inorganic and organic. Nowadays, the synthetic organic polymers are prepared by organic membrane.

Mostly, the parting process of membranes is based on pressure (nanofiltration, microfiltration, ultrafiltration, and reverse osmosis) is almost exclusively made of organic polymers (synthetic). Cellulose acetate, Polyethylene (PE), polypropylene, and polytetrafluoroethylene (PTFE), are few examples of zeolites, metals, silica, and ceramics, are all examples of inorganic membrane materials [13]. Thermally and Chemically stable are commonly used in various industries such as separation of hydrogen, ultrafiltration, and microfiltration [14].

Polymeric membrane is widely used as a treatment method because of the ineffectiveness of traditional treatment systems. Polymeric membrane has been described in properties which are Ultrafiltration (UF), Nanofiltration (NF), and Reverse Osmosis (RO). Which provides major benefits in the form of lesser osmotic pressure differential, higher permeate flux, molecular weight compounds (> 300), higher retention of multivalent salts, and adequacy of assets as well as operating and maintenance costs [15]. Generally, polymeric membranes are studied in the way of COD retention, dye persistence, permeate flux, and salt eliminations. The properties of various textile wastewater reclamation and membrane properties have been thoroughly investigated. The conclusions demonstrated that polymeric membranes are an appropriate therapy for the purification of textile wastewater, with a tolerable failure rate [16]. Nevertheless, using an appropriate previous treatment to avoid contamination and significant harm to the module is essential to maintain the efficiency of polymeric membranes at acceptable operating expenses [17].

The focus of this chapter is to re-examine and critically analyze the working mode of various techniques control over industrial effluent, precisely discharged from the dyeing process, another the testing facility, in contrast to membranes of polymer. To combine the different conditions of polymeric membranes for the purifications of industrial wastewater. A summary of the presently offered transport models is also conducted to attain a better perspective of the transport properties of dyes and salts in polymeric membranes. In order to serve the most effective ways to mitigate fouling, a basic understanding of fouling processes and reported strategies for fouling control are also addressed. In addition, the ultimate fate of polymeric membrane in the textile industry is discussed to develop cost-effective polymeric membrane, especially for the textile wastewater treatment.

2 Textile Wastewater

Textile manufacture is a complicated process that incorporates multiple phases and chemicals for the creation of various products [18–21]. The study employed a sample of wastewater from a textile industry that uses a variety of dyes (both reactive and direct) as well as chemicals such as salts, detergents, auxiliaries (such as emulsifiers and surfactants), and caustic soda [22]. Their quantities are determined by the type of process that produces the various effluents. The wastewater is usually produced by dyeing, washing, and other processes. These steps need a considerable amount of water and chemicals.

2.1 Characteristics of Textile Wastewater

The volumes of waste water are significantly heavily potentially and colored discharged in the form of hazardous effluents from the textile industries. The textile industry in China used roughly 8650 million m^3/year of fresh water and released 1840 million m^3/year of wastewater, ranking third among significant sectors in China [23]. Approximately 830 million m^3/year of textile industry effluent is generated in India each year, with an additional 640 million m^3/year of wastewater discharged. Treating this wastewater and releasing it into receiving water bodies might cost over a billion dollars per year [24]. The textile industries are generating waste water during the dyeing and washing process. Production of 8000 kg fabric per day in textile and typical textile industry uses 1.6 m (million) liters of ground water from this, 30–40% is used for dyeing process, 60–70% for washing level, and 10–50% for unused dyes. The total produced waste water has been directly discharged into ground without adding any anti-contamination [21, 25, 26]. Textile industry mechanisms such as bleaching, printing, scouring, and accomplish produce large amounts of wastewater, up to 1–10 million liters per day. In addition, more than 500 tons of dyestuffs have been produced from the only textile industries. The high dye content, higher temperature, chemical and biological demand of oxygen, pH, total organic carbon and dissolved solids, TS, TSS, SS, phosphate, chlorides, and other characteristics of Textile wastewater [27–29]. Generally, the textile waste water is containing chlorobenzenes, nitrates, sulfates, phenols, turbidity, VOCs (volatile organic compounds), salts, alkalinity, acid, bases, mordants, bleaching, surfactants, dioxin, electrical conductivity, fixing, and finishing agents, as well as several metals like as, Cr, Cd, Ni, Pb, Cu, Sb, and Zn are all found in Textile wastewater [25, 26, 30, 31]. Furthermore, the chemical composition and nature of Textile wastewater is mostly depending on the used chemicals as well as mechanisms used by the Textile industry.

2.2 Nature of Textile Effluents

On average, significant quantities of water, salts, and chemically active dyes are used in textile refining processes for each kilogram's cotton production. As a result, it generates a wastewater in huge amounts that contains stable pollutants. Depending on the textile, textile effluents usually contain a variety of colors, solvents, detergents, and multivalent salts. Textile effluents usually include a variety of detergents, salts, solvents, and dyes as per their nature of releasing in the textile procedure (bleaching, printing, dyeing, scouring, and finishing). Table 1 depicts the usual components of wastewater in the finishing and dyeing processes effluents, which includes the number of components in varying concentrations [5, 32].

A range of values is given for each of the parameters involved, verifying the high variability of the dyeing wastewater [33]. In addition of surfactants into these elements, are used to minimization of water surface tension during processing. Although, they exclusively account for a limited portion of the wastewater. The non-ionic surfactants such as alkyl phenol ethoxylates in textile wastewater should be managed in proper manner so, alkyl phenol can be biodegraded because it is more toxic as compared to ethoxylated [34]. Reactive dyes are used in huge amounts and owing to the vast usage of chemically reactive dyes for cellulosic fiber is still economic limitations of other dyes [35]. In general, reactive dyes are chemically reacting among the substrate of fiber for making a stable bond [36]. The chemically active dyes are met to react with water instead of the functional groups are present on textile fabric after that hydrogen-based dyes are produced. By the degree of fiber mixture and dye's fixation, the total exhaust amount effluent can be analyzed [37]. In the nut shell, all dye groups have the common environmental problem. In conclusion, it is necessary to eliminate all or maximum amounts of the effluent before releasing it directly to environment. On the other hand, Inorganic salt is applied to the process of dyeing to improve dye absorption by the fabric. The most popular inorganic salt is sodium chloride (NaCl), which is commonly applied in the process of dyeing. Sodium sulfate (Na_2SO_4) is the divalent salts, that are used in the place of NaCl during the process. In few regions, there is major ecological issue due to the high attention of salt in the waste stream and salinization of the soil. As a result, it is important to remember that the treatment of wastewater system is not just a way to deal with a problem with the environment, it is also a way to recycle precious rinsed water and reduce the amount of waste that is discharged.

3 Various Technologies for the Treatment of Water Waste in Textile

Numerous studies have been described a variety of biological, Physico-chemical, and advanced oxidation approaches for the treatments of wastewater products in textile are shown in Fig. 2. To improve or accelerate the of textile wastewater products

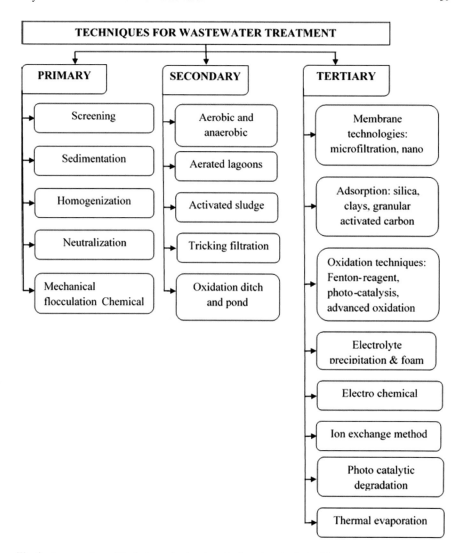

Fig. 2 An overview of techniques for treatment of wastewater in textile

treatment, these treatments can be utilized separately or in combination.

Despite various methods (physical, chemical, biological, etc.), subcategories of physical methods for treatment of textile wastewater are the most commonly used methods.

3.1 Physical Methods for Treatment of Wastewater in Textile

Physical methods are used for textile wastewater treatments; the approaches are based on the mass transfer process [38]. These procedures have a degradation range of 86.8–99%, with the adsorption procedure being the most effective. These are capable of easily decomposing practically any color or dye combination [38]. These strategies are frequently used because of their simplicity and efficacy. When compared to biological and chemical methods for textile wastewater treatment procedures, this process requires the least amount of chemicals along with several qualities in terms of less cost, ease of design, and simplicity of operation, high efficiency, and lack of hazardous substance effect [39]. This process is more predictable than the other two textile wastewater treatment procedures since it does not deal with living organisms [38]. Basically, three types of physical treatments were adopted can be shown in Fig. 3.

(a) **Adsorption**

In this method, adsorbate or ion molecules (liquid or gas) are concerned on adsorbent's (solid) surface, which is known as a surface phenomenon. The two types of adsorption are physisorption and chemisorption. These classifications are based on how dye molecules or other components bind to the adsorbent surface [40]. Hydrogen bonding, vander-wall interactions, electrostatic interactions, hydrophobic interactions, and other forms of interactions may exist in the process of adsorption in the molecules' dye [41]. In order to be more successful in adsorption, adsorbents commonly have a porous structure. This shape increases the total surface area to allow liquid more quickly. This technology helps to remove dyes from textile wastewater in a simple and cost-effective manner [42]. This approach has various advantages, including immense treatment efficiency and the ability to be reused. The performance of most adsorbents is affected by various parameters such as starting dye concentration, contact time, temperature, pH, and adsorbent quantity. A good adsorbent has

Fig. 3 Illustration of physical treatment techniques

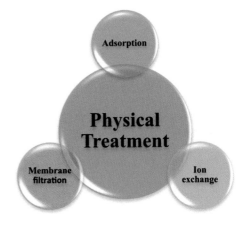

a rough surface like porous formation (which leads toward highly surface area) and a short adsorption symmetry traditional period for eliminating dye wastes in less time [43, 44]. The most important component is adsorbent in the adsorption process. The most significant characteristics of an adsorbent are its surface area, adsorption capacity, and less adsorption period [45, 46]. The effluents that can be removed from textile wastewater are frequently done with zeolites, activated carbon, alumina, and silica gel [40]. Well-structured activated carbons are represented by its porous structure, high surface area, and excellent adsorption capabilities against numerous types of contaminants [47]. Generally, Adsorption increases when pH rises [48].

(b)   **Ion exchange**

The ion conversion approach has generated a lot of curiosity. Due to its convenient applications such as water treatments, removal of heavy and toxic metals, and effluent, high efficiency, low prices, and interesting features in the textile wastewater. One of the most effective procedures is the removal of dyes with the strong interactions among charged dyes and available functional groups on ion exchange resins. The formation of strong interactions among solutes and resins leads to successful parting in this technique [49, 50]. The solutes were separated with differing surface charges, cation exchangers, or anion exchangers as resins are utilized [51]. Marin et al. confirmed that the 98.6% Acid Orange 10 Dye (conc: 10^{-2} M) was removed from the textile waste water with the help of Amberlite IRA 400 anion-exchange resin [52]. Various effective ion exchangers have been also mentioned by Samsami et al. [53].

(c)   **Membrane filtration**

Membrane separation is a method of filtering and separating certain compounds in wastewater by using the membrane's micropores and selective permeability. Presently, membrane separation processes based upon pressure on membrane, i.e., microfiltration, ultrafiltration, nanofiltration, and reverse osmosis are most commonly used for the treatment of textile effluent.

4   Membrane Fabrication Methods

The membrane preparation process is influenced by different variables, including the polymer used and the membrane's desired structure. Membrane fabrication techniques consist of phase inversion, track-etching, interfacial polymerization, stretching, and electrospinning. Let's take a closer look at these methods.

I.   **Phase Inversion**

Phase inversion is a method in which an apparently homogenous polymeric solution is managed to transition to solid from a liquid state [54]. This is the most prevalent approach, and it aids in the construction of polymeric porous membranes with complex structures [55]. The selections of polymer, as well as the additive materials used in the polymer blend, influences the chemical interaction, membrane structure, and characteristics, by using this process [56, 57]. In phase inversion process, phase transformations are the options of thermally induced phase separation, immersion precipitation, vapor-driven, and evaporation-driven phase separation. However, immersion precipitation is widely utilized process with diverse morphologies in the synthesis of polymeric membranes. Sidney Loeb and Srinivasa Sourirajan developed the technique, which is also popular as the Loeb-Sourirajan method [58]. They employed this approach for seawater treatments, which promote the expansion of the cellulose acetate membrane at very first. This is the well-known way of making polymeric membranes. Furthermore, the total time is taken in the coagulation or precipitation period in the precipitation bath of the polymer during the process of membrane creation for the casting solution-based film to dissolve and then remixed.

The total time for solvent-nonsolvent diffusion in solution precipitation bath interface among the casting solution [59]. In 1975 research described the straight-line plot between the coagulation time and the square of membrane thickness. Kang et al. discovered the equation as follows in 1993 [60]:

$$X = 2(D_e t)^{1/2} \tag{1}$$

By squaring both sides, the equation can be written as

$$X = 2(D_e t) \tag{2}$$

where X = membrane thickness (mm); t = the coagulation/precipitation time (s); and D_e = coefficient of effective diffusion (mm^2/s).

Zheng et al. elaborate an expression among the coagulation/precipitation time and thickness of a membrane (wet) [61]:

$$D_a = X^2/t \tag{3}$$

where D_a = kinetic parameter (mm^2/s).

II. Electrospinning

Electrospinning is a relatively recent method of fabricating porous membranes for a variety of utilization including desalination and filtration. In this method, high amount of potential is required for the grounded collector and polymer solution droplet. The electrostatic potential is strong sufficient which overcomes the surface tension of the droplet and results in the formation of a charged liquid jet. Controllable aspect ratios are a unique property of these fibrous membranes:

$$\text{Aspect ratio} = L/d \text{ (where } L = \text{length of fiber; } d = \text{diameter of fiber)}$$

The porosity, hydrophobic nature, surface design, and distribution of pore size electrospun fibrous membranes must be determined or finalized employing fiber size, shape, and morphology. The effects of humidity and nonsolvent among polystyrene (PS) electrospun fiber size and morphology were investigated by Pai et al. [62]. They discovered to increase in PS fiber diameter from 0.9 to 3.93 mm, electrospun from 30% by weight PS/DMF solution, and a relative humidity (RH) rise from 11 to 43%. Because RH impacts the rate of solidification of electrospun fibers, the fibers electrospun above % RH had favorable smooth surfaces. The fibers exhibit wrinkled surfaces below % RH, resulting in smaller diameter fibers. Again, Lin et al. reported the alteration of solvent composition and polystyrene solution concentration, nanoporous and micro structures in PS fiber can be formed. When electrospun with the solution in tetrahydrofuran (THF) 30% by weight PS solution in the fibers have densely packed nanopores with ribbon-like and densely packed nanopores shape [63]. By adding tetrahydrofuran (THF) and dimethylformamide (DMF), the fiber surface becomes free nanopores. The formation of beaded structure nanofibers occurs when the absorption of PS in DMF/THF is taken as less.

III. Stretching

In the 1970s, a method for stretching polymer membranes was invented. Extrusion is used to make microporous membranes for microfiltration, ultrafiltration, and membrane distillation after the stretching procedure. In this technique a thin sheet forms by extruding polymer and making it porous via stretching and then increasing the temperature up to melting point in order to make it more porous. The crystalline sections of the polymer give strength, while the amorphous sections form a porous structure, resulting in extremely crystalline polymers using this approach.

Stretching is performed in two parts: cold stretching first, then hot stretching.

1. In cold stretching: For nucleate micropores in the precursor film.
2. In hot stretching: For control/increase membrane's end pore shape.

The mixture of (dissimilar molecular weights) pore density and uniformity, Polypropylene (PP) on crystallinity, and tensile characteristics was examined by Tabatabaei et al. [64]. For the large molecular-weight species increases the membranes of homogeneous, better connection, and distribution of pores. Kim et al. examined the PP hollow fibers as a precursor and the effect of annealing on crystallinity size before stretching [65].

IV. Track-etching

The energetic heavy metal ions are commonly used to irradiate a nonporous polymeric film, to get result in form of linear spoiled tracks around the exposed polymeric surface. This approach is defined as its ability to precisely adjust the membrane's pore density, size, and distribution. These parameters are self-contained and can be adjusted throughout a large number of values, ranging from nanometer to micrometer. From these characteristics developed a straightforward link between membrane structure (pore shape and size) and qualities of water transport.

Since 1970s, polycarbonate (PC) track-etched-based membranes are available. Fleischer et al. contain basic information on particle track generation, the track-etching recipes, track production mechanism, and prospective applications [66]. Komaki and Tsujimura examined the impact of etching rate of porous polyethylene naphthalate (PET) films on the hole density and diameter and increased linearly with etching time [67]. They bombarded polyimide films with various heavy ions (Ag, I, Cu, and Br) before treating them with gamma rays in the presence of oxygen [68]. Gamma rays' exposure during existence of oxygen, the formation of etched tracks accelerates. The film is bombarded with the lighter ions and found the effect was more pronounced. Starosta et al. used a cyclotron to irradiate heavy ion beams with PET films and then etched with a NaOH solution [69]. Pore sizes ranged from 0.1 to 0.5 mm. The most commonly utilized polymers are PET and PC for track-etching because of their numerous characteristic resistances of acids, organic solvents, and their mechanical qualities. Additionally, PVDF (polyvinylidene fluoride) and its copolymers have been used as fluorinated polymers [70]. However, the pore formation takes a long period because of its strong oxidizer resistance.

V.   Interfacial Polymerization

NF (Nanofiltration) is a process that defines the RO membranes and thin-film composite (TFC) are built by combining polymerization (IP), which is the most useful approach for commercial creation. Cadotte et al. introduced the first TFC membranes manufactured by using this approach for the RO applications, as a result of this discovery the membrane performance used for RO applications [71, 72]. The TFC membrane has been successfully synthesized with the independent qualification of the surface and microporous substrate layer by the IP method [73].

The structural morphology, solvent type, monomer concentration, reaction period, and constituent of the barrier membrane layer are all affected and considered by many parameters such as posttreatment conditions. A membrane of TFC polyurea showed greater water flux and salt rejection than an intrinsic membrane of skinned asymmetric cellulose acetate, which is confirmed by the IP technique's authenticity. Most RO and NF membranes are made by the IP technique have a thin sheet of polyamide (PA) on surface of the membrane support. Trimesoyl chloride (TMC) and M-phenylenediamine (MPD) are the most frequent reactive monomers are employed to generate the functionalized PA layer on RO/NF membranes. Li et al. reported TFC and RO membranes with MPD are prepared by two new chemicals which are tri- and tetra-functional biphenyl acid chlorides: 3,30,5,50-biphenyl tetraacyl chloride and 3,40,5-biphenyl triacyl chloride [74].

On the basis of recent studies, the monomers have been utilized for the synthesis of TFC membranes by using the IP approach. However, the monomers consist of more polar functional groups and functional groups, that produce membranes have a smooth surface or are more hydrophilic in nature. The increased hydrophilicity aids in the improvement of the membranes' antifouling properties. In addition, great work has been put toward avoiding chlorine from oxidizing membranes of PA TFC. Moreover, the chemical alterations of the PA layer with diamine moieties can significantly improve the membranes' chlorine resistance, with aromatic, aliphatic diamines and

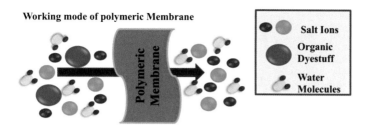

Fig. 4 Illustration of working mode of polymeric membrane filtration

cycloaliphatic having the most effective chlorine resistance. The development of polymeric membranes TFC polyester, polyesteramide, and TFC-PA by using the IP method.

4.1 Polymeric Membrane Filtration

To eliminate the color, salinity, and COD from the textile wastewater the filtration of polymeric membrane is most favorable cutting-edge treatment [75, 76]. The process involves passing textile wastewater into a membrane with small pores, that allow larger solutes than the holes on membrane and the larger solutes to be stuck on it after that the passing solution is free from solutes. Filter cake is formed by the trapped solutes (see Fig. 4).

These layers must be cleaned on a regular basis in order to facilitate the filtration process. Membranes can be classified depending on several aspects, including the size of their holes. Its performance is typically defined by rejection and penetrates flux [50].

Phase inversion, is a fabrication method that provides an anisotropic material that may subsequently be utilized as a substrate for a film composite membrane, and the most important for the textile wastewater treatment. Table 2 provides a closer look at recently used polymeric materials for manufacture of commercial membranes. According to Wagner [77], the total worldwide usage of membranes is generally as follows:

- polysulfone UF and MF membranes: 5–7%
- composite NF membranes: 3–5%
- composite RO membranes: 85%
- other membranes (polyacrylonitrile, ceramic materials, and cellulose system): 3–5%.

Table 2 Polymeric materials and related membrane processes [78]

S. No.	Polymer	Morphology		Barrier thickness (mm)	Membrane process
		Cross-section	Barrier type		
1.	Perfluorosulfonic acid polymer	Isotropic	Nonporous	50–500	ED
2.	Polyetherimides	Anisotropic	Mesoporous	~ 0.1	UF
3.	Polyethersulfones	Anisotropic	Mesoporous Macroporous	~ 0.1	UF
		Isotropic		50–300	MF
4.	Polyacrylonitrile	Anisotropic	Mesoporous	~ 0.1	UF
5.	Cellulose acetates	Anisotropic	Nonporous Mesoporous Macroporous	~ 0.1	GS, RO
		Anisotropic		~ 0.1	UF
		Isotropic		50–300	MF
6.	Cellulose nitrate	Isotropic	Macroporous	100–500	MF
7.	Cellulose, regenerated	Anisotropic	Mesoporous	~ 0.1	UF, D
8.	Polyethylene terephthalate	Isotropic track-etched	Macroporous	6–35	MF
9.	Polyphenilene oxide	Anisotropic	Nonporous	~ 0.1	GS
10.	Poly(styrene-co-divinylbenzene), sulfonated or aminated	Isotropic	Nonporous	100–500	ED
11.	Polytetrafluoroethylene	Isotropic	Macroporous Nonporous	50–500	MF
				~ 0.1	GS
12.	Polycarbonates, aromatic	Anisotropic	Nonporous Macroporous	~ 0.1	GS
		Isotropic track-etched		6–35	MF
13.	Polyamide, aromatic, in situ synthesized	Anisotropic/composite	Nonporous	~ 0.05	RO, NF
14.	Polyamide, aromatic	Anisotropic	Mesoporous	~ 0.1	UF
15.	Polyamide, aliphatic	Isotropic	Macroporous	100–500	MF

(continued)

Table 2 (continued)

S. No.	Polymer	Morphology		Barrier thickness (mm)	Membrane process
		Cross-section	Barrier type		
16.	Polyether, aliphatic crosslinked, in situ synthesized	Anisotropic/composite	Nonporous	~ 0.05	RO, NF
17.	Polypropylene	Isotropic	Macroporous	~ 0.1 < 1–10	MF
18.	Polyimides	Anisotropic	Nonporous	~ 0.1	GS, NF
19.	Polyethylene	Isotropic	Macroporous	50–500	MF
20.	Polyvinyl alcohol, crosslinked	Anisotropic	Nonporous	< 1–10	PV
21.	Polyvinylidene fluoride	Anisotropic Isotropic	Mesoporous Macroporous	~ 0.1 50–300	UF MF
22.	Polysulfones	Anisotropic	Nonporous	~ 0.1	GS
23.	Polysiloxanes	Anisotropic/composite	Nonporous	~ 0.1	GS, PV, NF

Abbreviations RO—reverse osmosis; MF—microfiltration; PV—pervaporation; NF—nanofiltration; GS—gas separation; ED—electrodialysis; D—dialysis; UF—ultrafiltration

4.1.1 Microfiltration (MF)

Microporous membranes are used in this membrane technique to filter pollutants from the fluid. The pore size range of a common microfiltration membrane, which is between 0.1 and 10 micrometers (mm), can be used to classify it. The removal of suspended solid particles is one of the uses of this method [77].

4.1.2 Ultrafiltration (UF)

Ultrafiltration membranes are found in the diameters of pores ranging from 0.1 to 0.001 microns, the reverse osmosis and nanofiltration are more cost-effective because they require less pressure. Large pore size causes low rejection but it is an inexpensive way to reject organic dyes from textile wastewater [77]. Despite the fact that MEUF has been proposed as a sustainable model for the retention of heavy metal ions, and organic pollutants in various studies, it has yet to be tested on a large amount. In this method, in the aqueous solution with pollutants, the surfactants were added, which were very concentrated to order to compare with CMC (critical micelle concentration). Therefore, the produced micelles of surfactant molecules dissolve the organic and inorganic pollutants. The non-dispersed solutes-containing micelles are then eliminated throughout by ultrafiltration membrane treatment. They are practically pollutant-free after passing through the membrane [79].

4.1.3 Reverse Osmosis (RO)

Reverse osmosis (RO) is a technique for reducing and separating dissolved solids, nitrate, bacteria, organics, pyrogens, submicron colloidal material, and color from water using RO wound membranes [80]. This technique has been worked in the textile effluents elimination of salts and dyes from solutions, resulting in nearly refined water [81]. The benefits of this technology or membrane-based process include achieving separation and concentration none changing the state nor using the chemical as well as thermal energy. These characteristics result in acceptable recovery applications and efficient process of energy. Some of the uses of RO systems include the distillery wasted wash, groundwater treatment, beverage industry, recovery of chemicals such as phenol, wastewater reclamation, and seawater reverse osmosis (SWRO) treatment [80].

4.1.4 Nanofiltration (NF)

Nanofiltration (NF) is a recently developed membrane method that may be used to treat and decontaminate a variety of wastewaters [82]. Textile effluents have been treated with NF. NF has currently been demonstrated as a viable treatment method that can be used to replace or supplement several traditional separation procedures, specifically in the elimination of textile effluents [83]. Its aperture is merely a few nanometers wide, with a retention molecular weight of 80–1000 Da. Textile dye effluents can be treated using a mix of adsorption and nanofiltration. Because the adsorption step comes before the nanofiltration stage, the concentration polarization during the filtration process is reduced, increasing the process output. Lower the weight (molecular) of organic molecules, reactive hydrolyzed dyes, large monovalent ions, divalent ions, and auxiliaries of dyeing are all retained by nanofiltration membranes as shown in Fig. 5 [50, 84]. Reverse osmosis (RO) and Ultrafiltration (UF) have been illustrated in this technique. Kurt et al. examined the removal of textile effluents, salts, and COD from the industrial sewages using a commercial NF-270, Dow Filmtec, with 200–300 MWCO (molecular weight cutoff) (NF membrane).

The NF-270 can remove approx. 100% of dyes in continuous mode, eliminate maximum to 95% of COD, and 76% of salts [85]. Furthermore, lower fouling tendency of NF membrane and also lesser the expending of energy than RO and UF [86–88]. An outline of polymeric NF membranes that are commercially available and used for textile effluent treatment can be shown in Table 3.

Polymers are commonly utilized as flotation aids, flocculants, and coagulants in addition to being frequently attained for purification of water in membrane technology. Polymers are appealing materials for membrane construction due to their distinctive qualities that are low cost, flexibility, and enhanced controllability

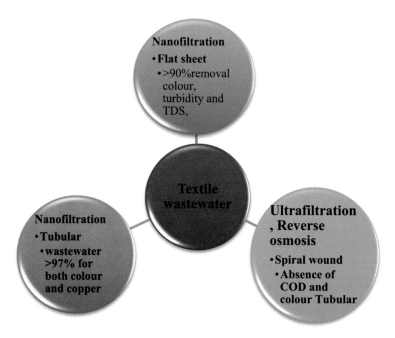

Fig. 5 Illustrations of membrane use for textile wastewater treatment. *Symbols* COD—carbon organic dissolved; TDS—total dissolved solids

Table 3 Various applications for commercially available NF membranes in textile wastage [7]

S. No.	Membrane (company)	Polymer material	MWCO (Da)	Process methods	Remarks
1	ATF 50 (Adv. Membr. Tech)	Spiral wound (TFC[c] of PIPd on PSf[e])	340	The wastewater of Industries was examined mainly two types as (a) COD = 14,200 mg L^{-1} with pH 10.2; (b) COD = 5430 mg L^{-1} with pH 5.5. The transmembrane pressure taken in between 0.2 and 1.1 MPa along with the temperature was taken as 25–40 °C [89]	R_{COD} = 80.9% for pH 5.5 wastewater R_{COD} = 95% for pH 10.2 wastewater
2	Desal 5 DK (Osmonics)	Flat sheet (TFCc –PA$_b$)	150–300	The concentration of dye was taken 1 gL^{-1} and prepared a synthetic solution of dye without incorporating the auxiliary components. This synthesis was performed with 10 bar of pressure, 25 °C temperature, and pH 6 without and without stirring the solution [16]	P_{ave} = 41.1 L/(m^2 h) $R_{dye, ave}$ = 100% for direct red 80 at Reynolds no. of 4100
3	1073 DK (Osmonics)	Spiral wound (PA)[b]	300	The concentrations of dye varying from 400 to 500 mg L^{-1} at the temperature 60 °C with pressure were taken 25 bars. The solution contains Na_2SO_4 (15 g L^{-1}), NaCl (10 g L^{-1}) and $CaCl_2$ (10 g L^{-1}) [90]	P_{ave} = 60.25 L/(m^2 h) $R_{dye, ave}$ = 94.5%

(continued)

Table 3 (continued)

S. No.	Membrane (company)	Polymer material	MWCO (Da)	Process methods	Remarks
4	2540 DK (Osmonics)	Spiral wound (NA)*	NA*	The COD 1576 g L^{-1}, conductivity 3.5 μS/cm and the color > 500 Hz of Industrial wastewater and the transmembrane pressure was 25 bars taken along that the cross-flow velocity of 1.66 ms^{-1} was used during the experiments [91]	$P_{ave} = \sim 60$ L/(m^2 h) $R_{salt} = 60$–80%
5	4040F DL (Osmonics)	Spiral wound (NA)*	150–300	The three stage treatment system used for the process of dyeing and finishing plant of textile wastewater are NF, UF and Sand filtration. The pH 7.8, COD 142 mg L^{-1}, TSS 12 mg L^{-1} and the conductivity 3950 μS/cm are the mean values of these parameters. NF and UF modules are worked at 9 and 0.4 bar, respectively [92]	$R_{COD} = >93\%$ $R_{TSS} = >60\%$ R conductivity $= 40.5\%$
6	31 MPS (Weizmann)	Spiral wound (NA)*	NA*	The concentrations of dye varying from 400 to 500 mg L^{-1} at the temperature 60 °C with pressure were taken 25 bars. The solution contains Na_2SO_4 (15 g L^{-1}), NaCl (10 g L^{-1}) and $CaCl_2$ (10 g L^{-1}) [90]	$P_{ave} = 66.25$ L/(m2.h) $R_{dye, ave} = 94.9\%$

(continued)

Table 3 (continued)

S. No.	Membrane (company)	Polymer material	MWCO (Da)	Process methods	Remarks
7	70 NF (Dow/Film Tec)	Flat sheet (PA)[b]	250	The mainly two types of exhausted dye baths from the wool dyeing process were used in this experimental process, i.e., (1) Metal complete dye bath, (2) Acid dye bath. The pressure of the transmembrane were taken at 10 bar [93]	$P_{SA} = 33$ L/(m^2 h) $P_{MC} = 32$ L/(m^2 h)
8	NF45 (Dow/Film Tec)	Spiral wound (PA)[b]	200	The concentrations of dye varying from 400 to 500 mg L^{-1} at the temperature 60 °C with pressure were taken 25 bars. The solution contains Na$_2$SO$_4$ (15 g L^{-1}), NaCl (10 g L^{-1}) and CaCl$_2$ (10 g L^{-1}) [90]	$P_{ave} = 39.2$ L/(m^2 h) $R_{dye,\ ave} = 92\%$
9	NTR 7450 (Nitto–Denko)	Flat sheet (sPES)[f]	600–800	The experiment conducted with the 0–60 bar pressure and the cross-flow velocity was 0–0.75 ms^{-1} for reactive blue 2 or reactive orange 16 (15 g L^{-1}), Na$_2$SO$_4$ (56 g L^{-1}), surfactant-EDTA (0.2 g L^{-1}), Na$_2$SO$_3$ (1 g L^{-1}) and NaOH (2.5 g L^{-1}) contained synthetic dye bath solution [94]	$P = 64$ L/(m^2 h) $R_{dye} = 92.1\%$ $R_{salt} = 87.3\%$ at operating pressure of 20 bar

(continued)

Table 3 (continued)

S. No.	Membrane (company)	Polymer material	MWCO (Da)	Process methods	Remarks
10	TFC–SR2 (fluid system)	Flat sheet (TFC[c] of PSf[e])	200–400	The wastewater sample solutions prepared by 92-1583 ppm of reactive black 5 and NaCl (10–80 g L^{-1}) and these solutions are filtered under 100–500 Kpa pressure and the cross-flow velocity were taken 3–5 L/min [95]	$P_{ave} = 45.05$ L/(m^2 h) $R_{dye, ave} = 97.71\%$
11	UTC 20 (Toray Ind)	Flat sheet (PA)[b]	180	Reactive orange 16 and reactive black 2 taken 1 g L^{-1} dye were examined with the addition of NaOH (2.5 g L^{-1}), Na$_2$SO$_3$ (1 g L^{-1}), EDTA (0.2 g L^{-1}), Na$_2$SO$_4$ (11 g L^{-1}), Na$_2$SO$_4$ (19 g L^{-1}) [93]	$R_{dye, RO16} = \sim99\%$ $R_{dye, RB2} = >99.3\%$

* Not available, [b] Polyamide, [c] Thin-film composite, [d] Piperazineamide, [e] Polysulfone, [f] Sulfonated polyethersulfone, P_{ave} Average permeability, R Rejection

[96]. The most commonly used polymers are polyethylene, polypropylene, cellulose acetate, polysulfone, polyethersulfone, poly-acrylonitrile, polycarbonates, and poly (vinylidene fluoride) in membrane technology. Table 4 shows an outline of current studies on textile wastewater treatment utilizing membranes of polymeric material. Membrane fouling has been successfully controlled with polymeric materials in a variety of systematic filtration. On the membrane's surface, the addition of grafting polymer chains is a typical method of improving the membrane's antifouling capabilities [97].

Tian et al., synthesized an MF membrane for the adsorption of higher molecular weight metal ions via electrospinning method along with surface modification by PMAA (polymethacrylic acid) [98]. The adsorption selectivity of this membrane for Hg^{2+} is quite high. By employing a saturated ethylene dinitrilotetra acetic acid solution, the deposited heavy ions are removed easily from the surface of membrane and can be reused for the adsorption of metal ions. Gao et al., investigated the efficiency of anaerobic membrane bioreactor containing the artificial sewage protein 30% and

Table 4 Water treatment solution by using polymeric membranes in previous studies

Polymer	Structure	Process	Description	Refs.
Cellulose acetate (CA)		MF	• The Electrospun CA was functionalized with polymethacrylic acid • Main function: heavy metal ions adsorption, i.e., Cu^{2+}, Cd^{2+}, Hg^{2+}, Cr^{2+} and Pb^{2+} • Selectively used for the adsorption of mercury (Hg^{2+})	[98]
		RO	• The rejection rate approx. 99% was observed • Selectively used for the elimination of ammonium ions	[99]
Poly(vinylidene fluoride)		UF	• The immobilization method is used for the modification of membranes surface by poly(vinyl alcohol) polymeric material • Used for improvement of hydrophilicity antifouling membrane's performance • Decrease membrane roughness	[100]
		UF	• By modification of poly (vinylidene fluoride) were studied • The treatment of Congo red and reactive black 5 dye solutions were reported • Modified PVDF membranes showed (COD) rejection, lower membrane fouling and moderate color removal for purification, and separation of water	[101]
Polysulfone		MF	• THF is used for the modification of surface • The rejection of sodium alginate got improved	[102]
Polyethersulfone		NF	• This polymer claimed for the reduction of the size of plant and overall cost reduction • Focused on the removal of radioactive atoms such as uranium	[103]

(continued)

Table 4 (continued)

Polymer	Structure	Process	Description	Refs.
Polypropylene		MF/UF	• The improvements of the performance of polyethersulfone polymeric membrane the additives were used such as linear Pluronic 31R1, star-like Tetronic 904, and Polyvinylpyrrolidone • As hydrophilicity increased the fouling property was decreased • As the roughness of membrane increased fouling resistivity was increased	[104]
Polypropylene	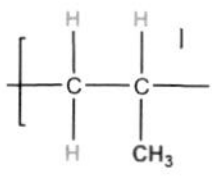	MF	• Treated with three different kinds of polymer (biopolymers), i.e., cellulose, potato starch, and chitosan • Used for adsorption of Cu^{2+} ions Hernandez	[105]
Polyethylene	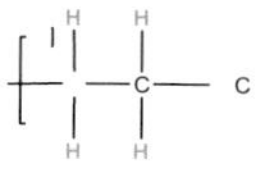	MF	• Chemically treated by a silicon micropillar array with diameter of 2 mm and the height of pillar was mm • Utilized for rejection of water/oil separation and *Escherichia coli* bacteria • The polymer increases the retention capacity and flux	[106]
Polyetherimide	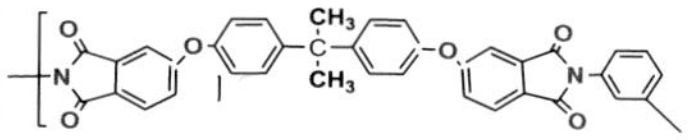	NF	• m-phenylene diamine (MPD) and trimesoyl chloride (TMC) polymerized by interfacial polymerization techniques • Promising tool for eliminating an azo-based dye from textile effluents such as reactive black 5	[107]
Polyarylate	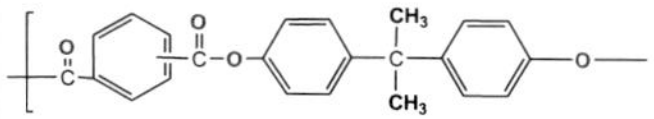	NF	• 5,50,6,60-tetrahydroxy-3,3,30,30-tetramethyl-1,10-spirobisindane (TTSBI) and isophthaloyl dichloride (IPC) as monomers polymerized by interfacial polymerization technique • It leads the treatment of high salinity and high acidic dyes from textile wastewater	[108]
Polyacrylonitrile	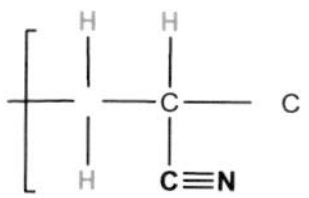	UF	• Modified by sodium hydroxide having high surface potential provides better function • Efficiency of this polymeric material polyacrylonitrile membrane was demonstrated in treatment of textile effluent, and selective dyes	[109]

(continued)

Table 4 (continued)

Polymer	Structure	Process	Description	Refs.
Polyamide		NF	• The polymeric material polyamide act as an active layer used for making a thin-film composite membrane • In this, the presence of salt leads to a further decrease in flux • Used for dye removal	[16]

indicate the PEI is much better than the cPVDF also explain the reaffirm is an important material in fouling [110]. Polyether-block- amide used as a coating agent on poly (vinylidene fluride) shows very slower fouling as compared to uncoated polyetherimide. Because hydrophobic membranes are sensitive to fouling, incorporating polymers that are hydrophilic in nature into them, for example poly (vinylidene fluoride) membranes, is a straightforward way to improve interaction of water in nature [111]. During the phase separation, before incorporating into the polymer (vinylidene fluoride) the pyrrolidone and polyvinyl alcohol which is hydrophilic polymer can be added for making of membrane solution. Despite their beneficial factor on membrane fouling but also, they are incompatible with hydrophobic matrices, which can lead to serious issues such as the development of interfacial micro-voids [112]. Srivastava investigated the large reduction in COD, the modified PVDF membranes were able to successfully remove over 70% of RB5 dye and over 97% of CR dye from the pure solutions. PVDF UF membranes among 60% SAN content have outstanding antifouling behavior (83% FRR) and tailored PVDF membranes can be again used for various continuous works with the recovery of substantial flux using only for water cleaning. Nevertheless, it is all modified membranes showed lesser membrane fouling, reduction in COD, minor membrane fouling, and mild color removal for purification and separation of textile wastewater effluents [101]. S. Ma et al., fabricated a novel catalytic (POM) membrane by grafting an ionic liquid polymeric membrane onto a membrane of polypropylene and after that complexing via polyoxometalate. Under visible light, the membrane was successfully employed to degrade dyes in water. Firstly, Photoinduced grafting of poly(1-(4-vinylbenzyl)-3-methylimidazolium chloride) (PVBMC) is used for the modification of PP nonwoven fabric membrane (PP NWF) after that anchored by POM. This membrane is capable of degradation of 95% of AO7 in 0.02 g L^{-1} of the solution in 2 h with the help of two fluorescent lamps (55 W). A membrane disk with a diameter of 47 mm may degrade AO7 with the percent of 70 in 250 mL of 0.02 g L^{-1} solution in 10 min only by using cycled filtration. The catalytic membrane shows great stability as well as great reusability, supported by long-term stability tests and cyclic usages [113].

Akbari et al., reported in their work the membrane generally demonstrated good rejection for dyes of anionic such as acid red 4, acid orange 10, direct red 80, direct yellow 8, and reactive orange 16, are owing to its comparatively less cutoff, but the dyes of cationic were persisting in excess of 95% regardless of pH and concentration employed. Direct yellow 80 and red 8, in particular, were fully maintained, resulting in permeate that could be reused. However, due to its sensitivity to fouling, the membrane's flow dropped. The osmotic pressure caused a further drop in flux in the presence of salt. Furthermore, more decrement in flux was found in the presence of salt [98].

Yung et al., examined the retention ratio of NaCl is very low in the membrane of PAR NF, indicating for dye/NaCl fractionation separation selectivity was great. As a result, the acid-tolerant PAR NF membrane has a lot of promise for treating higher salinity and acidic textile dye effluents efficiently. Interfacial polymerization of IPC (isophthaloyl dichloride) and TTSBI (5,50,6,60-tetrahydroxy-3,3,30,30-tetramethyl-1,10-spirobisindane) produced an active layer of ultrathin PAR on the surface of

SWCNT/PES support membrane. Herein, a promising technique is PAR NF is found for the treatment of high salinity and the acidic textile dyes effluents with great efficiency [108].

Gabriel et al., reported that the extraction of azo-based dyes and Reactive Black 5 from textile wastewater the NF thin-film composite membranes with a polymeric materials polyetherimide (PEI) are used. PEI is initially dissolved in the solvent which is N-methyl-2 pyrollidone (NMP) after a while it is mixed with nonsolvent which is acetone transformed to a membrane that indicates the phase inversion process found in this procedure. Then the IP process was used for changing the membrane, in which the m-phenylene diamine (MPD) and trimesoyl chloride (TMC) as acyl chloride, the precursors amine was taken and immersed into two immiscible solvents such as hexane and water. The PEI-TFC membrane was produced by addition of TFC selective layer, and it was able to separate the dyes in textile wastewater for ex. RB5, 150 ppm with the rejection rate of 92%, which is proved by various characterization techniques [114].

5 Conclusion and Future Scope

It is intricate to figure out a wide-ranging conclusion on the efficiency and the feasibility of membrane in favor of treatment of the wastewater in textile industries. This chapter mainly focused on the physical treatment of textile wastewater such as membrane filtration, Ion exchange, and adsorption. Among all textile effluents removal methods, membrane filtration (physical method) can be the best way to eliminate colors and other chemical substances. The polymeric membranes have been confirmed to be effective in dealing with highly colored textile effluent that is also highly saturated with monovalent and/or divalent ions, according to various researches carried out thus far. The interconnections between polymeric membrane structure, synthesis, performance, and surface properties have been examined in this chapter. We saw that the production of membranes for water treatment has made significant development. In general, it can be stated that polymeric membranes provide much more positives than conventional treatment methods and various membrane technologies. The development and optimization of polymeric membranes with various methods have been mentioned in this chapter. Moreover, thorough relationship understanding among surface properties, structure of porous and the membrane performance in textile wastewater treatment processes. At the commercial scale, the disadvantage of using these technologies is membrane fouling, which leads to the lesser permeate flux, and membrane cleaning is required repeatedly.

Polymeric membrane processes can be more efficient if the right polymeric materials are chosen, taking into account factors like permeability, selectivity, chemical, and mechanical resistance. It has been concluded in comparison of RO, NF, UF, and MF membrane filtration processes that indicate the NF is more effective. Although the NF membrane cannot achieve the retention behavior than RO membrane, dyeing tests showed that treated water can be used instead of fresh water in textile operations. Furthermore, NF operates under lesser demanding conditions (approximately pressure 6.5–7.0 bar) than RO, which allows for even more cost reductions. We look forward to the more advanced technologies are developed so that wastewater may be handled cheaply as possible at both the pilot and industrial scales.

References

1. Petersen L, Heynen M, Pellicciotti F (2016) Freshwater resources: past, present, future. In: International encyclopedia of geography: people, the earth, environment and technology, pp 1–12
2. Moussavi G, Mahmoudi M (2009) Removal of azo and anthraquinone reactive dyes from industrial wastewaters using MgO nanoparticles. J Hazard Mater 168(2–3):806–812
3. Liu C, Guo Y, Wei X, Wang C, Qu M, Schubert DW, Zhang C (2020) An outstanding antichlorine and antibacterial membrane with quaternary ammonium salts of alkenes via in situ polymerization for textile wastewater treatment. Chem Eng J 384:123306
4. Lau WJ, Ismail AF (2009) Polymeric nanofiltration membranes for textile dye wastewater treatment: preparation, performance evaluation, transport modelling, and fouling control—a review. Desalination, 245(1–3):321–348
5. Marcucci M, Ciabatti I, Matteucci A, Vernaglione G (2003) Membrane technologies applied to textile wastewater treatment. Ann NY Acad Sci 984:53–64
6. Soyekwo F, Liu C, Wen H, Hu Y (2020) Construction of an electroneutral zinc incorporated polymer network nanocomposite membrane with enhanced selectivity for salt/dye separation. Chem Eng J 380:122560
7. Cecille L, Toussaint JC (1989) Future industrial prospects of membrane processes
8. Quist-Jensen CA, Macedonio F, Drioli E (2015) Membrane technology for water production in agriculture: desalination and wastewater reuse. Desalination 364:17–32
9. Stoquart C, Servais P, Bérubé PR, Barbeau B (2012) Hybrid membrane processes using activated carbon treatment for drinking water: a review. J Membr Sci 411:1–12
10. Ravanchi MT, Kaghazchi T, Kargari A (2009) Application of membrane separation processes in petrochemical industry: a review. Desalination 235(1–3):199–244
11. Fane AT, Wang R, Jia Y (2011) Membrane technology: past, present and future. In: Membrane and desalination technologies. Humana Press, Totowa, NJ, pp 1–45
12. Obotey Ezugbe E, Rathilal S (2020) Membrane technologies in wastewater treatment: a review. Membranes 10(5):89
13. Aliyu UM, Rathilal S, Isa YM (2018) Membrane desalination technologies in water treatment: a review. Water Practice Technol 13(4):738–752
14. Kayvani Fard A, McKay G, Buekenhoudt A, Al Sulaiti H, Motmans F, Khraisheh M, Atieh M (2018) Inorganic membranes: preparation and application for water treatment and desalination. Materials 11(1):74
15. Hilal N, Al-Zoubi H, Darwish NA, Mohamma AW, Arabi MA (2004) A comprehensive review of nanofiltration membranes: treatment, pretreatment, modelling, and atomic force microscopy. Desalination 170(3):281–308
16. Akbari A, Remigy JC, Aptel P (2002) Treatment of textile dye effluent using a polyamide-based nanofiltration membrane. Chem Eng Process 41(7):601–609
17. Marcucci M, Ciardelli G, Matteucci A, Ranieri L, Russo M (2002) Experimental campaigns on textile wastewater for reuse by means of different membrane processes. Desalination 149(1–3):137–143
18. Garg N, Garg A, Mukherji S (2020) Eco-friendly decolorization and degradation of reactive yellow 145 textile dye by *Pseudomonas aeruginosa* and *Thiosphaera pantotropha*. J Environ Manage 263:110383

19. Bener S, Bulca Ö, Palas B, Tekin G, Atalay S, Ersöz G (2019) Electrocoagulation process for the treatment of real textile wastewater: effect of operative conditions on the organic carbon removal and kinetic study. Process Saf Environ Prot 129:47–54

20. Fazal T, Razzaq A, Javed F, Hafeez A, Rashid N, Amjad US, Rehman MSU, Faisal A, Rehman F (2020) Integrating adsorption and photocatalysis: a cost effective strategy for textile wastewater treatment using hybrid biochar-TiO$_2$ composite. J Hazard Mater 390:121623

21. Cai H, Liang J, Ning XA, Lai X, Li Y (2020) Algal toxicity induced by effluents from textile-dyeing wastewater treatment plants. J Environ Sci 91:199–208

22. Ellouze E, Tahri N, Amar RB (2012) Enhancement of textile wastewater treatment process using nanofiltration. Desalination 286:16–23

23. Li K, Liu Q, Fang F, Wu X, Xin J, Sun S, Wei Y, Ruan R, Chen P, Wang Y, Addy M (2020) Influence of nanofiltration concentrate recirculation on performance and economic feasibility of a pilot-scale membrane bioreactor-nanofiltration hybrid process for textile wastewater treatment with high water recovery. J Clean Prod 261:121067

24. Jegatheesan V, Pramanik BK, Chen J, Navaratna D, Chang CY, Shu L (2016) Treatment of textile wastewater with membrane bioreactor: a critical review. Biores Technol 204:202–212

25. Kumar A, Sharma G, Naushad M, Ala'a H, Garcia-Penas A, Mola GT, Si G, Stadler FJ (2020) Bio-inspired and biomaterials-based hybrid photocatalysts for environmental detoxification: a review. Chem Eng J 382:122937

26. Rosa JM, Tambourgi EB, Vanalle RM, Gamarra FMC, Santana JCC, Araújo MC (2020) Application of continuous H2O2/UV advanced oxidative process as an option to reduce the consumption of inputs, costs and environmental impacts of textile effluents. J Clean Prod 246:119012

27. Chandanshive V, Kadam S, Rane N, Jeon BH, Jadhav J, Govindwar S (2020) In situ textile wastewater treatment in high rate transpiration system furrows planted with aquatic macrophytes and floating phytobeds. Chemosphere 252:126513

28. Hussain Z, Arslan M, Shabir G, Malik MH, Mohsin M, Iqbal S, Afzal M (2019) Remediation of textile bleaching effluent by bacterial augmented horizontal flow and vertical flow constructed wetlands: a comparison at pilot scale. Sci Total Environ 685:370–379

29. Maksoud MA, Elgarahy AM, Farrell C, Ala'a H, Rooney DW, Osman AI (2020) Insight on water remediation application using magnetic nanomaterials and biosorbents. Coord Chem Rev 403:213096

30. Hubadillah SK, Othman MHD, Tai ZS, Jamalludin MR, Yusuf NK, Ahmad A, Harun Z (2020) Novel hydroxyapatite-based bio-ceramic hollow fiber membrane derived from waste cow bone for textile wastewater treatment. Chem Eng J 379:122396

31. Sun Y, Cheng S, Lin Z, Yang J, Li C, Gu R (2020) Combination of plasma oxidation process with microbial fuel cell for mineralizing methylene blue with high energy efficiency. J Hazard Mater 384:121307

32. Barredo-Damas S, Alcaina-Miranda MI, Iborra-Clar MI, Bes-Pia A, Mendoza-Roca JA, Iborra-Clar A (2006) Study of the UF process as pretreatment of NF membranes for textile wastewater reuse. Desalination 200(1–3):745–747

33. Suksaroj C, Heran M, Allegre C, Persin F (2005) Treatment of textile plant effluent by nanofiltration and/or reverse osmosis for water reuse. Desalination 178(1–3):333–341

34. Vandevivere PC, Bianchi R, Verstraete W (1998) Treatment and reuse of wastewater from the textile wet-processing industry: review of emerging technologies. J Chem Technol Biotechnol Int Res Process Environ AND Clean Technol 72(4):289–302

35. Gatewood BM (1996) Evaluation of aftertreatments for reusing reactive dyes. Textile Chemist and Colorist 28(1):38–42

36. Srivastva SR (1979). Recent processes of textile bleaching. Dyeing Finishing Small Business, Delhi

37. O'Neill C, Hawkes FR, Hawkes DL, Lourenço ND, Pinheiro HM, Delée W (1999) Colour in textile effluents–sources, measurement, discharge consents and simulation: a review. J Chem Technol Biotechnol Int Res Process Environ Clean Technol 74(11):1009–1018

38. Katheresan V, Kansedo J, Lau SY (2018) Efficiency of various recent wastewater dye removal methods: a review. J Environ Chem Eng 6(4):4676–4697
39. Khan NA, Bhadra BN, Jhung SH (2018) Heteropoly acid-loaded ionic liquid@ metal-organic frameworks: effective and reusable adsorbents for the desulfurization of a liquid model fuel. Chem Eng J 334:2215–2221
40. Gupta VK (2009) Application of low-cost adsorbents for dye removal–a review. J Environ Manage 90(8):2313–2342
41. Kumar R, Rashid J, Barakat MA (2014) Synthesis and characterization of a starch–AlOOH–FeS 2 nanocomposite for the adsorption of Congo red dye from aqueous solution. RSC Adv 4(72):38334–38340
42. Kyzas GZ, Lazaridis NK, Mitropoulos AC (2012) Removal of dyes from aqueous solutions with untreated coffee residues as potential low-cost adsorbents: equilibrium, reuse and thermodynamic approach. Chem Eng J 189:148–159
43. Boer JH (1970) Physical and chemical aspects of adsorbents and catalysts: dedicated to JH de Boer on the occasion of his retirement from the technological university, Delft. Academic Press, The Netherlands
44. Tien C (1994) Adsorption calculations and modeling. Butterworth-Heinemann
45. Robinson T, McMullan G, Marchant R, Nigam P (2001) Remediation of dyes in textile effluent: a critical review on current treatment technologies with a proposed alternative. Biores Technol 77(3):247–255
46. Ballav N, Das R, Giri S, Muliwa AM, Pillay K, Maity A (2018) L-cysteine doped polypyrrole (PPy@ L-Cyst): a super adsorbent for the rapid removal of Hg+2 and efficient catalytic activity of the spent adsorbent for reuse. Chem Eng J 345:621–630
47. Tan IAW, Ahmad AL, Hameed BH (2008) Adsorption of basic dye on high-surface-area activated carbon prepared from coconut husk: equilibrium, kinetic and thermodynamic studies. J Hazard Mater 154(1–3):337–346
48. Sivaraj R, Namasivayam C, Kadirvelu K (2001) Orange peel as an adsorbent in the removal of acid violet 17 (acid dye) from aqueous solutions. Waste Manage 21(1):105–110
49. Greluk M, Hubicki Z (2013) Evaluation of polystyrene anion exchange resin for removal of reactive dyes from aqueous solutions. Chem Eng Res Des 91(7):1343–1351
50. Ahmad A, Mohd-Setapar SH, Chuong CS, Khatoon A, Wani WA, Kumar R, Rafatullah M (2015) Recent advances in new generation dye removal technologies: novel search for approaches to reprocess wastewater. RSC Adv 5(39):30801–30818
51. Xu T (2005) Ion exchange membranes: state of their development and perspective. J Membr Sci 263(1–2):1–29
52. Marin NM, Pascu LF, Demba A, Nita-Lazar M, Badea IA, Aboul-Enein HY (2019) Removal of the acid orange 10 by ion exchange and microbiological methods. Int J Environ Sci Technol 16(10):6357–6366
53. Samsami S, Mohamadi M, Sarrafzadeh MH, Rene ER, Firoozbahr M (2020) Recent advances in the treatment of dye-containing wastewater from textile industries: overview and perspectives. In: Process safety and environmental protection
54. Giorno L, Mazzei R, Drioli E (2009) Biochemical membrane reactors in industrial processes. In: Membrane operations: innovative separations and transformations, pp 397–409
55. Mulder M, Winterton J (1991). Chapter 1: introduction. In: Basic principle of membrane technology. Kluwer Academic Publisher, Dordrecht/Boston/London, pp 1–15
56. Marchese J, Ponce M, Ochoa NA, Prádanos P, Palacio L, Hernández A (2003) Fouling behaviour of polyethersulfone UF membranes made with different PVP. J Membr Sci 211(1):1–11
57. Chuang WY, Young TH, Chiu WY, Lin CY (2000) The effect of polymeric additives on the structure and permeability of poly (vinyl alcohol) asymmetric membranes. Polymer 41(15):5633–5641
58. Loeb S, Sourirajan S (1960) Sea-water demineralization by means of a semipermeable membrane: UCLA water resources center report WRCC-34. Los Angeles, CA

59. Strathmann H, Kock K, Amar P, Baker RW (1975) The formation mechanism of asymmetric membranes. Desalination 16(2):179–203
60. Kang YS, Kim UY (1993) Asymmetric membrane formation via immersion precipitation method: II. A membrane formation scheme. Korea Polymer Journal 9:9–15
61. Zheng QZ, Wang P, Yang YN (2006) Rheological and thermodynamic variation in polysulfone solution by PEG introduction and its effect on kinetics of membrane formation via phase-inversion process. J Membr Sci 279(1–2):230–237
62. Pai CL, Boyce MC, Rutledge GC (2009) Morphology of porous and wrinkled fibers of polystyrene electrospun from dimethylformamide. Macromolecules 42(6):2102–2114
63. Lin J, Ding B, Yu J (2010) Direct fabrication of highly nanoporous polystyrene fibers via electrospinning. ACS Appl Mater Interfaces 2(2):521–528
64. Tabatabaei SH, Carreau PJ, Ajji A (2008) Microporous membranes obtained from polypropylene blend films by stretching. J Membr Sci 325(2):772–782
65. Kim J, Kim SS, Park M, Jang M (2008) Effects of precursor properties on the preparation of polyethylene hollow fiber membranes by stretching. J Membr Sci 318(1–2):201–209
66. Fleischer RL, Price PB, Walker RM, Walker RM (1975) Nuclear tracks in solids: principles and applications. University of California Press
67. Komaki Y, Tsujimura S (1976) Growth of fine holes in polyethylenenaphthalate film irradiated by fission fragments. J Appl Phys 47(4):1355–1358
68. Komaki Y, Ishikawa N, Sakurai T (1995) Effects of gamma rays on etching of heavy ion tracks in polyimide. Radiat Meas 24(2):193–196
69. Starosta W, Wawszczak D, Sartowska B, Buczkowski M (1999) Investigations of heavy ion tracks in polyethylene naphthalate films. Radiat Meas 31(1–6):149–152
70. Shirkova VV, Tretyakova SP (1997) Physical and chemical basis for the manufacturing of fluoropolymer track membranes. Radiat Meas 28(1–6):791–798
71. Cadotte JE, Petersen RJ, Larson RE, Erickson EE (1980) A new thin-film composite seawater reverse osmosis membrane. Desalination 32:25–31
72. Cadotte J, Forester R, Kim M, Petersen R, Stocker T (1988) Nanofiltration membranes broaden the use of membrane separation technology. Desalination 70(1–3):77–88
73. Lau WJ, Ismail AF, Misdan N, Kassim MA (2012) A recent progress in thin film composite membrane: a review. Desalination 287:190–199
74. Li L, Zhang S, Zhang X, Zheng G (2007) Polyamide thin film composite membranes prepared from 3,4′,5-biphenyl triacyl chloride, 3,3′,5,5′-biphenyl tetraacyl chloride and m-phenylenediamine. J Membr Sci 289(1–2):258–267
75. Zheng Y, Yao G, Cheng Q, Yu S, Liu M, Gao C (2013) Positively charged thin-film composite hollow fiber nanofiltration membrane for the removal of cationic dyes through submerged filtration. Desalination 328:42–50
76. Rambabu K, Bharath G, Monash P, Velu S, Banat F, Naushad M, Arthanareeswaran G, Show PL (2019) Effective treatment of dye polluted wastewater using nanoporous CaCl₂ modified polyethersulfone membrane. Process Saf Environ Prot 124:266–278
77. Cheryan M (1998) Ultrafiltration and microfiltration handbook. CRC Press
78. Ulbricht M (2006) Advanced functional polymer membranes. Poly 47(7): 2217–2262
79. Zaghbani N, Hafiane A, Dhahbi M (2007) Separation of methylene blue from aqueous solution by micellar enhanced ultrafiltration. Sep Purif Technol 55(1):117–124
80. Garud RM, Kore SV, Kore VS, Kulkarni GS (2011) A short review on process and applications of reverse osmosis. Univ J Environ Res Technol 1(3)
81. Ciardelli G, Corsi L, Marcucci M (2001) Membrane separation for wastewater reuse in the textile industry. Resour Conserv Recycl 31(2):189–197
82. Lin J, Tang CY, Ye W, Sun SP, Hamdan SH, Volodin A, Van Haesendonck C, Sotto A, Luis P, Van der Bruggen B (2015) Unraveling flux behavior of superhydrophilic loose nanofiltration membranes during textile wastewater treatment. J Membr Sci 493:690–702
83. Gao J, Thong Z, Wang KY, Chung TS (2017) Fabrication of loose inner-selective polyethersulfone (PES) hollow fibers by one-step spinning process for nanofiltration (NF) of textile dyes. J Membr Sci 541:413–424

84. Amini M, Arami M, Mahmoodi NM, Akbari A (2011) Dye removal from colored textile wastewater using acrylic grafted nanomembrane. Desalination 267(1):107–113
85. Kurt E, Koseoglu-Imer DY, Dizge N, Chellam S, Koyuncu I (2012) Pilot-scale evaluation of nanofiltration and reverse osmosis for process reuse of segregated textile dyewash wastewater. Desalination 302:24–32
86. Gunawan FM, Mangindaan D, Khoiruddin K, Wenten IG (2019) Nanofiltration membrane cross-linked by m-phenylenediamine for dye removal from textile wastewater. Polym Adv Technol 30(2):360–367
87. Sutedja A, Josephine CA, Mangindaan D (2017) Polysulfone thin film composite nanofiltration membranes for removal of textile dyes wastewater. IOP Conf Ser Earth Environ Sci 109(1):012042
88. Karisma D, Febrianto G, Mangindaan D (2017) Removal of dyes from textile wastewater by using nanofiltration polyetherimide membrane. IOP Conf Ser Earth Environ Sci 109(1):012012
89. Chen G, Chai X, Po-Lock Y, Mi Y (1997) Treatment of textile desizing wastewater by pilot scale nanofiltration membrane separation. J Membr Sci 127(1):93–99
90. Lopes CN, Petrus JCC, Riella HG (2005) Color and COD retention by nanofiltration membranes. Desalination 172(1):77–83
91. Bes-Piá A, Iborra-Clar MI, Iborra-Clar A, Mendoza-Roca JA, Cuartas-Uribe B, Alcaina-Miranda MI (2005) Nanofiltration of textile industry wastewater using a physicochemical process as a pre-treatment. Desalination 178(1–3):343–349
92. Marcucci M, Nosenzo G, Capannelli G, Ciabatti I, Corrieri D, Ciardelli G (2001) Treatment and reuse of textile effluents based on new ultrafiltration and other membrane technologies. Desalination 138(1–3):75–82
93. Van der Bruggen B, Cornelis G, Vandecasteele C, Devreese I (2005) Fouling of nanofiltration and ultrafiltration membranes applied for wastewater regeneration in the textile industry. Desalination 175(1):111–119
94. Van der Bruggen B, Daems B, Wilms D, Vandecasteele C (2001) Mechanisms of retention and flux decline for the nanofiltration of dye baths from the textile industry. Sep Purif Technol 22:519–528
95. Tang C, Chen V (2002) Nanofiltration of textile wastewater for water reuse. Desalination 143(1):11–20
96. Ghaemi N, Madaeni SS, Daraei P, Rajabi H, Zinadini S, Alizadeh A, Ghouzivand S (2015) Polyethersulfone membrane enhanced with iron oxide nanoparticles for copper removal from water: application of new functionalized Fe_3O_4 nanoparticles. Chem Eng J 263:101–112
97. Shahkaramipour N, Ramanan SN, Fister D, Park E, Venna SR, Sun H, Lin H (2017) Facile grafting of zwitterions onto the membrane surface to enhance antifouling properties for wastewater reuse. Ind Eng Chem Res 56(32):9202–9212
98. Tian Y, Wu M, Liu R, Li Y, Wang D, Tan J, Wu R, Huang Y (2011) Electrospun membrane of cellulose acetate for heavy metal ion adsorption in water treatment. Carbohydrate Polymers 83(2):743–748; Bodalo A, Gomez JL, Gomez E, Leon G, Tejera M (2005) Ammonium removal from aqueous solutions by reverse osmosis using cellulose acetate membranes. Desalination 184(1–3):149–155
99. Bodalo A, Gomez JL, Gomez E, Leon G, Tejera M (2005) Ammonium removal from aqueous solutions by reverse osmosis using cellulose acetate membranes. Desalination 184(1–3):149–155
100. Du JR, Peldszus S, Huck PM, Feng X (2009) Modification of poly (vinylidene fluoride) ultrafiltration membranes with poly (vinyl alcohol) for fouling control in drinking water treatment. Water Res 43(18):4559–4568
101. Srivastava HP, Arthanareeswaran G, Anantharaman N, Starov VM (2011) Performance of modified poly (vinylidene fluoride) membrane for textile wastewater ultrafiltration. Desalination 282:87–94
102. Hwang J, Choi J, Kim JM, Kang SW (2016) Water treatment by polysulfone membrane modified with tetrahydrofuran and water pressure. Macromol Res 24(11):1020–1023

103. Torkabad MG, Keshtkar AR, Safdari SJ (2017) Comparison of polyethersulfone and polyamide nanofiltration membranes for uranium removal from aqueous solution. Prog Nucl Energy 94:93–100

104. Abdel-Karim A, Gad-Allah TA, El-Kalliny AS, Ahmed SI, Souaya ER, Badawy MI, Ulbricht M (2017) Fabrication of modified polyethersulfone membranes for wastewater treatment by submerged membrane bioreactor. Sep Purif Technol 175:36–46

105. Hernández-Aguirre OA, Nunez-Pineda A, Tapia-Tapia M, Gomez Espinosa RM (2016) Surface modification of polypropylene membrane using biopolymers with potential applications for metal ion removal. J Chem

106. Fan F, Wang L, Jiang W, Chen B, Liu H (2016) A novel polyethylene microfiltration membrane with highly permeable ordered 'wine bottle'shaped through-pore structure fabricated via imprint and thermal field induction. J Phys D Appl Phys 49(12):125501

107. Febrianto G, Karisma D, Mangindaan D (2019) Polyetherimide nanofiltration membranes modified by interfacial polymerization for treatment of textile dyes wastewater. IOP Conf Ser Mater Sci Eng 622(1):012019

108. Lu Y, Fang W, Kong J, Zhang F, Wang Z, Teng X, Zhu Y, Jin J (2020) A microporous polymer ultrathin membrane for the highly efficient removal of dyes from acidic saline solutions. J Membr Sci 603:118027

109. Dutta M, Bhattacharjee S, De S (2020) Separation of reactive dyes from textile effluent by hydrolyzed polyacrylonitrile hollow fiber ultrafiltration quantifying the transport of multi-component species through charged membrane pores. Separation Purif Technol 234:116063

110. Gao DW, Zhang T, Tang CYY, Wu WM, Wong CY, Lee YH, Yeh TH, Criddle CS (2010) Membrane fouling in an anaerobic membrane bioreactor: differences in relative abundance of bacterial species in the membrane foulant layer and in suspension. J Membr Sci 364(1–2):331–338

111. Kang GD, Cao YM (2014) Application and modification of poly (vinylidene fluoride)(PVDF) membranes–a review. J Membr Sci 463:145–165

112. Li N, Xiao C, An S, Hu X (2010) Preparation and properties of PVDF/PVA hollow fiber membranes. Desalination 250(2):530–537

113. Ma S, Meng J, Li J, Zhang Y, Ni L (2014) Synthesis of catalytic polypropylene membranes enabling visible-light-driven photocatalytic degradation of dyes in water. J Membr Sci 453:221–229

114. Karisma D, Febrianto G, Mangindaan D (2018) Polyetherimide thin film composite (PEI-TFC) membranes for nanofiltration treatment of dyes wastewater. IOP Conf Ser Earth Environ Sci 195(1):012057. IOP Publishing

Dye Removal Using Polymer Composites as Adsorbents

Rwiddhi Sarkhel, Shubhalakshmi Sengupta, Papita Das, and Avijit Bhowal

1 Introduction

The utilization and regeneration of the sustainable resources for the development of industrial technology has become a great concern worldwide. Water pollution has become a major worldwide environmental issue harming the sustainability of the environment [4, 27]. Water pollution disrupts the supply chain in industrial belts posing a serious threat to the environmental feasibility [53]. During last few decades, there has been a lot of research for the minimization of toxic pollutants from the environment to make it healthy and sustainable. Highly toxic compounds have been released into the environment directly or indirectly. These compounds maybe dyes, pesticides, polycyclic aromatic hydrocarbons, radionuclides, and heavy metals [14]. They are widely used in industries like paper mills, leather, textile, pharmaceuticals, and cosmetics [44, 57, 58].

It has been reviewed that the main causes of water pollution are caused by the excessive discharge of textile wastewater which can pose harmful effects on the human mankind like cancer, skin allergies, headaches, different mutations, etc. due to its complex aromatic structure [17]. The removal of organic dyes from the ecosystem is found to be a challenging task, so effective approaches are developed to mitigate

R. Sarkhel · P. Das (✉) · A. Bhowal
School of Advanced Studies in Industrial Pollution Control Engineering, Jadavpur University, Jadavpur, Kolkata, West Bengal 700032, India
e-mail: papitasaha@gmail.com

R. Sarkhel · S. Sengupta
Department of Sciences and Humanities, Vignan's Foundation for Science, Technology and Research (VFSTR), Deemed to be University, Vadlamudi, Guntur, Andhra Pradesh 522213, India

P. Das · A. Bhowal
Department of Chemical Engineering, Jadavpur University, 188, Raja S. C. Mullick Road, Kolkata 700032, India

© The Author(s), under exclusive license to Springer Nature Singapore Pte Ltd. 2022
A. Khadir and S. S. Muthu (eds.), *Polymer Technology in Dye-containing Wastewater*, Sustainable Textiles: Production, Processing, Manufacturing & Chemistry, https://doi.org/10.1007/978-981-19-0886-6_4

the destruction of natural resources. The minimization of dye wastes can reduce the risk of water pollution [42]. Due to the complex molecular structure of dyes, they can resist degradation caused by light, physical or biological pathways using micro-organisms [8]. More than 10,000 different dyes have been utilized in pharmaceutical, leather, cosmetics, textile, and paper industries [38]. Approximately 2% of these textile industry dyes directly end up as effluent in water sources. Therefore, the amount of dye concentration estimated in wastewater varies in the range of 5-200 mg/L [2].

Different techniques are available for the removal of dye which includes microbial as well as physical methods like advanced oxidation (by photodegradation), coagulation, flocculation, membrane separation, adsorption, microbial degradation [7, 18].

The complete depletion of dye molecule can be readily obtained by the process of oxidation, either by photochemical process or simple chemical method [20]. The main goal of any advanced oxidation technique is to synthesize the hydroxyl molecule (HO*) for better interaction through hydrogen bonding and utilizing it as a strong oxidizing agent to degrade the contaminant in the aqueous solution. Photocatalytic degradation method is one of the most attractive and promising techniques among the various approaches due to its reliability on the utilization of irradiation energy, generation to less harmful by-products, simplicity, and eco-friendly. Coagulation and flocculation are the processes utilized in water treatment to remove pollutants like oil, grease, dyes, metals maintaining COD, and water concentration. This can lead to development of highly cost-effective polymer composites [65]. Membrane separation process is widely used nowadays to remove the contaminants in a physical manner by the utilization of a membrane thus yielding a high environmental stability and the removal capacity is also high by involving high pressure-driven methods such as ultrafiltration, nanofiltration, reverse osmosis [32]. The stability of the membrane is at high permeation flux with 99% in neutral or alkaline environments [52]. Adsorption process is the most widely used technology for the dye removal process among all these methods due to its cost-effectiveness, high stability, low toxicity [6, 13].

The second important aspect is designing and commercial utilization of affordable and novel material for the industrial wastewater treatment has gained a significant growth in a few years. Nanocomposite materials have proven as a suitable alternative in order to overcome the drawbacks of monolithic and micro composites [48].

Nano-structured polymer composites help in improving the water quality thus enhancing environmental stability [56, 66]. Research on polymer composites helps in effective enhancement of wastewater treatment removing the micropollutants and focusing on the challenges of future research. Research on the Polymer composite was done due to their attractive properties like better adsorption and removal characteristics of metal ions, dyes, and other toxic pollutants [10]. Polymer composites consist of organic or inorganic compounds to provide allied advantageous properties such as stiffness, low density, high thermal resistance, chemical stability [40].

Polymer composites blended with natural fibers like cellulose obtained from different Agro waste like sugarcane bagasse, peanut shell proved to be an attractive option for the researchers. The polymer composites blended with graphene, carbon, or clay-based were majorly used in water treatment as well as desalination technologies [33, 59]. The development of a reinforced composite polymer by the addition of different linked monomers into it, which are derived from easily available and user-friendly agricultural wastes makes it productive and enhances sustainability [26]. The generation of the polymer reinforced composites in wastewater remediation and involving it in dye removal techniques is gaining worldwide interest for researchers and young scientists. Utilization of the industrial wastes used for the preparation of polymer composites reinforced with polymer, clay, cellulose/chitin, and carbon can functionalize the fiber surfaces and reduce the absorption of moisture making them compatible [29].

In this chapter, we have discussed various types of polymer composites that have been used for dye removal purposes. Recent works in this field have also been discussed here.

2   Various Types of Polymer Composites in Dye Removal

Different types of Polymers and polymeric composites are used for dye removal purposes (Fig. 1). The polymer composites are filled with fillers like clay, metal, carbon, or fibers.

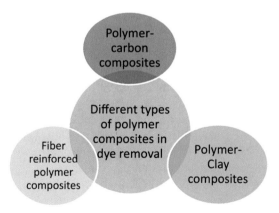

Fig. 1 Diagram illustrating different types of polymer reinforced composites

3 Polymer Composites

Polymer-polymer composites are mainly used in the process of dye removal since it has many attractive and advantageous properties like mechanical strength, high thermal stability, flexibility, durability, and high surface area with high surface-to-volume ratio. These properties make the polymer capable of doping with the other polymer composites by the synthesis involving functionalization by the method of crosslinking and blending with different fillers. The fillers maybe comprised of different Agrowaste like Sugarcane bagasse, peanut shells, rice husk ash, etc. The polymer-polymer composites have attracted the scientists and researchers all over the world because of their properties. So, these composites are well utilized in the purpose of wastewater treatment and in the desalination methods. The main advantage of these types of composites is wider range of pollutants can be utilized for the Bioremediation treatment, but the disadvantage is high production cost [11].

Polymers are often used as fillers in polymers, or one or more polymers are synthesized to form a new polymeric material where different fillers are reinforced to form a composite system. Elkady and co-workers developed a copolymer were developed using Styrene and acrylonitrile by the in-situ polymerization method and using nanofibers as filler and used it for dye removal (by basic Crystal violet dye) through the adsorption batch process. The polymer composite was produced by the technique of electrospinning [19]. The composite morphology is shown in (Fig. 2).

The surface morphological characteristics were identified by the SEM technique indicating a uniformity for the carboxylated nanofibers. The adsorptivity of the process was also been affected by the structural modifications with the isolated microbes. The adsorption capacity was noted at about 30 min as 67.11 mg/g. This is because of the toxicological and structural changes interpreted in the copolymer with the functional groups of carboxyl acid and amino acid bringing about immunological changes in the polymer by the high adsorptive power in the removal of basic dyes like Crystal violet and Methylene blue dye [34]. Synthesis of cyclodextrin composites also proved to be an exceptional group of polymer-polymer composites with cost-effective power, high feasibility, and effectiveness in the process of dye removal due to their good physicochemical properties [31, 68]. The efficacy for dye removal was increased by crosslinking the cyclodextrins with the polydopamine for the high thermal stability and structural feasibility immobilized with starch to utilize it in wastewater and biodegradation treatment using fungi and cyanobacteria due to their microporous nature. The β cyclodextrin plays a significant role in the adsorption process of dye removal using microbes due to the intraparticle diffusion and interlinking forces of attraction in dyes with the polymer doped composites [63].

Similarly, polymer composites like PVA/Cellulose, PVA/chitin were generated from the fillers combined with the monomers. The fillers were utilized as the natural fibers like cellulose. Cellulose was extracted from the Agrowaste like Sugarcane bagasse, Peanut shell waste. The cellulose was also grafted with 2-acrylamidomethylpropane generating efficient polymer composites for dye and heavy metal removal involving micro-organisms like *Aspergillus sp.* For the preparation of

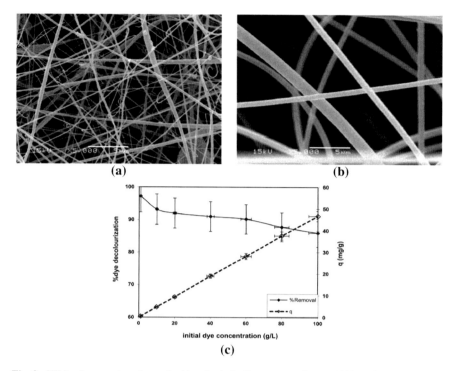

Fig. 2 SEM micrographs of **a** poly (Acrylonitrile-Styrene copolymer (AN-co-ST)) nanofibers; and **b** modified poly (AN-co-ST) nanofibers; **c** effect of initial dye concentration on dye removal capacity and percentage decolorization onto the chemically modified nanofiber (reproduced from [20], open-access article)

PVA/chitosan or PVA/Cellulose polymer, initially, PVA polymeric membrane was produced by adding 4 g PVA with 100 mL water, and continuously stirred for two hours. After the solution turns transparent, some amount of chitosan or cellulose in a 1:2 ratio is added to form a homogenous solution by the process of ultra-sonification. Finally, the polymer composites were obtained. And then the weights were measured, analyzed for the dye removal with Methyl orange dye solution under dark conditions with UV irradiation and the adsorption-desorption study was evaluated through UV-vis spectroscopy. Then the dye-loaded composites were used for the Biodegradation treatment using microbes like fungi and bacteria maintaining an optimum alkaline pH and temperature of about 35 °C [47]. Also, polymer composites with diethylen-etriamine with polyacrylonitrile were produced by the membrane separation and electrospinning technique. The composites were characterized by SEM and TGA. SEM confirmed the presence of roughness in the structure due to the incorporation of diethylamine with the presence of microbes which affected the dye removal process [3]. The advantages and disadvantages of Polymer doped polymer composites are represented in (Table 1).

Table 1 Interpretation of advantages and disadvantages of polymer-polymer composites

Advantages	Disadvantages
(1) Polymer composites are very cost-effective to use	(1) High cost of production
(2) High dye removal efficacy and adsorptive capacity	(2) High labor cost
(3) Structural feasibility increases due to highly porous nature	(3) High viscous nature inherits low electrical conductivity

As mentioned earlier various types of fillers are reinforced in polymer matrices to form polymer composites which can be used in wastewater treatment. Some of these types of filler reinforced composites are discussed in detail in the following sections.

4 Polymer-Carbon Composites

Carbon when effectively involved with Polymer composites has been significantly used in the water treatment involving Biodegradation processes and adsorption with dye removal process exists in different molecular properties with advantages like high mechanical and chemical properties, high thermal stability, and cost-effective technique. Polymer doped carbon composites are generated with different forms of carbon like carbon nanotubes (CNT), Activated carbon, Graphene and can be utilized in various industrial and commercial applications. The advantages of the composites enable them for the high removal efficiency of varied micropollutants like dye, heavy metals from wastewater. The polymer doped carbon composites have attracted the scientists due to the excellent properties of high surface area and high solubility persists. The generation process of polymer-carbon composites is represented in (Fig. 3).

Carbon nanotubes (CNT) are the cylindrical structure of carbon materials with a one-dimensional graphitic structure having high surface area making it promizable to be utilized in Bioremediation process involving dye and oil removal. They are of two types multiwalled carbon tubes and single-walled carbon nanotubes. But the carbon nanotubes possess a low solubility technique due to the absence of surface functional groups which resulted in low adsorption performance. The carbon nanotubes show

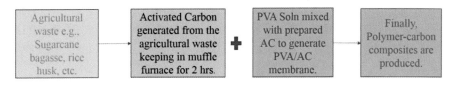

Fig. 3 Synthesis process of polymer-carbon composites, a generalized scheme using polyvinyl alcohol (PVA) as matrix and activated carbon as filler [9]

an excellent functionalization property for the degradation process. Graphene when interlinked with the multiwalled nanotube structure grafted with a PVA polymer enhances the functional property due to crosslinking by sp^2 hybridization [50]. The polymer composites in the combination of activated carbon and Graphene possess a high stability for the bioremediation by the utilization of advanced treatment technologies like Photodegradation and ozonation (in presence of ozone) and microorganisms. Activated Carbon preparation from the agricultural wastes has foreseen the environmental problems and thus increases the mechanical stability of the carbon nanocomposites with the reinforcement of activated carbon derived from Agrowaste like Sugarcane bagasse, Rice husk ash, Peanut, Coconut waste, Tea waste, etc.

The composite research on carbon nanotubes attracted attention due to the improved characteristics by surface modifications in the adsorption performance involving dye removal enhancing the thermal and mechanical stability. Fabrication of carbon nanotubes by surface functionalization is a facile approach. A recent literature study was done where carbon nanotubes were grafted with poly (sodium-p-styrene sulfonate) where the CNT was initially coated with a layer of dopamine and then polymerized to utilize in water treatment involving biodegradation by *Azolla* cyanobacteria then in the removal of a basic Methylene blue dye. The enhanced dye removal showed a great adsorptive capacity at 25 min with 174 mg/g [61]. Polyaniline when encapsulated with multiwalled CNT's by the in-situ polymerization process possesses a high removal capacity of about 884.84 mg/g with alizarin yellow dye [60]. This suggested that the adsorption study interpreted pseudo second-order Langmuir isotherm and thus the polymer doped carbon CNT composite may be hence utilized as a cost-effective adsorbent for the removal of various dyes.

PVA/AC polymer doped composite is synthesized by the process of 4 g of PVA (Polyvinyl Alcohol) powder was mixed with 100 mL of distilled water and kept for stirring for 2 h to form a transparent solution [47]. Then 2 g of activated carbon was generated from Agrowaste Sugarcane bagasse and carbonized in muffle furnace to obtain the activated carbon then was added in 10 ml of prepared PVA solution and stirred for about 1 h. Hence, the obtained solution was ultrasonicated for 30 min. Then cast on a sterile petri dish by the solution casting polymerization technique and utilized in dye removal process and finally dye-loaded composites were used in biodegradation using bacterial strains [47].

Polymer doped graphene composites have gained a lot of interest due to their effective physicochemical properties and their three-dimensional structure which enhances the porous nature, high thermal resistance, and high electrical conductivity. Polymer-Graphene composites can be structured and fabricated onto membranous films to enhance the bioremediation and dye removal process, also may be used in ultrafiltration membrane separation techniques [28]. Sodium alginate beads were grafted with Graphene to form a composite with excellent selectivity and permeavity [12]. In a study Poly (N, N-2-ethyl aminoethyl methacrylate) was cast with graphene oxide for the removal of methylene orange dyes [15]. A nanoporous fibrous membranous composite was developed using multiwalled carbon nanotube encapsulated with polymer for dye removal by Methylene blue and Congo red because of its low permeation flux and low pressure-driven process carried out by electrospinning and

Table 2 Representation of advantages and disadvantages of polymer-carbon composites

Advantages	Disadvantages
(1) These polymer composites are easy to use and cost-effective	(1) Ability of dispersion is poor
(2) High surface area and porosity	(2) Production process is very critical with the advanced treatment methods
(3) High adsorptive capacity	(3) Lack of design guidance

spraying technique [51]. The polymer composites doped with graphene possess good adsorption properties focusing on magnetic adsorbents with biopolymers. Biopolymers were used where iron oxide played a significant role in the magnetic property. The dye removal by polymer-graphene composites was subjected to photocatalytic processes by the isolated micro-organisms of fungi [69]. The polymer doped graphene composite is most commonly used composite because of its high adsorption and removal property due to the functionalization and high porosity which enables it in the utilization of Bioremediation process. The advantages and Disadvantages of Polymer doped Carbon composites are interpreted in (Table 2).

5 Polymer-Clay Composites

Some treatment technologies for the water treatment processes involving Bioremediation and Adsorption process are favored due to their attractive properties such as high effectiveness in the removal process, easy operation technique, and high efficiency. Polymer doped clay composites are one of them which are emerging technologies utilized in the wastewater processes with flexible nature and nature of adsorbents. Clays constitute natural materials with low cost used as natural efficacious adsorbents for the removal of many micropollutants like oil, heavy metals, and dyes. The advantage of these types of composites is their high porosity and high surface-to-volume ratio. But their disadvantage is their wettability process is very poor and slow. Polymer-clay composites can be produced by different associative methods like flocculation, exfoliation, coating, and intercalation (explained in Fig. 4).

A polymer cationic clay composite was grafted with polyvinyl pyridine (polymer) and montmorillonite (cationic clay) as a novel adsorbent for the removal of dyes like Methylene blue with an effective pH response. The high elimination of organic pollutants has been due to the high presence of leachate ions and high zeta potential. At an increase of pH level, the desorption of pollutants takes place using these types of composites [21]. A similar study was conducted using montmorillonite and sodium 2-acrylamido-2-methylpropane sulfonate and N isopropylacrylamide with a graft polymerization using a sulfate free technique. This enhanced the dye removal property by Methylene blue dye and the bioremediation process with these dye-loaded composites. The composites have been so efficient that they can be reused three to

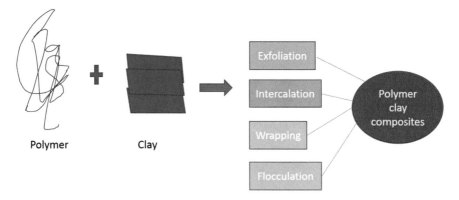

Fig. 4 Diagram representing the formation of polymer-clay composites by different methods

four times with no removal loss [5]. The dye has an adsorption capacity of 1875 mg/g representing the highest removal involving mercuric ions due to its highly porous nature and chelation property [64]. Also, a polymer-clay composite produced from montmorillonite with polyvinyl alcohol and poly (4-styrenesulfonic acid-comaleic acid having highly efficient and cost-effective adsorbents [36]. A hydrogel composite with polymer doped clay composite comprising polymers with montmorillonite was exfoliated using bis [2-(methacryloyloxyethyl] phosphate as a crosslinker used for the elimination of methyl red, methylene blue, and crystal violet from wastewater with maximum adsorption capacity of 113, 155, and 176 mg/g. The desorption technique was exhibited in the Ethanol solution [39]. From the recent literature survey, anionic clay-based composite of magnesium aluminum coated with polydopamine was produced for the desalination and water treatment for the bioremediation purposes by using microbes. Also, utilizing removal of copper ions as well as dye removal by Methylene blue and Methyl orange with an adsorption capacity of 198.73 mg/g [11]. The good adsorption capacity was due to the interaction of hydroxyl molecules through hydrogen bonding with removal of dye from wastewater enhancing the adsorption and thus this composite can be effectively used for the bioremediation purposes too [16, 62]. The schematic representation of synthesis of acrylamide and N-isopropyl acrylamide-montmorillonite composite and its use for methylene blue elimination from aqueous media is represented in (Fig. 5).

The advantages and disadvantages of Polymer-clay composites are discussed in (Table 3).

6   Fiber-Reinforced Polymer Composites

Fiber-reinforced composites have been nowadays the most prominent option to be used in case of water treatment since they are very lightweight and have high tensile strength due to the reinforcement of fillers into them. The fillers maybe of metal,

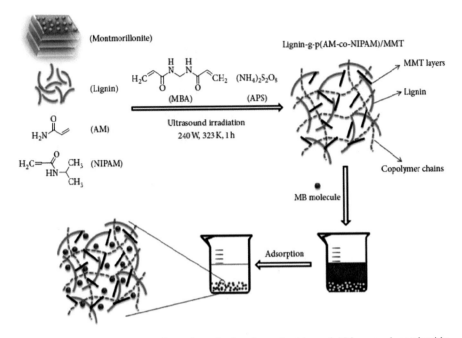

Fig. 5 A schematic representation of synthesis of acrylamide and N-isopropyl acrylamide-montmorillonite composite, and its use for methylene blue elimination from aqueous media (reproduced from Berber et al. [11], open-access article)

Table 3 Representation of advantages and disadvantages of polymer-clay composites

Advantages	Disadvantages
(1) Utilization of cost-effective adsorbents	(1) High labor cost
(2) High adsorption capacity due to more interaction of hydroxyl molecules	(2) Structure of composites was complex
(3) High efficacy for bioremediation purposes	(3) Production costs for developing these composites are very high

ceramic, carbon, and natural fibers like cellulose. These composites are a heterogeneous mixture of two materials, one embedded into the other. Fiber-reinforced polymer composites have high performance due to their crosslinking nature of cellulosic fibers with excellent structural properties. These properties make the polymer capable of reinforcing with fillers by the synthesis involving functionalization by the method of crosslinking and blending. The fillers maybe comprised of cellulose extracted from different Agrowaste like Sugarcane bagasse, peanut shells, rice husk ash, etc. The synthesis of fiber-reinforced polymer composite has been interpreted in (Fig. 6).

Polyvinyl composites reinforced with nanocellulose, chitin, carbon were synthesized. These composites were cost-effective, eco-friendly, and reusable. The

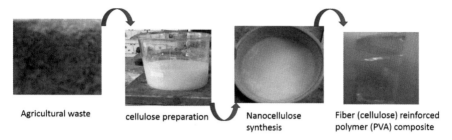

Agricultural waste | cellulose preparation | Nanocellulose synthesis | Fiber (cellulose) reinforced polymer (PVA) composite

Fig. 6 Synthesis of fiber-reinforced polymer composite membrane, a pictorial representation with nanocellulose as filler and PVA as matrix [37]

crosslinking component was the maleic acid to generate the biopolymer composite film. The composite film was evaluated for the dye removal process involving Methylene blue dye, and the effect of parameters like pH, temperature, concentration was studied for the batch process. The adsorption capacity was extremely high as 467.5 mg/g, confirmed from the FTIR and SEM analysis. SEM analysis showed a rough surface after the incorporation of methylene blue dye into it. FTIR evaluated structural bonds linked to the fillers with different functional characteristics [37].

PVA/Chitin and PVA/nanocellulose was prepared by initially preparing PVA film, dissolving 10 g of PVA granules with 70 g of water, and kept stirring until the solution becomes homogenous at 85 °C for about 2 h. Then nanocellulose was prepared by ultrasonicating the prepared cellulose mixture for 15 min. Now, finally, the chitin or nanocellulose is mixed with maleic acid and PVA in a weight ratio of 1:5 and stirring it for about 3 h. at 500 rpm. HCL was added to the mixture of maleic acid since it behaves as a catalyst. After the solution turns homogenous, pour into a sterile petri dish and keep it under a low heat for the generation of the fiber-reinforced polymeric composite membrane [22]. Finally, the polymer composites were utilized for the bioremediation process involving dye removal and the dye-loaded composites can also be used for the biodegradation process. The isotherm and adsorption studies were done to investigate the adsorption-desorption, and kinetic parameters at an alkaline pH range about 8.5 so that the composites can be regenerated again for further use [43]. The advantages and disadvantages of Fiber-reinforced polymer composites are discussed in (Table 4). Various properties of all these polymer reinforced composites have been explained in (Table 5).

Table 4 Advantages and disadvantages of fiber-reinforced polymer composites

Advantages	Disadvantages
(1) They have high strength and good rigidity	(1) They have low electrical conductivity
(2) The adsorption properties with respect to dye removal are very high	(2) The density of these composites is low so adsorption resistance is high
(3) They have high thermal resistance, so durability is high and thus can be used for bioremediation purposes	(3) They are fragile and hard to repair

7 Dye Removal Properties for the Different Polymer Composites Used in the Study

See Table 5.

Table 5 Representing SWOT analysis of the composites

Types of composites	Strengths	Weaknesses	Opportunities	Threats
(1) Polymer-polymer composites	(1) Very lightweight (2) Cost-effective (3) Environment friendly	(1) Low density (2) Low performance on mechanical properties	(1) Utilization of natural fibers as fillers (2) Design may be changed when needed	(1) High production costs (2) Disposal is not easy
(2) Polymer-carbon composites	(1) Cost-effective (2) High surface-to-volume ratio (3) Manufacturing processes benefits bioremediation techniques through dye elimination and using microbes (4) Rise of market economy	(1) Untrained labor (2) Dispersion ability is low (3) Advanced treatment technologies are critical (4) No loss with recycling	(1) High adsorption capacity (2) High porosity (3) Use of sustainable renewable resources (4) New applications are produced	(1) Low energy density (2) Regeneration is critical (3) Little applied R&D
(3) Polymer-clay composites	(1) High electrical conductivity (2) Weight loss in composites is high (3) High adsorption	(1) Thermal resistance is not very high (2) Mechanical properties are low (3) Fatigue properties develop	(1) New and emerging market technologies (2) Sustainability with green materials (3) Surface area is high	(1) Lack of production chains (2) High labor cost (3) High demand for dimensional analysis with biomass cropping
(4) Fiber-reinforced composites	(1) High porosity (2) High strength to volume ratio (3) Adsorption capacity is high thus enabling high removal efficiency	(1) High production cost (2) Low electrical conductivity (3) No magnetic property	(1) Cost-efficient (2) Easily available (3) Durability is high, and high thermal resistance so reinforcement is strong	(1) They are hard to repair due to fragility and porosity (2) Low corrosion resistance (3) Maintenance and production cost is very high

8 Recent Works of Dye Removal Using Polymer Composites

In the recent literature survey, for the purpose of photocatalytic process, a hybrid composite was produced. Polyoxometalate reinforced with polymer was prepared at optimum temperature conditions by the photocatalysis technique. The polyoxometalate was reinforced into the polymeric resin and then was studied under light intensity to interpret the dye removal capacity. The prepared composites reinforced into the polymer matrix revealed high adsorption properties, and the photocatalytic ability was studied under UV irradiation on the Eosin dye [25]. It was noted that the removal efficiency was about 98 and 93%, respectively using these polyoxometalate hybrid composites, whereas it was 94% using silicon reinforced with polyoxometalate composites. The photocatalytic characteristics were studied to increase the quantum yield for the dye decolorization which is crosslinked with the ligands of functional groups occupying oxygen molecules [45].

The characteristics of the polymer were evaluated by the FTIR analysis. After the photopolymerization process, the photocatalytic properties of the prepared polyoxometalate reinforced with the polymeric resin were reserved thus enabling efficient dye removal using Eosin dye. The composites were generated due to the immobilization process studied under UV light. The composite efficiency was interpreted utilizing UV spectroscopy with a wavelength of 405 nm maintaining a pH of 8. The development of these composites leads to high yield and offers new possibilities to remove contaminants from wastewater enhancing the photocatalytic efficacy. Thus, these polymer composites can be regenerated, recycled, and utilized for bioremediation removing pollutants like dye and for other purposes to enhance environmental sustainability.

The development of polymer composites of clay with the green composite polyethylene was done for removing methylene blue dye from wastewater. The different optimum parameters were maintained during the experimental process such as pH (9), concentration of the dye at 1×10^{-5} M with an agitation speed of 1440 rpm. The removal of methylene blue dye from the polymer composites was due to the interaction of the molecules through hydrogen bonding increasing the surface area and porosity [23]. The adsorption and kinetic parameters like temperature, pH, and dye concentrations were investigated for the composites having high adsorptive capacity stated that the adsorption process depicted pseudo second-order.

The three different concentrations were maintained as 5×10^{-6}, $10 \times 10^{-6,}$ and 25×10^{-6} M, with room temperature and the values of pH were 5.5, 7, and 9, respectively. The sample was used for the adsorption experiments evaluated through UV-vis spectroscopy and was characterized by SEM, TGA. The literature survey revealed from the results increasing methylene blue dye concentration enhances the adsorption capacity maintaining a pH value of 5.5–9. Due to the molecular interaction and hydrogen bonding, the kinetic energy increases. The adsorption isotherm models for the best fit curve were obtained by pseudo first-order, pseudo second-order, mass transfer diffusion, and intraparticle diffusion. The activation energies, isotherm studies evaluating kinetic as well as thermodynamic parameters were discussed. The Gibbs's free energy was found to be $-$ 70.64 kJ/mol, Entropy $-$ 70.64 J/mol/K, and the activation energy as 12.37 kJ/mol maintaining a room temperature. The results revealed that the reaction was spontaneous and exothermic and thus this type of composites can be utilized in water treatment and providing green, sustainable environment [49]. Formation of the polymer-clay composite and its dye removal characteristics has been shown in (Figs. 7 and 8).

The polymer composites of polyaniline reinforced with hexaferrite and PVA was prepared and examined for dye removal analysis by Reactive black dye. Composite hydrogels were produced by the process of oxidation of aniline hydrochloride mixed with ammonium sulfate with 5% PVA in the presence of hexaferrite particles. These composites show a good adsorptive and magnetic power due to the presence of ferrite molecules with crosslinking of functional bonds incorporated into the hydrogel composites. The mixture prepared was initially kept for thawing at $-$ 5 °C for 5 days for the polymerization to take place [54]. After the thawing process, the composites were produced by the lyophilization process. The innovation for the electromagnetic property of the polymer composites attracted different researchers and scientists worldwide. These polymer composites were prepared due to their advantageous properties such as high chemical strength, high thermal stability, cost-effectiveness, high electrical conductivity, and environmental feasibility.

The porosity of the composites becomes high due to the presence of hexaferrite molecules thus enabling high efficacy for dye removal utilizing the hydrogels [11]. The freezing attitude of the hydrogels reinforced with polyaniline was incorporated because of the magnetic hexaferrite molecules which exhibit easy separation of

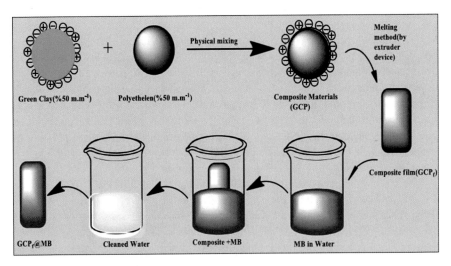

Fig. 7 Formation of GCPf (clay polyethylene composite) and its removal of methylene blue from water solution (reproduced from [49], open-access article)

Fig. 8 Removal of methylene blue by clay polyethylene composites from water in the laboratory (reproduced from [49], open-access article)

adsorbents from the fluid. The adsorptive removal capacity was about 99% by Reactive black dye, this is due to the reason of cohesive forces present in the hydrogels by the reinforcement of polyaniline and hexaferrite in presence of PVA, thus can be utilized in bioremediation and water treatment. The deformation of composites results in effective magnetic anisotropy and low coerciveness [46].

An eco-friendly bio adsorbent polymer composite was generated by the process of graft polymerization with high removal capacity of azo dyes from solutions. 1,1-diallyl-4-carboxypiperidin-1-ium bromide was incorporated into biomaterials like cellulose, chitosan. Different parameters which affected the adsorption property of dye are temperature, pH, adsorbent dosage, and solution concentration. The biopolymer composites served as efficient adsorbents with excellent amphoteric approach which enabled the composites in the bioremediation, desalination, and water treatment study. The bio adsorbent composite served as a pH-responsive material for the removal of azo dyes such as indigo and Congo red from the solutions [35, 67]. The polymer composites helped in the decontamination of dye-containing water solutions. The surface morphological properties and functionalization characteristics were determined by SEM, FTIR, TGA, and VSM which revealed the roughness nature of the composites after dye degradation. The adsorption isotherm models like Langmuir isotherm and pseudo second-order kinetic model were investigated to obtain the best fit model for the experimental analysis and use it further. The polymer composites were regenerated with a basic solution of HCL or NaOH [41]. The adsorption capacity for the azo dyes was 840.33 mg/g and 909.1 mg/g, respectively. This is due to the entrapment and immobilization of dye molecules into the polymer bio adsorbents with cost-effective materials as an environment-friendly technique that enables the sustainability with regeneration, and ease in magnetic separation for the polymer composites.

9 Future Perspectives

The chapter foresees the generation of different types of polymer composites and their application in dye removal as well as for bioremediation in water treatment and desalination. The demand for clean water has been increased drastically in past decades due to increased marine pollution. Different methods were being employed for the treatment of above problems [1, 24]. The removal processes mainly involved adsorption techniques and separation processes (bioremediation or phytoremediation).

A vast variety of polymer reinforced composites for dye removal have been exhibited and utilized in lab-scale experimentation techniques, therefore exhibiting a positive outcome. Thus, challenges can be overcome by employing more applications to these composites in removing dyes and other micropollutants. Elimination of dye and weight loss of the composites results due to the polymerization techniques and electrochemical processes [30, 55].

10 Conclusion

As discussed in this chapter, the polymer reinforced composites made of all synthetic and natural polymers have remarkable, and attractive properties of adsorption and bioremediation when combined with different adsorbents like polymer, clay, cellulose, and carbon. The recent works related to the dye removal study utilizing the polymer composites have also been discussed in this chapter. These composites are prepared for the sustainability of the environment since they are cost-effective, and environment friendly. The production of these polymers yields a high cost, but they have easy fabrication, have high chemical and thermal strength, and application process. Using Agro waste in the preparation of natural fibers as a filler in developing the polymer-polymer composites and in the generation of activated carbon for polymer-carbon composites exhibit high efficacy. A lot of research has been done to make the polymers durable, highly efficient, and have high adsorption properties to be utilized in dye removal from industrial wastewaters.

Acknowledgment The author acknowledges people of Jadavpur University for their valuable guidance and support at each level of the research. The work was supported by RUSA 2.0 (financial assistance) from Jadavpur University.

Conflict of Interest There is no conflict of interest by the authors.

References

1. Abdullah N, Tajuddin M, Yosuf H (2018) Carbon based polymer nanocomposites for dye removal:32493262
2. Ahmad R, Kumar R (2010) Adsorption studies of hazardous malachite green onto treated ginger waste. J Environ Manage 91(4):1032–1038
3. Almasian A, Olya ME, Mahmoodi NM (2015) Preparation and adsorption behavior of diethylenetriamine/polyacrylonitrile composite nanofibers for a direct dye removal. J Fib Polym 16(9):1925–1934
4. Ao C, Zhao J, Xia T, Huang B, Wang Q, Gai J, Chen Z, Zhang W, Lu C (2021) Multifunctional La (OH)$_3$@cellulose nanofibrous membranes for efficient oil/water separation and selective removal of dyes. Sep Purif Technol 254:117603
5. Atta AM, Al-Lohedan H, AL Othman HL, Abdel AA (2015) Characterization of reactive amphiphilic montmorillonite nanogels and its application for removal of toxic cationic dye and heavy metals water pollutants. J Ind Eng Chem 31:374–384
6. Bagha AT, Nikkar H, Mahmoodi N, Markazi M, Menger FM (2011) The sorption of cationic dyes onto kaolin: kinetic, isotherm and thermodynamic studies. Desalination 266(1–3):274–280
7. Barka N, Qourzal S, Assabbane A, Nounah A, Ait-Ichou Y (2010) Photocatalytic degradation of an azo reactive dye, reactive yellow 84, in water using an industrial titanium dioxide coated media. Arab J Chem 3(4):279–283
8. Bedin KC, Souza IPAF, Cazetta AL, Spessato L, Ronix A, Almeida VC (2018) CO$_2$ spherical activated carbon as a new adsorbent for methylene blue removal: kinetic, equilibrium and thermodynamic studies. J Mol Liq 269:132–139
9. Behera SK, Kim JH, Guo X, Park HS (2008) Adsorption equilibrium and kinetics of polyvinyl alcohol from aqueous solution on powdered activated carbon. J Hazard Mater 153:1207–1214
10. Berber MR (2020) Current advances of polymer composites for water treatment and desalination. J Chem 19:7608423
11. Bober P, Zasońska BA, Humpoliček P, Kucekova Z, Varga M, Horak D, Babayan V, Kazantseva N, Prokeš J, Stejskal J (2020) Polyaniline-maghemite based dispersion: electrical, magnetic properties and their cytotoxicity. J Synth Met 214:23–29
12. Cao K, Jiang K, Zhao J (2014) Enhanced water permeation through sodium alginate membranes by incorporating graphene oxides. J Mem Sci 469:272–283
13. Crini G. Non-conventional low-cost adsorbents for dye removal: a review. Biores Technol 97(9):1061–1085
14. Diez MC (2010) Biological aspects involved in the degradation of organic pollutants. J Soil Sci Pl Nutri 10(3):244–267
15. Dong L, Fan W, Tong X, Zhang H, Chen M, Zhao Y (2018) A CO$_2$-responsive graphene oxide/polymer composite nanofiltration membrane for water purification. J Mat Chem A 6(16):6785–6791
16. Dou J, Huang Q, Huang H (2019) Mussel-inspired preparation of layered double hydroxides-based polymer composites for removal of copper ions. J Coll Inter Sci 533:416–427
17. Du JJ, Yuan YP, Sun JX, Peng FM, Jiang X, Qiu LG, Xie AJ, Shen YH, Zhu JH (2011) New photocatalysts based on MIL-53metal–organic frameworks for the decolorization of methylene blue dye. J Hazard Mater 190:945–951
18. Elahmadi MF, Bensalah F, Gadri A (2009) Treatment of aqueous wastes contaminated with Congo red dye by electrochemical oxidation and ozonation processes. J Hazar Mat 168(2–3):1163–1169
19. Elkady M, El-Assar M, Hassan H (2016) Adsorption profile of basic dye onto novel fabricated carboxylated functionalized co-polymer nanofibers. Polymers 8(5):177
20. Forgacs E, Cserháti T, Oros G (2004) Removal of synthetic dyes from wastewaters: a review. Environ Int 30:953–971
21. Gardi I, Mishael YG (2018) Designing a regenerable stimuli responsive grafted polymer-clay sorbent for filtration of water pollutants. Sci Technol Adv Mat 19(1):588–598

22. Gatabi MP, Moghaddam HM (2016) Point of zero charge of maghemite decorated multiwalled carbon nanotubes fabricated by chemical precipitation method. J Mol Liq 216:117–125

23. Ghaedi M, Hossainian H, Montazerozohori M, Shokrollahi A, Shojaipour F, Soylak M, Purkait MK (2011) A novel acorn-based adsorbent for the removal of brilliant green. Desalination 281:220–233

24. Ghaemi N, Madaeni SS, Daraei P, Rajabi H, Shojaeimehr T, Rahimpour F (2015) PES mixed matrix nanofiltration membrane embedded with polymer wrapped MWCNT: fabrication and performance optimization in dye removal by RSM. J Hazard Mater 298:111–121

25. Ghali M, Brahmi C, Benltifa M, Dumur F, Duval S (2019) New hybrid polyoxometalate/polymer composites for photodegradation of eosin dye. J Poly Sci 57(14):1538–1549

26. Ghosh SK, Pal S, Ray S (2013) Study of microbes having potentiability for biodegradation of plastics. Int J Environ Sci Pollut 20:4339–4355

27. Halder JN, Islam MN (2015) Water pollution and its impact on the human health. J Environ Hum 2:36–46

28. Jayakaran P, Nirmala GS, Govindarajan L (2019) Qualitative and quantitative analysis of graphene-based adsorbents in wastewater treatment. Int J Chem Eng:9872502:17

29. Kalia S, Vasishta S, Kaith BS (2011) Cellulose nanofibers reinforced bioplastics and their applications, handbook of bioplas bio computer engineering application. Wiley-Scrivener Publishing, NY, USA

30. Li Z, Li Y, Zhu RS (2015) Synthesis and characterization of mesoporous carbon nanofibers and its adsorption for dye in wastewater. Adv Powder Technol 27:591–598

31. Li Y, Zhou Y, Lei J, Pu S (2019) Cyclodextrin modified filter paper for removal of cationic dyes/Cu ions from aqueous solutions. Wat Sci Tech 78(12):2553–2563

32. Lin AG, Liu PY, Liu G, Zhang GZ (2006) Ind Water Treat 26:5–8

33. Liu X, Ma R, Wang X (2019) Graphene oxide-based materials for efficient removal of heavy metal ions from aqueous solution: a review. Environ Pollut 252:62–73

34. Liu Q, Li Y, Chen H (2020). Superior adsorption capacity of functionalized straw adsorbent for dyes and heavy-metal Ions. J Hazard Mats 382:121040

35. Mahida VP, Patel MP (2016) Removal of heavy metal ions from aqueous solution by superabsorbent poly (NIPAAm/DAPB/AA) amphoteric nanohydrogel. Desalin Water Treat 57:13733–13746

36. Medhat F, Pakizeh M (2018) Preparation and characterization of a nano clay/PVA/PSf nanocomposite membrane for removal of pharmaceuticals from water. Appl Clay Sci 162:326–338

37. Mok CF, Ching YC, Muhamad F, Osman NAA, Dai Hai N, Hassan CRC (2020) Adsorption of dyes using poly (vinyl alcohol) (PVA) and PVA-based polymer composite adsorbents: a review. J Polym Env 28:775–793

38. Mondal S (2008) Methods of dye removal from dye 399 house effluent—an overview. Environ Eng Sci 25(3):383–396

39. Nakhjiri MT, Bagheri GM, Kurdtabar M (2018) Effect of bis [2-(methacryloyloxy)ethyl] phosphate as a crosslinker on poly (AAm-co-AMPS)/Na-MMT hydrogel nanocomposite as potential adsorbent for dyes: kinetic, isotherm and thermodynamic study. J Poly Res 25(11):244

40. Pang H, Wu H, Wang X, Hu B, Wang X (2019) Recent advances in composites of graphene and layered double hydroxides for water remediation: a review. Chem Asian J 14(15):2542–2555

41. Patel SR, Patel R, Patel MP (2020) Eco-friendly bio adsorbent-based polymer composites as a pH-responsive material for selective removal of anionic and azo dyes from aqueous solutions, J Macromol Sci Part A:1827957

42. Paulino AT, Guilherme MR, Reis AV, Campese GM, Muniz EC, Nozaki J (2006) Removal of methylene blue dye from an aqueous media using superabsorbent hydrogel supported on modified polysaccharide. J Colloid Interface Sci 301:55–62

43. Pirbazari A, Pargami N (2015) Surfactant-coated tea waste: preparation, characterization and its application for methylene blue adsorption from aqueous solution. J Environ Anal Toxicol 5(5):1

44. Raffi F, Hall JD, Cernigila CE (1997) Mutagenicity of azo dyes used in foods, drugs and cosmetics before and after reduction by clostridium species from the human intestinal tract. Food Che Toxicol 35:897–901
45. Rajeshwar K, Osugi ME, Chanmanee W, Chenthamarakshan CR, Zanoni MV, Kajitvichyanukul P, Krishnan-Ayer J (2008) J Photochem Photobiol C Photchem Rev 9(4):171
46. Sanchez PA, Stolbov OV, Kantorovich SS, Raikher YL (2019) Modeling the magnetostriction effect in elastomers with magnetically soft and hard particles. Soft Matter 15:7145–7158
47. Sarkhel R, Sengupta S, Das P, Bhowal A (2020) Comparative biodegradation study of polymer from plastic bottle waste using novel isolated bacteria and fungi from marine source. J Poly Res 27(16)
48. Schmidt D, Shah D, Giannelis EP (2002) New advances in polymer/layered silicate nanocomposites. Curr Opin Solid State Mater 6(3):205–212
49. Sen F, Demirbas O, Calimli MH, Aygun A, Alma MH, Nas MS (2018) The dye removal from aqueous solution using polymer composite films. J Appl Wat Sci 8:206
50. Serp P, Corrias M, Kalck P (2003) Carbon nanotubes and nanofibers in catalysis. Appl Catal A 253(2):337–358
51. Shi J, Wu T, Teng K, Wang W, Shan M, Xu Z (2016) Simultaneous electrospinning and spraying toward branch-like nanofibrous membranes functionalized with carboxylated MWCNTs for dye removal. Mater Lett 26(9):166
52. Song CW, Wang TH, Pan YQ (2006) Sep Purif Technol 51:80–84
53. Soni S, Bajpai PK, Mittal J, Arora C (2020) Utilization of cobalt doped iron based MOF for enhanced removal and recovery of methylene blue dye from wastewater. J Mol Liq 314:113642
54. Stejskal J, Bober P, Trchova M, Kovalcik A, Hodan J, Hromadkova J, Prokeš J (2017) Polyaniline cryogels supported with poly (vinyl alcohol): soft and conducting. Macromolecules 50:972–978
55. Sun Y, Wang G, Dong Q, Qian B, Meng Y, Qiu J (2014) Electrolysis removal of methyl orange dye from water by electrospun activated carbon fibers modified with carbon nanotubes. J Chem Eng 253:73–77
56. Takka S, Gürel A (2010) Evaluation of chitosan/alginate beads using experimental design: formulation and in vitro characterization. AAPS Pharm Sci Tech 11:460–466
57. Vandeviveri PC, Bianchi R, Verstraete W (1998) Treatment and reuse of wastewater from the textile wet-processing industry: review of emerging technologies. J Chem Tech Bio 72:289–302
58. Verma P, Madamwar D (2003) Decolorization of synthetic dyes by a newly isolated strain of *Serratia marcescens*. World J Microbi Biotech 19:615–618
59. Wang X, Chen L, Wang L (2019) Synthesis of novel nanomaterials and their application in efficient removal of radionuclides. Sci China Chem 62(8):933–967
60. Wu K, Yu J, Jiang X (2018) Multi-walled carbon nanotubes modified by polyaniline for the removal of alizarin yellow from aqueous solutions. Adsorp Sci Tech 36(1–2):198–214
61. Xie Y, He C, Liu L (2015) Carbon nanotube-based polymer nanocomposites: biomimic preparation and organic dye adsorption applications. RSC Adv 5(100):82503–82512
62. Yang XJ, Zhang P, Li P (2019) Layered double hydroxide/polyacrylamide nanocomposite hydrogels: green preparation, rheology and application in methyl orange removal from aqueous solution. J Mol Liq 280:128–134
63. Yue X, Jiang F, Zhang D, Lin H, Chen Y (2017) Preparation of adsorbent based on cotton fiber for removal of dyes. J Fib Polym 18(11):2102–2110
64. Zeng H, Wang L, Zhang D, Wang F, Sharma VK, Wang C (2019) Amido-functionalized carboxymethyl chitosan/montmorillonite composite for highly efficient and cost-effective mercury removal from aqueous solution. J Coll Int Sci 554:479–487
65. Zeng YB, Yang CZ, Zhang JD, Pu WH (2007) J Hazard Mater 147:991–996
66. Zhang P, Lo I, O'Connor D, Pehkonen S, Cheng H, Hou D (2017) High efficiency removal of methylene blue using SDS surface modified $ZnFe_2O_4$ nanoparticles. J Colloid Interface Sci 508:39–48
67. Zhen Y, Ning Z, Shaopeng Z, Yayi D, Xuntong Z, Jiachun S, Weiben Y, Yuping W, Jianqiang C (2015) A pH-and temperature-responsive magnetic composites adsorbent for targeted removal of nonylphenol. ACS Appl Mater Interfaces 7:24446–24457

68. Zhou Y, Hu Y, Huang W, Chen G, Cui C, Lu J (2018) A novel amphoteric β-cyclodextrin-based adsorbent for simultaneous removal of cationic/anionic dyes and bisphenol a. J Chem Eng 341:47–57

69. Zinadini S, Zinatizadeh AA, Rahimi M, Vatanpour V, Zangeneh H (2014) Preparation of a novel antifouling mixed matrix PES membrane by embedding graphene oxide nanoplates. J Membr Sci 453:292–301

Polymer-Derived Ceramic Adsorbent for Removal of Dyes from Water

Parimal Chandra Bhomick, Akito I. Sema, and Aola Supong

1 Introduction

The global population has been increasing at an exponential rate over the past decade. This has led to the increase in the industries to cope up with the demands of the growing population in terms of food, cloth, shelter, and energy. Thus, industrialization is a vital sector for the welfare and development of a country. But, this growth in the industries has resulted in the deterioration of the environment to a larger scale especially the water bodies. According to the CDP (Carbon Disclosure Project) Global Water Report 2018, companies in the seven sectors (food, textile, energy, industry, chemicals, pharmaceuticals, and mining) account for and wield influence over 70% of the world's freshwater use and pollution [10]. Among these sectors, the textile industry is one of the largest consumers of freshwater and also generates tons of wastewater [30]. It is reported that for average-sized textile mills about 100–200 L of water is utilized for one kilogram of fabric processed in a single day [42] and generates around 2700 MLD of wastewater in India alone by woolen industries [32].

Textile wastewater is the result of a magnitude of processes involved in the textile industry—de-sizing, scouring, bleaching, mercerizing, dyeing, printing, and finishing stages—and it utilizes numerous synthetic dyes. As the uptake of these synthetic dyes by fabrics is very poor, large amounts of highly colored water are often discharged as wastewater. These wastewater effluents contain a multifarious load of pollutants—waxes, ammonia, disinfectants, insecticides, surfactants, oils, anti-static

P. C. Bhomick (✉)
Department of Chemistry, Immanuel College, Lengrijan, Dimapur 797112, India
e-mail: paribhomick15589@gmail.com

A. I. Sema
Department of Chemistry, National Institute of Technology, Nagaland, Dimapur 797103, India

P. C. Bhomick · A. Supong
Department of Chemistry, Nagaland University, Lumami 798627, India

A. Khadir and S. S. Muthu (eds.), *Polymer Technology in Dye-containing Wastewater*,
Sustainable Textiles: Production, Processing, Manufacturing & Chemistry,
https://doi.org/10.1007/978-981-19-0886-6_5

105

agents, solvents, organic stabilizers, colors, metal ions, salts, sulfide, and formalde-hyde—which differ from mill to mill, country to country, fashion trend, fabric weight and type, season, and chemicals applied [41]. These pollutants directly devitalize the quality—pH, biological oxygen demand (BOD), chemical oxygen demand (COD), hardness, electrical conductivity, and total alkalinity—of water [28]. All these relate mostly to the dyeing and finishing processes and add to the larger portion of the total wastewater, as they are mostly released to the environment without prior treatment. Globally, an estimated 80% of all industrial and municipal wastewater is released into the environment without any prior treatment, with detrimental effects on human health and ecosystems [38].

It is reported that in India the textile industry due to the high demand for polyester and cotton, consumes around 80% of the total global production of 130,000 tons of dyestuff [18]. Dyes even in trace quantity can affect the quality of water as no one will drink colored water even in the least diluted form; they even hinder munic-ipal treatment processes such as ultraviolent decontamination [18]. Dyes can affect aquatic plants since they diminish sunlight transmission through water. Likewise, dyes may pose a danger to amphibian life and aquatic life; their presence in water results in the increase of turbidity, which obstructs the synthesis of food and oxygen by aquatic plants such as algae [6]. Most of the dyes used in the textile industry have complicated and stable aromatic structures, which make their biodegradation ineffective both chemically and photolytically. Even if they degrade, these degraded products might be mutagenic or teratogenic, carcinogenic, and may cause serious damage to people [1]. For example, failure of the kidneys, regenerative framework, liver, mind, and focal sensory system [28].

Though dyes have a wider application yet because of the risk associated with dyes and toxic derivatives, the treatment of wastewater from dyes becomes signif-icantly important and also challenging. Over the years, numerous methods have been used to remove dyes such as biological treatment, ultrafiltration, photocatal-ysis, electrochemical processes, and adsorption. Adsorption over other techniques is advantageous due to its low cost, ease of operation, and high removal efficiency and it is regarded as the most effective process of advanced wastewater treatment which industries employ to reduce/treat wastewater effluents [40].

Though various adsorbents—clays, porous carbon, biomass, zeolites—are exten-sively investigated for removal of pollutants, substantial works are dedicated currently toward the development of new adsorbent material or to improve the perfor-mance of existing adsorbents. In this regard, polymer-derived ceramics (PDCs) offer many advantages as they can be easily prepared by preceramic polymers (PP) as the molecular composition/architecture of PP can be altered easily to obtain various forms via plastic forming technique. PDCs are great nano-porous adsorbent mate-rials with large surface area, tunable porosity, good corrosion resistance, excellent mechanical strength, and high thermal stability due to which PDCs can be signifi-cantly employed for advanced adsorption and catalysis. These advance and enhanced properties of PDCs makes them an interesting candidate for applications in the fields of adsorption and catalyst support.

Since, "water is a basic human need, required for drinking and to support sanitation and hygiene, sustaining life and health. In fact, both water and sanitation are human rights" [36]. Thus, in this chapter, our focus is to highlight the different types of PDCs with reference to organosilicon polymers especially polysiloxane, as the preceramic polymer precursor, its synthesis and its application for removal of dyes from water by both adsorption and as catalyst support for dye degradation. The idea behind this chapter is to understand how polymer-derived ceramics can serve as a treatment method for textile wastewater especially colored substances like dye which is considered as a significant pool of pollution.

2 Polymer-Derived Ceramics

Polymer-derived ceramics (PDCs) are a recent class of porous ceramic materials, which are synthesized by direct thermal conversion of preceramic polymer (PP) into ceramics in an inert atmosphere. They are a novel class of multifunctional nanostructured materials that are composed chiefly of silicon, carbon, nitrogen, boron, and oxygen. The properties of PDCs depend on various parameters: preceramic polymer composition, temperature, heating rate, heating time, and chamber atmosphere. The PP are of various classes however, organosilicon polymers—polysiloxanes, polycarbosilanes, and polysilazanes—are of significant interest and are widely investigated due to their versatile properties: heat resistance, high mechanical strength, and chemical stability [25]. These properties provide the PP-derived PDCs with high potential applications in high-temperature sensors, electrode materials, optical devices, gas separation, catalysis, and water filtration.

2.1 Preceramic Polymers

Though PDCs can be prepared by powder technology, which requires the presence of sintering additives that constrain technical applications, they are preferably and easily obtained through PP. The type of PDCs, depends on the types of preceramic polymers (PPs) from which it is derived and are classified as binary systems (Si_3N_4, SiC, BN, and AlN), ternary systems (SiCN, SiCO, and BCN), quaternary systems (SiCNO, SiBCN, SiBCO, SiAlCN, and SiAlCO), and even pentanary systems.

Among many available preceramic polymers, organosilicon polymers are widely used for the preparation of PDCs and are therefore selected for discussion under this chapter; some of the organosilicon polymers that are used as a precursor to derive PDCs are shown in Fig. 1.

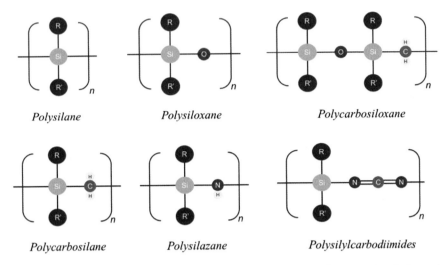

Fig. 1 Some examples of organosilicon preceramic polymers (R and R' are organic substituents)

2.1.1 Polysilanes

They are the simplest one-dimensional organosilicon polymers that are composed of a Si–Si backbone where organic substituents are attached. These organic side-chain substituents are responsible for the varied molecular weight of polysilanes and also affect their properties. Delocalization of σ electrons on—Si–Si—bonds (σ conjugation) produces varied properties in polysilanes such as photoconductivity, luminescence, and high thermal stability [13].

2.1.2 Polysiloxane (Silicones)

Polysiloxanes are made up of silicon–oxygen–silicon backbone with low intermolecular force. The reason behind the flexibility of the polymers which results in low glass transition temperature. They are a class of organosilicon polymers that have been used in various industries—textile, electronics, medicine—as preceramic polymers because of their high resistance to temperatures, low surface tension and energy, and high gas permeability.

2.1.3 Polycarbosilane

Polycarbosilanes are polymer with—Si–C—backbone. The carbon chain in this backbone varies such as methylene, vinylidene, phenylene—the reason behind the complexity of the polycarbosilane structure—through which different ceramic yields are obtained. But, in some cases the polycarbosilanes possess an alternate π-conjugated unit like phenylene, ethenylene, and a sil(an)ylene unit. Polycarbosilanes as a precursor are mostly used for SiC-based fibers because of their high ceramic yield [11].

2.1.4 Polycarbosiloxane

Polycarbosiloxanes are usually inexpensive and are widely used for sealing applications or as semiconductor encapsulant material. A variety of derivatives polycarbosiloxane are commercially available. Many of them have excellent chemical, physical, and electrical properties. The general synthesis method for the preparation of polycarbosiloxanes comprises the reaction of chloro(organo)silanes with water [11]. Additionally, hyperbranched polycarbosiloxanes synthesized via the Piers–Rubinsztajn (P–R) and hydrosilylation reactions unveil as a promising candidate for high-technology performance such as solar space and solar applications [39, 48].

2.1.5 Polysilazane

They are polymers that consist of alternating silicon–nitrogen bonds with carbon containing side groups. They are mostly used as the precursors for constructing SiCN ceramics. Polysilazanes have a versatile property such as high thermal stability, oxidation, and corrosion resistance due to which they are significantly used for heat exchange barriers or on steel against oxidation. Eventhough they were used as precursors to prepare silicon (carbo)nitrides and other related materials back in the 1950s, they are widely used even to this day as silylating agents and as precursors for the preparation of ceramic materials.

3 Preparation of PDCs

PDCs are traditionally prepared by powder technology; since it uses sintering suitable powders, this method constrains technical applications [11]. In recent years, the use of preceramic polymers especially silicon-based polymer as precursors for the synthesis of PDCs have been mushrooming as it provides several advantages—homogeneous distribution of the elements in PDCs, PDCs with high purity, synthesis at relatively low temperatures (800–1000 °C), ease of changing the molecular structure of the precursors, and ease of tailorability (producing ceramics with new elemental compositions)—compared to powder technology. Another advantage of PPs is that different elements—Si, C, N, and O can be easily grafted on the polymer backbone to get different organosilicon polymer, which various side chains on Si can be modified to give the final ceramic a versatile decomposition property. The final PDCs composition depends on the type and synthesis of PPs—organosilicon polymers. To prepare silicon-based polymer ceramics, the organosilicon polymers should possess certain properties such as an appropriate solubility for shaping, high molecular weight, cross-linking ability [13]. Thus, the preparation of preceramic polymers is the first vital phase to construct polymer-derived ceramics. This is because the properties and composition of PPs affect the final compositions of PDCs, which subsequently determine the ceramic properties.

3.1 Conversion of Molecular Precursors to Preceramic Polymers

In this process, normally, PP are prepared from molecular precursors, this opens a gateway for the researchers to manually tailor the architecture and cross-linking ability of the PPs, which will help to produce PDCs with high purity and versatility at lower processing temperature, and easily alterable to a multi-component hybrid material at molecular levels. In a typical synthesis, the preceramic polymer is first synthesized and shaped using conventional polymer-forming techniques such as polymer infiltration pyrolysis (PIP), injection molding, coating from solvent, extrusion, or resin transfer molding (RTM). Preceramic polymers especially organosilicon polymers are generally produced by polymerization of halosilanes as the molecular precursor.

For example, PPs such as Polysilanes are known to be synthesized by Wurtz type Coupling of halosilanes, by anionic polymerization reaction, or by catalytic dehydrogenation of silanes and by reduction of silanes [13]; Polycarbosilane can be synthesized by Kumada rearrangement [20]. This rearrangement is based on the thermal decomposition of a polymethylsilane—synthesized by the reaction of dimethyldichlorosilane with sodium at 120 °C in xylene—to yield the polycarbosilane [22]; Polycarbosilane can also be prepared by a sequential Grignard coupling reaction of (chloromethyl) triethoxysilane and vinyl magnesium bromide followed by LiAlH$_4$ reduction [15]. They can also be prepared by ring-opening polymerization [13]. Polysilazanes is synthesized by ammonolysis of chlorosilanes with ammonia or by ring-opening polymerization of cyclic polysilanane [31]. Polysiloxanes are prepared by polycondensation of linear silanes with terminated active functional groups (e.g., usually dimethyl-dichloro silane) and by ring-opening polymerization of cyclic silaethers [11], while Polysilylcarbodiimides can be synthesized by pyridine catalyzed polycondensation reaction of chlorosilanes with bis(trimethylsilylcarbodiimide) prepared via reaction hexamethyldisilazane, cyanoguanidine, and a catalytic amount of ammonium sulfate [19]. Figure 2 shows the route of formation of some polyorganosilicon polymers via halosilanes through various approaches.

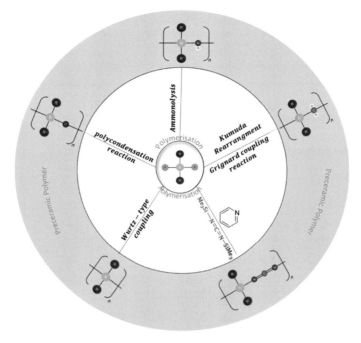

Fig. 2 Transformation of molecular precursors to preceramic polymer

3.2 Conversion of Preceramic Polymers to Polymer-Derived Ceramics

Polymer-derived ceramics can be easily prepared through controlled heat treatment of preceramic polymers. Since the physical state and rheological properties of preceramic polymers can be tailored and controlled, polymers can be shaped before or during the cross-linking steps followed by pyrolysis and annealing if required (Fig. 3). Preceramic polymers are converted into ceramic through pyrolysis at temperatures (generally > 800 °C) in an inert atmosphere; pyrolysis results in the release of organic small volatile species such as H_2, CH_4, or other volatile compounds, from the polymer matrix due to the breaking of C–H bonds. This route allows the preparation of unusual ceramic structures with tuned mesoporosity, which cannot be obtained with any other method. It takes into consideration the advantage of the chemistry of molecular and polymeric precursors and their ability to be coupled with forming methods to form into ceramics after pyrolysis [21]. For example, polycarbosilanes can be directly converted to SiC by pyrolysis; [24] prepared SiC foams with a tunable porosity (Fig. 4a, b). They synthesized high molecular weight polycarbosilanes via the thermal backbone rearrangement of polydimethylsilane granules, and then they were modified with epoxy by dissolving in epoxy-tetrahydrofuran solution to generate the controlled foaming sites in the PCS. SiC foams (Fig. 4c) were derived

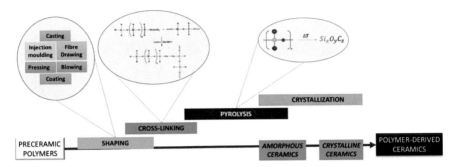

Fig. 3 Processes associated with the conversion of preceramic polymers to Polymer-Derived Ceramics (*taking polysiloxane as the preceramic polymer example*)

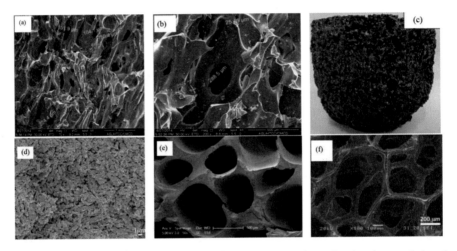

Fig. 4 **a, b** SEM images of SiC foam converted from PCS powder, **c** showing shape and size of synthesized SiC form, Dia = 50 mm, ht = 80 mm (Reproduced from [24] with permission Elsevier, Copyright 2017), **d** SEM images of the M–SiOC pyrolyzed at 1000 °C (Reproduced from [46] with permission Elsevier, Copyright 2011), **e** SEM image of fracture surfaces of pyrolyzed SiOC Ceramics [37], **f** SEM images of SiOC ceramic foams showing struts formed during the solvent evaporation [35]

from the pyrolysis of polycarbosilanes at 1000 °C and ceramization over 1200 °C in a vacuum furnace at a heating rate of 1–2 °C/min by maintaining the furnace pressure at 100 mbar [24].

Of various polymer-derived ceramics, Silicon carbide oxide (SiCO) and silicon oxycarbide (SiOC) ceramics have gained immense attention because of their wide application in the field of adsorption studies especially due to their surface area, pore-volume, high-temperature stability, and thermal–mechanical durability.

As the presence of porous structure is significant to employ as an adsorbent or catalyst support, the formation of this porous structure is created by direct foam methods, replica, sacrificial template, sacrificial filler, freeze casting, preceramic polymer mixtures, preceramic polymer decomposition, etc. For example, Yu et al. [43], fabricated a mesoporous SiOC ceramic by pyrolysis of polymeric gels prepared from mixtures of cross-linkable methyl-terminated polydimethylsiloxane(PDMS) as pore-former and allylhydridopolycarbosilane as a preceramic precursor. The polymeric gel was then pyrolyzed at 900 °C in an argon atmosphere for 2 h giving a specific surface area and a total pore volume of 87.83 m^2g^{-1} and 0.195 cm^3g^{-1}, respectively. They also prepared macroporous ceramics by mixing PDMS with polysiloxane, followed by cross-linking and then pyrolyzing it at 900 °C. [27] have followed a soft templating approach using polyurethane sponge as a porous template infiltrated with silicone resin pyrolyzed at 1400 °C in an argon atmosphere to synthesize SiC nanowires-filled cellular SiCO ceramics [27].

Yuan et al. [46] fabricated ordered mesoporous SiOC composites using liquid poly(hydridomethylsiloxane) as preceramic polymer and mesoporous carbon as a direct template by replica technique via double nano casting. They mixed poly(hydridomethylsiloxane) and mesoporous carbon CMK-3 with tetrahydrofuran to form a gel-like mixture (composite monoliths), which were cross-linked at 150 °C for 20 h in a muffle furnace under humid air. The sample was then pyrolyzed at 1000–1200 °C in a horizontal ceramic tube furnace for 2 h in an argon atmosphere to obtain the SiOC-carbon composite monoliths. Finally, the ordered mesoporous SiOC monoliths were produced by removing the carbon template by thermally treating the composite monolith at 1000°C in an ammonia atmosphere for 10 h.

Shibuya et al. [34] prepared porous SiOC ceramics by pyrolysis of methylsilicone resin with the help of sacrificial microbeads of polymethyl methacrylate (PMMA). The precursors were melt mixed at 90 °C for 5 min and were uniaxially pressed at 200 °C for 1–2 h at a pressure of 20 MPa. While, cross-linking and elimination of PMMA were performed by heating the samples in vacuum at 400 °C for 1 h and were then pyrolyzed at 800, 1000, and 1200 °C in vacuum, with heating for 2 h, to produce SiOC ceramics. [29], tailored a porous SiOC aerogel by using a linear polyhydridomethyl siloxane cross-linked with divinylbenzene (DVB) via hydrosilylation reaction in presence of a Platinum divinyltetramethyldisiloxane complex as Pt catalyst in presence of acetone as solvent. The mixture was autoclaved at a temperature of 150 °C for cross-linking reactions to form a gel. The wet gels were supercritically dried and pyrolyzed at 1000 °C in an alumina tubular furnace at a rate of 5 °C/min for 1 h to obtain SiOC aerogel with a surface area of 180 m^2g^{-1} and pore volume of 1.09 cm^3g^{-1} [29]. SiOC microcellular foams were prepared by [7] using a preceramic polymer (MK silicone resin) and polymethylmetacrylate (PMMA) microbeads as sacrificial filler. MK silicone resin and PMMA were dry mixed at a 2:8 ratio using a ball mill and then warm pressed at 150 °C for cross-linking and to put the PP in the PMMA voids. The sample obtained was then treated in the air at 300 °C for 2 h to burn out the PMMA microbeads and for cross-linking, which was then pyrolyzed at 1100 °C for 2 h at nitrogen atmosphere and etched in an HF solution and heated to produce the SiOC foam [7].

SiOC ceramic foams were prepared via replica method using polyurethane foam waste and a linear polysiloxane—polyhydridomethylsiloxane(PHMS)— containing Si–H as well as Si–CH3 moieties as a preceramic polymer crossed linked using a cyclic siloxane containing allyl groups 2,4,6,8-tetramethyl-2,4,6,8-tetravinylcyclotetrasiloxane (TMTVS) via hydrosilylation reactions and pyrolyzed at 1100–1300 °C. The sample was then etched with HF, washed and oven-dried to form the final ceramic foams with a maximum surface area of 167 m^2g^{-1} and total pore volume of 0.068 cm^3g^{-1} [35] (Table 1).

4 PDCs for Removal of Dyes From Textile Wastewater

PDCs especially derived from poly(organo)silanes are widely used as an adsorbent and as catalyst support for removal of various organic matter from textile water.

Gou et al. [14], prepared porous silicon oxycarbide (SiOC) ceramic membranes with high corrosion resistivity and with tunable pore structures (pore size in the range of 50 nm–2 μm and porosity of 35–40%) for industrial wastewater treatment. They reported that the polymer-derived membrane on treating with 20% H_2SO_4 lost only a minor fraction of the weight i.e., 0.07 wt% and a negligible decrease in water permeate flux. Bruzzoniti et al. [9] successfully synthesized porous silicon carbide-based polymer-derived ceramics, SiOC ceramic aerogels as adsorbents to remove methyl blue from water. SiOC aerogel found to have a SSA of 163 m^2g^{-1} and total pore volume of 0.723 cm^3g^{-1}. It showed a maximum capacity for methyl blue removal of 44.2 mgg^{-1} conducted at pH 5. They reported that the maximum adsorption capacity of SiOC aerogels were much higher than that of commercial mesoporous silica, SBA-15 (41.2 mgg^{-1}) with respect to its SSA (490 m^2g^{-1}). Their group elaborated that the higher adsorption of methyl blue could be because of the multifunctionality of the SiOC aerogels surface which led to the adsorption of MB through strong adsorbate-adsorbent ionic interactions and also through van der Waals interactions with sp^2 free carbon phase present in the Si-based PDCs due to pyrolysis process.

Hojamberdiev et al. [16], reported the synthesis of polymer-derived ceramic composites to remove organic dyes from contaminated water by both adsorption and photocatalysis. They prepared mesoporous polymer-derived ceramic composites (SiOC/TiO_2 and SiOC/N-doped TiO_2) by incorporating TiO_2 and N-doped TiO_2 powders into vinyl-functionalized polysiloxane polymer having high S_{BET} values of ($336 m^2g^{-1}$ and $254 m^2g^{-1}$) and narrow pore size distributions. They found that higher adsorption and degradation for SiOC/N-doped TiO_2 ceramic as N-doped TiO_2 is more negative than that of pure TiO_2. This resulted in a higher rate constant of 27×10^{-3} min^{-1} mostly due to the change of the electronic structures and crystallinity of TiO_2 with nitrogen doping and also due to homogeneously dispersed TiO_2 nanoparticles, the large S_{BET} value also facilitates the adsorption, utilization of light and offers more active sites.

Table 1 Preparation of SiOC and SiOC using preceramic polymers through various processing methods

Process	Preceramic polymer	Atmosphere	Temperature °C	Surface area (m^{2}g^{-1})	Total pore volume (cm^{3}g^{-1})	Refs.
Hard template	Polycarbosilane	Nitrogen	1200	720	0.8	[33]
Soft template	Silicone resin and polyurethane sponge	Argon	1000–1400	0.5–45	na	[27]
freeze casting	Blending of three types of polysiloxanes	Nitrogen	1300	4.3–121.9	0.023–0.473	[37]
Hard template	Silicone resin and rice husk	Argon	1100–1200	480.35–766.61	0.43–0.84	[26]
Hard template	Polysiloxane and wood powder	Argon	1000–1300	91.4–463.1	0.063–0.311	[25]
	Polydimethylsilane powder	Nitrogen	1100–1200			[22]
Replica, double casting	Poly(hydridomethylsiloxane) and CMK-2	Argon	1000–1200	602–616	0.50–0.62*	[46]
Sacrificial filler	Methylsilicone resin and polymethyl methacrylate (PMMA)	Vaccum	800–1200	–	–	[34]
Aerogel	Linear polyhydridomethyl siloxane and divinylbenzene	–	1000	180	1.09	[29]
Etching	MK silicone resin and polymethylmethacrylate (PMMA)	Nitrogen	1100	26–37	–	[7]
Etching	Polysiloxane and polyurethane foam waste	Argon	1000–1300	4–147	0.002–0.068*	[35]

*mesopore volume

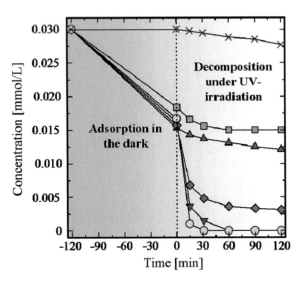

Fig. 5 Change of MB concentration ($C_0 = 0.03$ mM) by reaction with the SiOC ceramic and SiOC/20% ZnO nanocomposite synthesized at 700 °C for 2 h as functions of adsorption time in the dark and photodegradation time under UV illumination (Reproduced from [17] with permission Elsevier, Copyright 2011)

Hojamberdiev et al. [17] synthesized a mesoporous polymer-derived ceramics nanocomposite using ZnO and vinyl-functionalized polysiloxane as the preceramic polymer with S_{BET} of 220 m^{2}g^{-1} and total pore volume of 0.34 mlg^{-1} at 20% ZnO. They studied the adsorption and photoactivity of the ceramic nanocomposite toward methyl blue removal and degradation. They carried out methyl blue was adsorption onto the composite in the dark and photodegradation under UV light. The equilibrium adsorption capacity was found to be in the range 0.0067–0.0074 mmol/l for the nanocomposite. The MB degradation rate of the SiOC/ZnO nanocomposite containing 20 wt% ZnO showed 100% removal of MB at just 30 minutes. An increase in ZnO dosage significantly increased the removal of MB to 100%. The reaction of SiOC ceramic toward dye adsorption and degradation is shown in Fig. 5.

In another study by Yu et al. [43], they used methyl-terminated polydimethyl-siloxane (PDMS) as pore-former and allylhydridopolycarbosilane (AHPCS) as a preceramic precursor to produce mesoporous SiC(O) ceramic. The latter was used to investigate the adsorption of Rhodamine B (RhB) dye from water. They found temperature to affect the morphology of the ceramics, which was a result of different polymer-polymer miscibility. The samples also were having high thermal stability till 1400 °C as they were shown to retain their mesoporous structure. They reported that the ceramic to have a specific surface area and a total pore volume of 87.83 m^{2}g^{-1} and 0.195 cm^{3}g^{-1}, respectively. The mesoporous ceramic was reported to have a higher R hB removal efficiency of more than 90% within 20 min, which indicates the adsorbent's outstanding application for removal of dye from textile wastewater.

Feng et al. [12] reported the ceramic nanocomposites made of allylhydridopoly-carbosilane (AHPCS) and Bi$_2$WO$_6$ nanopowders which were cross-linked to form a polymeric gel and then pyrolyzed (Fig. 6). The nanocomposite was found to have elevated photodegradation efficiency toward RhB. The ceramic nanocomposite was having both meso and macroporous features in it. They studied the degradation of

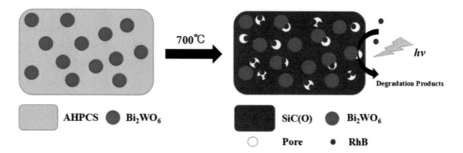

Fig. 6 Schematic illustration of the synthesis and photodegradation behavior of porous $Bi_2WO_6/SiC(O)$ ceramic nanocomposites (reproduced from [12] with permission Elsevier, copyright 2018)

RhB using both pure Bi_2WO_6 and the ceramic nanocomposite. They reported that the degradation efficiency to RhB solution (10 mg/L) was as high as around 90%, indicating that the $Bi_2WO_6/SiC(O)$ ceramic nanocomposites as a promising material for the purification of water contaminated with organic dyes. They also reported the effect of heating, Bi_2WO_6 content on the degradation behavior of the ceramic nanocomposites.

They explained the RhB photodegradation of the PDC nanocomposite, with the first step being the adsorption of RhB onto the composite surface, followed by the excitation of, Bi_2WO_6 on the nanocomposite by UV radiation to generate superoxide and hydroxyl radical and finally the degradation of Rhodamine B into smaller degradation products (Fig. 5).

Bruzzoniti et al. [8] reported the synthesis of SiC aerogel via cross-linkage of liquid allyllhydropolycarbosilane (SMP-10) with divinyl benzene (DVB) and SiC foams from PU foams impregnated with preceramic polymeric solution. Similarly, SiOC aerogel was synthesized from polyhydridomethylsiloxane and DVB at 900 °C under an inert atmosphere. The synthesized aerogels (SiC and SiOC) and SiC foams were used against methyl blue (MB) and rhodamine-B (RB) removal. Based on the experimental data, it was observed that both aerogels (SiC and SiOC) showed rapid sorption of dyes at 85% within the first hour and achieved maximum sorption at 5 h. However, the SiC foam was found to have maximum sorption of 70 and 30% for MB and RB indicating the dependency of SiC foam on the dye concentration. To further understand the complexity and performance of the SiOC aerogel under diverse interfering species, various textile wastewater and water samples from Po River were used in their study. It was found that the SiOC aerogels showed a quantitative adsorption capacity of 91.8% from textile wastewater and 93.4% for the river sample with no effect from the interfering species.

Awin et al. [2] reported the synthesis of polymer-derived ceramics/TiN composite having (BET) surface area of 0.9 m^2g^{-1}, i.e., predominantly macroporous with a pore diameter of approximately ~ 73 μm. They studied the potential application of the composite for the photodegradation of dyes taking MB as the model. The photocatalytic activity of the nanocomposite showed a degradation efficiency of 64% and the hydroxyl radicals were found to be the active species for dye degradation.

In another study, Yu et al. [44] investigated the efficiency of magnetic porous Ni-modified SiOC(H) ceramic nanocomposites with BET surface area of 253–344 m^2g^{-1} and pore volume of 0.134–0.185 cm^3g^{-1} toward adsorption of acid fuchsin. They found that the few composites despite their lower BET surface area shows a significantly enhanced adsorption performance describing factors such as microstructure and surface chemistry as the other essential factors for the dye adsorption. The equilibrium adsorption capacity for the composite was found to be between 129.8 and 123.8 mg g^{-1} at 120 min.

Bazarjani et al. [4] synthesized c-WO_{3-x}/$WO_3 \times H_2O$/[$-Si(O)CH_2-$]$_n$ nanocomposite using $WO_3 \times H_2O$ nanowhiskers dispersed in a homogeneous solution of mesoporous polycarbosilane-siloxane ([$-Si(O)CH_2-$] n) matrix. In their study, the nanocomposites displayed high photocatalytic activity toward the degradation of methyl blue ($\sim$ 70%) as compared to $WO_3 \times H_2O$ and ([$-Si(O)CH_2-$] n) species under visible light irradiation. The degradation mechanism of methyl blue by the nanocomposites was attributed to the diffusion of electrons and holes to the surfaces of $WO_3 \times H_2O$ nanowhiskers due to the presence of a high surface-to-bulk ratio of the nanowhiskers, thereby, reducing the recombination rate of photo-generated electrons and holes. The other mechanistic approach was predominant by the extensive electron reduction reactions of O_2 and the formation of the c-WO_{3-x}/$WO_3 \times H_2O$ nano heterostructures enabling high catalytic activity.

Pan et al. [25] reported the synthesis of a porous polymer-derived ceramic from polysiloxane preceramic polymer and wood biomass for dye adsorption application. The composite pyrolyzed at 1400 °C was found to have the highest BET surface of 463 m^2g^{-1} and pore volume of 0.311 cm^3g^{-1}. They reported that adsorption was affected by the pyrolysis temperature and maximum methyl blue adsorption capacity of 173.5 mgg^{-1} and followed Langmuir adsorption of monolayer adsorption. The highest MB adsorption was found to occur at an adsorbent dosage of 10 mg and pH of 6.6 and equilibrium was achieved at a time of 2 h.

Awin et al. [3] successfully synthesized TiO_2/SiOC nanocomposite to achieve visible light photocatalytic degradation of organic dye. Based on their characteristics result, the anatase TiO_2 was uniformly distributed on the amorphous SiOC matrix enhancing robust stability on the newly formed nanocomposites. Accordingly, the TiO_2/SiOC nanocomposite achieved BET surface area of 129 m^2g^{-1} with a pore volume of 0.22 cm^3g^{-1} and an average pore size of 6.8 nm. Based on their experimental analysis, the photocatalytic degradation of methyl blue under visible light showed high adsorption attributing to the high surface area/mesoporous and macroporous nature of the nanocomposite. The other mechanism was explained by the reduction in the bandgap (formation of heterojunction between the crystalline TiO_2 and the amorphous SiOC matrix) as compared to pure titania, thereby, enhancing the photocatalytic activity.

Hierarchically micro/macro-porous SiCN hybrid material was selectively applied for the adsorption of organic dyes by Meng et al., and co-workers [23]. In their study, polysilazane precursor, polydivinylbenzene microspheres (PDVB) as sacrificial templates and Nickelocene (Nicp$_2$) were used to synthesize the hybrid material. According to their study, the hybrid material synthesized at 600 °C (C1) showed a better BET surface area of 256.6 m^{2}g^{-1} and pore volume of 0.09 cm^{3}g^{-1} as compared to hybrid material at 700 °C (C2). It was observed that dyes with an azobenzene structure (methyl orange, methyl red, and congo red) showed no adsorption capability on C1. However, the bulky triphenyl structure (methyl blue, basic fuchsin and methyl violet) showed excellent adsorption on the hybrid material. The high adsorption capacity could potentially be attributed to the van der Waals interaction between the dyes aromatic structure and sp^2 hybridized carbon domain or via the electrostatic interaction between the hybrid SiCN matrix and the organic dye molecules. In addition, the regeneration and recyclability of the SiCN hybrid material exhibit high sorption capacity for methyl blue (MB), basic fuchsin (BF), methyl orange (MO), and acid fuchsin (AF). Further, the Langmuir isotherm model showed better fitting with a sorption capacity of 1327.7 mgg^{-1} for MB and 1084.5 mgg^{-1} for AF.

Bruzzoniti et al. [9] synthesized three different porous polymer-derived ceramics (SiC and SiOC aerogels and SiC foam) as an adsorbent for organic dyes removal from aqueous solution. Based on their experimental observation (1 ppm dye concertation, 48 h contact time), it unveiled that both SiC and SiOC aerogels showed 100% removal efficiency for methyl blue and rhodamine-B as well (MB and RB). However, while compared to the SiC foam, the removal percentage was 65.3% (MB) and 35.5% (RB), respectively. The lower sorption capacity of the SiC foam maybe attributed to van der Waals interaction between the sp^2 hybridized carbon present during the pyrolysis of the Si-based PDCs and the aromatic structure of the dye molecules. They also suggested the presence of electrostatic interaction between the cross-linked Si-OH moieties or the carbon-oxygen complex may also result to lower adsorption efficiency. Further, it was observed SiOC aerogel showed maximum adsorption capacity of 44.2 mgg^{-1} (MB) which may correspond to the free carbon and presence of monolithic aerogel structure, thereby increasing the van der Waals interaction with the organic dye.

Bhaskar co-workers [5] designed the nanocomposites made of t-ZrO$_2$ crystallized *in situ* in an amorphous precursor route of SiOC(N) matrix for visible light photocatalytic degradation of methyl blue. According to their research finding, the structural characterization of *t*-ZrO$_2$/SiOC(N) nanocomposite rendered the hierarchical porous nature of the nanocomposites. Additionally, the presence of oxygen and carbon vacancies in the zirconia lattice resulted in the reduction with a bandgap of 2.2 eV. The retention of *t*-ZrO$_2$ confined in the nanocomposite matrix exhibit high photodegradation removal rate (72%). The photocatalytic mechanism could be explained due to the formation of heterojunctions between the crystalline and amorphous material, easing a way for the visible light photocatalysis process with a lower bandgap.

Zeydanli et al. [47] employed the use of polymer-derived ceramics for the removal of various organic dyes from water. The maximum adsorption capacity was found to be 104.27, 185.87, and 110.74 mgg^{-1} for methyl blue, crystal violet, and Rhodamine

B, respectively. They also studied the thermal regeneration studies for MB, which showed that the efficiency decreased slightly from 100 to 99% after the third cycle.

Yu et al. [45] reported the in-situ synthesis of a template-free synthesis of Fe_3O_4/SiOC(H) nanocomposites. Herein, Fe_3O_4 nanoparticles were grafted homogeneously onto the cross-linked nanoporous SiOC(H) matrix via pyrolysis under argon atmosphere to synthesize the nanocomposite material. In their study, the structural characteristics reported a high micropore surface area of 301 m^2g^{-1}, micropore volume of 0.142 cm^3g^{-1}, and BET surface area up to 390 m^2g^{-1}. Further adsorption analysis displays that the in-situ Fe_3O_4/SiOC(H) nanocomposite effectively increases the catalytic degradation of xylene orange in aqueous solution using H_2O_2 as oxidant as compared with the commercial Fe_3O_4 nanospheres.

Zhang et al. [49] synthesized a novel mesoporous polymer-derived ceramic membrane demonstrating its excellent efficiency on rhodamine B removal. The porous polymer-derived ceramic membrane was prepared using the polydimethylsiloxane and Karstedt's catalyst followed by the addition of a solution mixture of 2,4,6,8-tetramethylcyclotetrasiloxane (D4H) and 2,4,6,8-tertramethyl-2,4,6,8-tetravinylcyclotetrasiloxane (D4V). Their study suggested the cyclosiloxane used as a precursor showed a high ceramic yield of 83% at 1000 °C possibly decreasing the shrinkage size and keeping the pristine skeletal structure of SiOC ceramic more intact. Their study observed the size exclusion mechanism was not ideal to explain the high removal rate of 95% of rhodamine B onto the SiOC membrane primarily due to the molecular size of RB and small pore size. The ideal mechanism may be attributed to van der Waals forces that exist between the sp^2 free carbon phase present in the SiOC network and the aromatic structure of RB. Additionally, the electrostatic ionic interaction between the positively charged surface of RB and the carbon/silicon-oxygen complex may also explain the mechanistic approach for the high removal rate of RB.

5 Conclusion and Future Outlook

Adsorbents for the removal of pollutants from water bodies have an age-old method; polymer-derived ceramics in this regard have been mushrooming recently as compared to other adsorbents like activated carbon, chiefly because of its tunability and high surface area for removal of pollutants from water. The application of polymer-derived ceramic materials in wastewater treatment serves as a promising material and holds great assurance. They are a novel class of multifunctional nanostructured materials and their promising nature attributes to their higher specific surface area, eco-friendly nature, variation in the types of chemical bonds and functional groups, high adsorption capabilities, tunable porosity, heat resistance, chemical stability, and excellent mechanical properties. This chapter deals with the preparation of PDCs and their application in the removal of dyes from water via adsorption and photodegradation. It is seen that PDCs hold a superior prospect as an adsorbent for the removal of organic dyes from textile wastewater. However, despite the wide range of PDCs available as an adsorbent prepared so far; there is still a need to fabricate methods and processing, which will innovate steps to develop PDCs with tunable pore size for wastewater treatment in the future.

References

1. Aljeboree AM, Alshirifi AN, Alkaim AF (2017) Kinetics and equilibrium study for the adsorption of textile dyes on coconut shell activated carbon. Arab J Chem 10:S3381–S3393. https://doi.org/10.1016/j.arabjc.2014.01.020
2. Awin EW, Lale A, Hari Kumar KC et al (2018a) Plasmon enhanced visible light photocatalytic activity in polymer-derived TiN/Si–O–C–N nanocomposites. Mater Des 157:87–96. https://doi.org/10.1016/j.matdes.2018.06.060
3. Awin EW, Lale A, Nair Hari Kumar KC et al (2018b) Novel precursor-derived meso-/macroporous TiO$_2$/SiOC nanocomposites with highly stable anatase nanophase providing visible light photocatalytic activity and superior adsorption of organic dyes. Materials (Basel) 11.https://doi.org/10.3390/ma11030362
4. Bazarjani MS, Hojamberdiev M, Morita K et al (2013) Visible light photocatalysis with c-WO$_3-$x/WO$_3$ × H$_2$O nanoheterostructures in situ formed in mesoporous polycarbosilane-siloxane polymer. J Am Chem Soc 135:4467–4475. https://doi.org/10.1021/ja3126678
5. Bhaskar S, Awin EW, Kumar KCH et al (2020) Design of nanoscaled heterojunctions in precursor-derived t-ZrO$_2$/SiOC(N) nanocomposites: transgressing the boundaries of catalytic activity from UV to visible light. Sci Rep 10:1–13. https://doi.org/10.1038/s41598-019-57394-8
6. Bhomick PC, Supong A, Baruah M et al (2018) Pine cone biomass as an efficient precursor for the synthesis of activated biocarbon for adsorption of anionic dye from aqueous solution: isotherm, kinetic, thermodynamic and regeneration studies. Sustain Chem Pharm 10:41–49. https://doi.org/10.1016/j.scp.2018.09.001
7. Biasetto L, Peña-Alonso R, Sorarù GD, Colombo P (2008) Etching of SiOC ceramic foams. Adv Appl Ceram 107:106–110. https://doi.org/10.1179/174367607X227963
8. Bruzzoniti MC, Appendini M, Onida B et al (2018a) Regenerable, innovative porous silicon-based polymer-derived ceramics for removal of methylene blue and rhodamine B from textile and environmental waters. Environ Sci Pollut Res 25:10619–10629. https://doi.org/10.1007/s11356-018-1367-x
9. Bruzzoniti MC, Appendini M, Rivoira L et al (2018b) Polymer-derived ceramic aerogels as sorbent materials for the removal of organic dyes from aqueous solutions. J Am Ceram Soc 101:821–830. https://doi.org/10.1111/jace.15241
10. CDP (2018) treading water, corporate responses to rising water challenges (CDP global water report). CDP Glob Water Rep 2018:84
11. Colombo P, Mera G, Riedel R, Sorarù GD (2010) Polymer-derived ceramics: 40 years of research and innovation in advanced ceramics. J Am Ceram Soc 93:1805–1837. https://doi.org/10.1111/j.1551-2916.2010.03876.x
12. Feng Y, Lai S, Yang L et al (2018) Polymer-derived porous Bi$_2$WO$_6$/SiC(O) ceramic nanocomposites with high photodegradation efficiency towards rhodamine B. Ceram Int 44:8562–8569. https://doi.org/10.1016/j.ceramint.2018.02.061
13. Fu S, Zhu M, Zhu Y (2019) Organosilicon polymer-derived ceramics: an overview. J Adv Ceram 8:457–478. https://doi.org/10.1007/s40145-019-0335-3
14. Guo F, Su D, Liu Y et al (2018) High acid resistant SiOC ceramic membranes for wastewater treatment. Ceram Int 44:13444–13448. https://doi.org/10.1016/j.ceramint.2018.04.149
15. He L, Zhang Z, Yang X et al (2015) Liquid polycarbosilanes: synthesis and evaluation as precursors for SiC ceramic. Polym Int 64:979–985. https://doi.org/10.1002/pi.4889

16. Hojamberdiev M, Prasad RM, Morita K et al (2012a) Template-free synthesis of polymer-derived mesoporous SiOC/TiO$_2$ and SiOC/N-doped TiO$_2$ ceramic composites for application in the removal of organic dyes from contaminated water. Appl Catal B Environ 115–116:303–313. https://doi.org/10.1016/j.apcatb.2011.12.036

17. Hojamberdiev M, Prasad RM, Morita K et al (2012b) Polymer-derived mesoporous SiOC/ZnO nanocomposite for the purification of water contaminated with organic dyes. Microporous Mesoporous Mater 151:330–338. https://doi.org/10.1016/j.micromeso.2011.10.015

18. Holkar CR, Jadhav AJ, Pinjari DV et al (2016) A critical review on textile wastewater treatments: possible approaches. J Environ Manage 182:351–366. https://doi.org/10.1016/j.jenvman.2016.07.090

19. Iwamoto Y, Völger W, Kroke E et al (2001) Crystallization behavior of amorphous silicon carbonitride ceramics derived from organometallic precursors. J Am Ceram Soc 84:2170–2178. https://doi.org/10.1111/j.1151-2916.2001.tb00983.x

20. Kim Y, Shin DG, Kim HR et al (2006) Kumada rearrangement of polydimethylsilane using a catalytic process. Key Eng Mater 317–318:85–88. https://doi.org/10.4028/www.scientific.net/kem.317-318.85

21. Lale A, Schmidt M, Mallmann MD et al (2018) Polymer-derived ceramics with engineered mesoporosity: from design to application in catalysis. Surf Coatings Technol 350:569–586. https://doi.org/10.1016/j.surfcoat.2018.07.061

22. Lodhe M, Babu N, Selvam A, Balasubramanian M (2015) Synthesis and characterization of high ceramic yield polycarbosilane precursor for SiC. J Adv Ceram 4:307–311. https://doi.org/10.1007/s40145-015-0165-x

23. Meng L, Zhang X, Tang et al Y (2015) Hierarchically porous silicon-carbon-nitrogen hybrid materials towards highly efficient and selective adsorption of organic dyes. Sci Rep 5.https://doi.org/10.1038/srep07910

24. Mishra MK, Kumar S, Ranjan A, Prasad NE (2018) Processing, properties and microstructure of SiC foam derived from epoxy-modified polycarbosilane. Ceram Int 44:1859–1867. https://doi.org/10.1016/j.ceramint.2017.10.123

25. Pan J, Ren J, Xie Y et al (2017) Porous SiOC composites fabricated from preceramic polymers and wood powders for efficient dye adsorption and removal. Res Chem Intermed 43:3813–3832. https://doi.org/10.1007/s11164-016-2850-y

26. Pan J, Shen W, Zhao Y et al (2020) Difunctional hierarchical porous SiOC composites from silicone resin and rice husk for efficient adsorption and as a catalyst support. Colloids Surfaces A Physicochem Eng Asp 584:124041. https://doi.org/10.1016/j.colsurfa.2019.124041

27. Pan JM, Yan XH, Cheng XN et al (2012) Preparation of SiC nanowires-filled cellular SiCO ceramics from polymeric precursor. Ceram Int 38:6823–6829. https://doi.org/10.1016/j.ceramint.2012.05.081

28. Pattnaik P, Dangayach GS, Bhardwaj AK (2018) A review on the sustainability of textile industries wastewater with and without treatment methodologies. Rev Environ Health 33:163–203. https://doi.org/10.1515/reveh-2018-0013

29. Pradeep VS, Ayana DG, Graczyk-Zajac M et al (2015) High rate capability of SiOC ceramic aerogels with tailored porosity as anode materials for li-ion batteries. Electrochim Acta 157:41–45. https://doi.org/10.1016/j.electacta.2015.01.088

30. Ramlow H, Machado RAF, Marangoni C (2017) Direct contact membrane distillation for textile wastewater treatment: a state of the art review. Water Sci Technol 76:2565–2579. https://doi.org/10.2166/wst.2017.449

31. Riedel R, Mera G, Hauser R, Klonczynski A (1988) Silicon-based polymer-derived ceramics: synthesis properties and applications—a review: dedicated to Prof. Dr. Fritz Aldinger on the occasion of his 65th birthday. J Ceram Soc Japan 114:425–444

32. Sarayu K, Sandhya S (2012) Current technologies for biological treatment of textile wastewater—a review. Appl Biochem Biotechnol 167:645–661. https://doi.org/10.1007/s12010-012-9716-6

33. Shi Y, Meng Y, Chen D et al (2006) Highly ordered mesoporous silicon carbide ceramics with large surface areas and high stability. Adv Funct Mater 16:561–567. https://doi.org/10.1002/adfm.200500643

34. Shibuya M, Takahashi T, Koyama K (2007) Microcellular ceramics by using silicone preceramic polymer and PMMA polymer sacrificial microbeads. Compos Sci Technol 67:119–124. https://doi.org/10.1016/j.compscitech.2006.03.022
35. Soraru GD, Campostrini R, Ejigu AA et al (2016) Processing and characterization of polymer derived SiOC foam with hierarchical porosity by HF etching. J Ceram Soc Japan 124:1023–1029. https://doi.org/10.2109/jcersj2.16072
36. UNESCO World Water Assessment Programme (2021) The United Nations world water development report 2021: valuing water. Paris
37. Vakifahmetoglu C, Zeydanli D, Innocentini MDDM et al (2017) Gradient-hierarchic-aligned porosity SiOC ceramics. Sci Rep 7:1–12. https://doi.org/10.1038/srep41049
38. WWAP UNWWAP (2017) The United Nations world water development report 2017. Wastewater: the untapped resource. UNESCO, Paris
39. Xu N, Stark EJ, Dvornic PR et al (2012) Hyperbranched polycarbosiloxanes and polysiloxanes with octafunctional polyhedral oligomeric silsesquioxane (POSS) branch points. Macromolecules 45:4730–4739. https://doi.org/10.1021/ma300470m
40. Yagub MT, Sen TK, Afroze S, Ang HM (2014) Dye and its removal from aqueous solution by adsorption: a review. Adv Colloid Interface Sci 209:172–184. https://doi.org/10.1016/j.cis.2014.04.002
41. Yaseen DA, Scholz M (2019) Textile dye wastewater characteristics and constituents of synthetic effluents: a critical review. Springer, Berlin Heidelberg
42. Yin H, Qiu P, Qian Y et al (2019) Textile wastewater treatment for water reuse: a case study. Processes 7.https://doi.org/10.3390/pr7010034
43. Yu Z, Feng Y, Li S, Pei Y (2016a) Influence of the polymer–polymer miscibility on the formation of mesoporous SiC(O) ceramics for highly efficient adsorption of organic dyes. J Eur Ceram Soc 36:3627–3635. https://doi.org/10.1016/j.jeurceramsoc.2016.02.003
44. Yu Z, Li S, Zhang P et al (2017) Polymer-derived mesoporous Ni/SiOC(H) ceramic nanocomposites for efficient removal of acid fuchsin. Ceram Int 43:4520–4526. https://doi.org/10.1016/j.ceramint.2016.12.104
45. Yu Z, Zhang P, Feng Y et al (2016b) Template-free synthesis of porous Fe_3O_4/SiOC(H) nanocomposites with enhanced catalytic activity. J Am Ceram Soc 10:1–10. https://doi.org/10.1111/jace.14305
46. Yuan X, Jin H, Yan X et al (2012) Synthesis of ordered mesoporous silicon oxycarbide monoliths via preceramic polymer nanocasting. Microporous Mesoporous Mater 147:252–258. https://doi.org/10.1016/j.micromeso.2011.06.025
47. Zeydanli D, Akman S, Vakifahmetoglu C (2018) Polymer-derived ceramic adsorbent for pollutant removal from water. J Am Ceram Soc 101:2258–2265. https://doi.org/10.1111/jace.15423
48. Zhang H, Xue L, Li J, Ma Q (2020a) Hyperbranched polycarbosiloxanes: Synthesis by piersrubinsztajn reaction and application as precursors to magnetoceramics. Polymers (Basel) 12.https://doi.org/10.3390/polym12030672
49. Zhang Z, Bao Y, Sun X et al (2020b) Mesoporous polymer-derived ceramic membranes for water purification via a self-sacrificed template. ACS Omega 5:11100–11105. https://doi.org/10.1021/acsomega.0c01021

Polymeric Hydrogels for Dye Adsorption

Magdalena Cristina Stanciu

1 Introduction

Water is the main constituent of Earth and it is vital for all the living organisms (plants, animals and humans). Unsafe chemical compounds, produced by different industries (paper, textile, plastics) are discharged into water and determine a great quantity of polluted wastewater. So, the removal of hazardous pollutants from water becomes crucial to maintain pure water resources.

Between different types of wastewaters, dyeing wastewater has a significant percentage due to the development of dyeing industry. Dyes are chemical substances which impart color to textiles, leather, paper, plastics, rubber [3, 4, 19, 73, 79, 180, 223, 255]. Nowadays, according to the Color Index, above 10,000 various dyes and pigments are used in the industry.

Their presence in big amount in the wastewater can reduce sunlight transmission in rivers and thus, affects the photosynthetic activities of aquatic species and determines the decrease of dissolved oxygen level in water. Also, dyes contain toxic compounds, like trace heavy metals (Cd, Cu, Pb, Zn, Co), amines and aromatics. What is serious is that dyes have mutagenic, teratogenic and carcinogenic effects and lead to malfunction of several humans' organs [166]. Dyes also cause: dermatitis, allergic conjunctivitis, rhinitis, skin irritation, asthma or different tissular changes. Unfortunately, these synthetic compounds are very hard to degrade (i.e. half-life time of many years), having a strong chemical stability at light, heat or oxidizing agents [46, 242].

Based on the source of made, dyes can be divided into natural and synthetic dyes. Natural dyes, extracted from natural resources (plants, invertebrates, minerals),

M. C. Stanciu (✉)
Natural Polymers, Bioactive and Biocompatible Materials Department, "Petru Poni" Institute of Macromolecular Chemistry, Grigore Ghica Voda Alley 41A, 700487 Iasi, Romania
e-mail: cstanciu@icmpp.ro

A. Khadir and S. S. Muthu (eds.), *Polymer Technology in Dye-containing Wastewater*,
Sustainable Textiles: Production, Processing, Manufacturing & Chemistry,
https://doi.org/10.1007/978-981-19-0886-6_6

have not been able to meet the market demand. Consequently, synthetic dyes were produced and replaced natural dyes, mainly in textile industry. Depending on their electric charge, synthetic dyes are ionic (cationic or anionic) and non-ionic. Anionic (reactive, direct, acid) and cationic (basic) dyes have a good solubility in water while non-ionic ones (disperse, vat) have not.

Because of the adverse effects and long-lasting stay in water, it is essential to employ treat methods and technologies for dyes removal from wastewater. Treatment methods consist of physical separation (filtration, coagulation/flocculation, adsorption), chemical processes (oxidation, ozonation) or biological degradation by using microbes or enzymes. Among all the techniques of dye elimination, the adsorption process is the best choice. It is a simple and cheap method, with a short analysis time and safe by-products. Due to its high efficiency and possibility of adsorbent regeneration for multiple reuses, it is one of the most popular and used treat methods for dyes removal [107, 193].

Activated carbon has an excellent adsorption ability to remove dyes in wastewater and its sorption is one of the best accessible control technologies [59]. The main drawback of activated carbon is its high price which determined the search for other efficient adsorbents. Several unconventional sorbents, having low cost, were reported for dyes elimination in water. Such adsorbents consist of: clay/zeolites materials (bentonite, kaolinite, smectite, montmorillonite), siliceous material (alunite, perlite, sepiolite, attapulgite), agricultural wastes like as peels, leaves, seeds (bagasse pith, wheat straw, rice husk, sawdust, bark, coconut shell, sugarcane bagasse, cotton fiber, coffee residues, different fruit peels), industrial waste products (waste carbon slurries, metal hydroxide sludge, fly ash, red mud), biomass (algae, fungi, bacteria) [45].

Polysaccharides are an important class of natural polymeric materials. They are stable, abundant, non-toxic, biodegradable and low cost. Polysaccharide-based materials, in different frameworks (hydrogels, membranes, beads/resins or films), have a wide range of applications, including wastewater treatment. The existence of numerous hydrophilic functional groups ($-OH$, $-COOH$, $-NH_2$, $-SO_3H$, $-CONH_2$) in their chemical structure affords the adsorption of the contaminants, like dyes or heavy metals [168].

2 Hydrogels

Hydrogels are hydrophilic polymeric 3D-network crosslinked structures that are able to retain in their swollen state a large amount of water or biological fluids through their pores [226, 20, 64, 196]. Crosslinking determines their insolubility in water and affords mechanical strength. In the dry state, hydrogels contract and restore their original volume. Generally, these polymers retain a significant fraction (20–100%) of water inside their tridimensional network, but there are hydrogels called superabsorbent that can imbibe water between 1000 and 100,000% [29, 160]. Hydrogels have been used in: regenerative medicine [66, 190], tissue engineering [9, 138, 237], drug delivery [48, 55, 92, 148, 244], wastewater treatment, wound healing

[183], agriculture and horticultural [67], as antibacterial [225] or antifouling [231] agents.

Hydrogels can be classified according to various criteria. Thus, gels can be physically or chemically crosslinked according to crosslinking mechanism. Several methods were described in the literature for obtaining physically crosslinked hydrogels: freeze-thawing [87, 167], stereocomplex formation [220, 221], ionic interaction [254], H-bonding [218], maturation (heat-induced aggregation) [11, 12]. Methods reported in the literature for achieving chemically crosslinked hydrogels, were: chemical crosslinking [8, 91], chemical [14] or radiation grafting [184], radical polymerization [224, 102, 100], condensation reaction [56, 57], enzymatic reaction [205, 253], high-energy radiation [118, 215, 247]. Based on polymeric composition, hydrogels can be: homo-polymeric, co-polymeric and interpenetrating polymeric hydrogel-IPN (semi or full). Consistent with network electrical charge, hydrogels are: non-ionic (neutral), ionic (cationic or anionic), amphoteric electrolyte (having both acidic and basic functional groups) and zwitterion (polybetaines) (with anionic and cationic groups in each unit). According to starting materials, hydrogels are classified into: natural, synthetic or hybrid (combinations of both natural and synthetic gels).

Neat hydrogels have a poor mechanical strength and thermal stability. The addition of nanofillers, such as: graphene oxide, carbon clay, bentonite, hydroxyapatite or montmorillonite into gel matrix determines the formation of new materials, called nanocomposite, with superior properties.

2.1 Techniques Used for Hydrogels Characterization

Adsorption capacity (q_e) and removal efficienscy (R) are the parameters used to evaluate sorption efficiency (Eqs. 1, 2).

$$q_e = \frac{(C_i - C_{eq})}{100 \times m}, \text{mmol/g} \tag{1}$$

$$R(\%) = \frac{(C_i - C_{eq})}{C_i} \times 100 \tag{2}$$

where C_i is the initial dye concentration while C_{eq} is the residual dye concentration at equilibrium (mM) and m is the mass (g) of dry gel.

Swelling behavior, Fourier transform infrared spectroscopy, scanning electron microscopy and X-ray diffraction are among the most employed methods to study structure and properties of the hydrogels.

2.1.1 Swelling Behavior

The swelling ratio is defined as the fractional increase in the weight of the hydrogel due to water absorption. The water uptake at equilibrium, R_w (g H_2O/g dry gel), represents the ratio between the amount of water retained by the gel and the weight of the dry gel (Eq. 3). The amount of water retained by the gel (g H_2O) is calculated as the difference between the weight of swollen gel (W_s) and that of initial dry material (W_d). R_w is determined by immersing a known amount of the dry gel in deionized water until the equilibrium is reached, followed by the centrifugation of the hydrogel.

$$R_w = \frac{(W_s - W_d)}{W_d} \tag{3}$$

The water retention at equilibrium depends on several factors: sorbent chemical structure and crosslinking degree, pH, temperature, ionic strength. The augmentation of crosslinking density will decrease the probability of water molecules penetration in the polymer matrix, so, will decrease R_w values.

2.1.2 Fourier Transform Infrared Spectroscopy (FTIR)

FTIR is a widely used technique employed for the detection of the functional groups of the hydrogels by using the infrared radiations. The polymers having functional groups can absorb infrared radiations, the result being the oscillation of their bonds. The infrared spectrum is obtained by plotting the measured infrared light intensity against the energy range, expressed in wavenumber. This method helps in determination of the success of the chemical reactions by following the shifting, appearance or disappearance of specific absorption bands, which each of them are characteristic to a functional group. Also, the occurrence of hydrogen bonds can be proved by using FTIR technique.

2.1.3 Scanning Electron Microscopy (SEM)

SEM is one of the most employed methods used to analyze the shape, size, crosslinking degree and porosity of the hydrogels. Morphology of dextran-based gels, bearing two types of pendant quaternary ammonium chloride groups, was investigated with the aid of scanning electron microscopy (Fig. 1) [207, 208].

The polymeric microparticles, inspected in dry state, were perfectly spherical, having a diameter between 100 and 220 μm. At a magnification up to 1000 (Fig. 1a,b), their surface looked to be flatten, while at a magnification of 10,000 (Fig. 1e), the surface was creased, the look being obtained during the drying process due to the contraction of the surface layer by gradual dehydration. The surface and the cross-section of microspheres was free of pores (Fig. 1a–e). These results definitely indicate the absence of porosity for the hydrogels in dry state, as has been already proved for

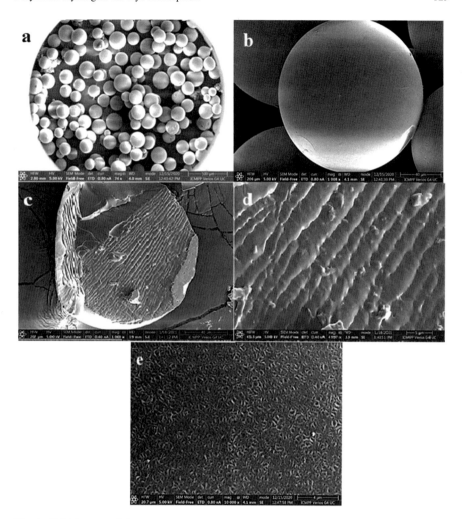

Fig. 1 SEM images of dextran-based microspheres, increased 74 times **a**, and 1000 times **b**; polysaccharide-based microparticles in cross-section, magnified 1000 times **c**, and 5000 times **d**; dextran hydrogel surface, increased 10,000 times **e**. *Reprinted by the permission from Elsevier,* Stanciu et al. [209]

other microparticles based on dextran which were obtained without porogen agents [81].

2.1.4   X-Ray Diffraction (XRD)

XRD is another technique frequently used for hydrogels characterization. This method consisted in the irradiation of the hydrogels with beam of monochromatic

X-rays. The resulted diffracted rays were examined depending on angle of diffraction, and the intensity of the diffracted rays offers details about the crystallinity percent, crystallite dimension and orientation, interplanar atomic spacing. As regards hydrogel nanocomposites, the addition of nanoparticles in the polymeric matrix determined the modification of XRD spectrum compared to pure gel which proves the success of the nanocomposite synthesis.

3 Natural Polysaccharides-Based Hydrogels

3.1 Chitosan and Chitin-Based Hydrogels

Chitin, the second most abundant natural biopolymer, is extracted from the exoskeleton of crustaceans (crabs, shrimps), mollusk cartilages and cell walls of fungi (Fig. 2a). It is composed of β-(1 → 4) linked N-acetyl-D-glucosamine units. Chitin is biocompatible, biodegradable, low cost and reusable. Some chitin-based derivatives were employed as adsorbents for the elimination of various pollutants from aqueous solution [26].

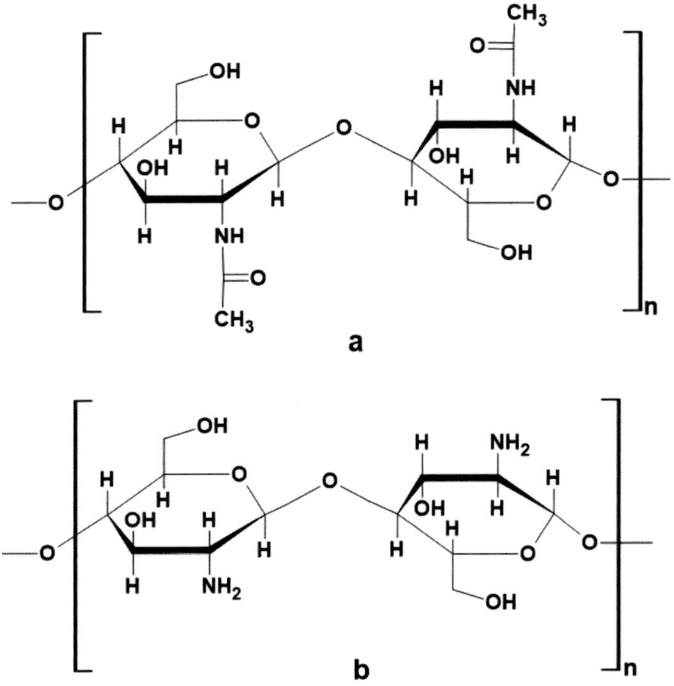

Fig. 2 Chemical structure of **a** chitin and **b** chitosan

Chitosan (poly-β-(1 $\rightarrow$ 4)-2-amino-2-deoxy-D-glucose) (Fig. 2b) is a linear polysaccharide composed of randomly distributed β-(1 $\rightarrow$ 4)-linked D-glucosamine and N-acetyl-D-glucosamine in its chemical structure [156, 179, 212]. Chitosan is produced by N-deacetylation of chitin [98, 124]. The intrinsic properties of chitosan and its derivatives (hydrophilicity, biocompatibility, biodegradability, bioadhesivity) afford their use in many domains, like: drug delivery [176, 171, 177, 112], biomedical research [99, 157, 23, 117, 189, 191], biotechnology [120], catalysis [82, 140], cosmetic industry [176], textiles [78], paper industry [176], enology [27].

Chitosan-based polymers are environment-friendly bio-absorbents in wastewater treatment due to their remarkable capacity of adsorption, low cost and the versatility of the manufacturing process (gels, beads, films, membranes, nanoparticles) [121, 123]. Chitosan can be derivatized due to the existence of amino group (C-2) along with both primary (C-6) and secondary (C-3) hydroxyl groups (Fig. 1a) which affords the obtaining of new polymers having improved properties, including adsorption capacity and a good endurance in severe media conditions. Hydrogels-based chitosan was mainly obtained by crosslinking and grafting reactions. The addition of supplementary functional groups on chitosan, as a result of its grafting reaction, augments the number of adsorption sites and therefore, the sorption capacity. Crosslinking reaction enhances the mechanical properties and consolidates the chemical stability of chitosan in acidic solutions. However, this reaction decreases the sorption process because of the binding of free amino groups of the polysaccharide with the crosslinker. Typical crosslinkers used with chitosan are: dialdehydes (glyoxal, formaldehyde, glutaraldehyde), epoxides (epichlorohydrin, ethylene glycol diglycidyl ether), isocyanates (hexamethylene diisocyanate) carboxylic acids, genipin. Also, chitosan-based gels can be obtained due to its ionotropic gelation with specific polyanions, like tripolyphosphate.

Chitosan-based sorbent, prepared by free radical crosslink copolymerization of acrylamide and acrylic acid on the polysaccharide in the presence of ceric ammonium nitrate/ascorbic acid (initiator) and N, N-methylene bisacrylamide (crosslinking agent), was used for Methyl orange removal [121]. Furthermore, the crosslinked copolymer proved a good antibacterial activity (*Pseudomonas aeruginosa, Escherichia coli, Staphylococcus aureus*), showing a 62% growth reduction in comparison with neat polysaccharide.

Poly(2-acrylamido-2-methylpropane sulfonic acid) grafted magnetic chitosan, was synthesized through free radical polymerization and used for cationic Methylene blue adsorption [239]. Magnetic nanoparticles of Fe_3O_4 and SiO_2, prepared by covering silica shell on the surface of Fe_3O_4, were inserted into chitosan matrix and polysaccharide microspheres resulted were grafted with 2-acrylamido-2-methylpropane sulfonic acid by free radical polymerization. The interactions between magnetic grafted chitosan and Methylene blue were both electrostatic and hydrophobic. Magnetic grafted chitosan microspheres could be quickly separated under magnetic field and efficiently regenerated under acid conditions.

Chitosan/montmorillonite intercalated composite, obtained by adding chitosan solution to montmorillonite suspension and then heated at 60 °C, was used for the

Reactive red 136 removal [131]. The sorption was both on the surface and inter-calation, and the reactive functional groups of the composite (hydroxyl, amide, amino, siloxane) were involved in the adsorption process. Toth isotherm model and Brouers-Weron-Sotolongo kinetic model fitted best with the experimental data. The composite-based chitosan proved great adsorption results still after 15 adsorption–desorption cycles.

Graphene oxide (GO) is a distinctive material that consisted of a single monomolecular layer of graphite with various oxygen-containing functionalities. GO shows exceptional properties such as: high specific surface area [211], high mechanical toughness, electrical and thermal conductivity, lack of corrosion [133, 159]. Several composites, in which multilayered graphene oxide sheets were mixed with natural or synthetic hydrogels, showed improved adsorption properties due to electrostatic and hydrogen bonding interactions with sorbates [88, 126, 127, 216]. The occurrence on graphene oxide basal planes of epoxide and hydroxyl groups and at its edges, of carboxyl and carbonyl groups led to an easy penetration of water molecules between nanosheets' layers which explains the high adsorbing capacity of these composites against dyes [18, 115, 210]. Dyes adsorption behavior of polysaccharide-based nanocomposites containing GO was described by many papers. Chitosan, chitin, starch, sodium alginate and xanthan are among the polysac-charides which were used for the synthesis of GO-based composites aiming to dyes sorption.

Hybrid gels composed of chitin and graphene oxide mixed in different ratios were employed for the removal of two dyes, namely Remazol Black and Neutral Red [80]. Sips and Redlich–Peterson isotherm models fitted well with the experimental sorption data. The optimum desorption pH values of Remazol Black and Neutral Red were 9.0 and 10.0 respectively.

A composite chitin-based hydrogel strengthened by tannic acid and having in its structure modified reduced graphene oxide was prepared via freezing–thawing [136]. The composite was used for the removal of Congo red dye. Tannic acid was utilized as reducing agent and surface modifier of graphene oxide. The hydroxyl groups of chitin and phenolic hydroxyl groups of tannic acid, adsorbed onto graphene oxide surface by π–π interactions, afforded the crosslinking reaction with epichlorohydrin and the obtained composite proved enhanced mechanical and adsorption properties.

A new chitosan/polyacrylate/graphene oxide composite hydrogel was tested for the removal of Food yellow 3 and Methylene blue dyes [35]. Polyacrylate, chitosan and deionized water were mixed into a reactor to obtain semi-soluble slurry like mixture over which graphene oxide was added and finally, this mixture was converted into a composite hydrogel by sol–gel conversion under acetic acid vapor. The swelling and mechanical properties of the composite hydrogel were improved by the presence of polyacrylic acid and graphene oxide, respectively. Dyes were retained on chitosan-based composite via hydrogen bonding, ionic bonding and covalent bonding.

The blending of β-chitosan with graphene oxide, previously functionalized with triethylenetetramine, was used as hybrid adsorbent for C.I. Reactive Blue 221 [43]. The presence of triethylenetetramine–graphene oxide in the hybrid polymer allowed both π–π interactions and electrostatic attraction with the reactive dyes and also preserved a good acid and alkali resistance of the sorbent.

The adsorption isotherm data for most of chitin/chitosan-based hydrogels fitted well with Freundlich or Langmuir models while the kinetic data obeyed the pseudo-second-order kinetics. Some of the latest dyes removal studies by using chitin/chitosan-based gels are summarized in Table 1.

3.2   Cellulose-Based Hydrogels

Cellulose, having the formula $(C_6H_{10}O_5)_n$, is the most abundant biomaterial on the globe. It contains glucose units which are linked by β-1,4 linkages and the degree of polymerization varies between several hundreds to ten thousands (Fig. 3).

Usually, cellulose is synthesized by plants, but in some cases it is produced by certain bacteria. Its chains are packed into microfibrils which are kept together due to intramolecular hydrogen bonds between the three hydroxyl reactive groups of each polysaccharide unit. This polysaccharide is biocompatible and biodegradable. With a rough and fibrous structure, neat cellulose is insoluble in water and it has high mechanical properties compared to other polysaccharides. Alkali/urea (thiourea), LiCl/dimethylacetamide, N-methyl morpholine-N-oxide are appropriate solvents for cellulose. The formation of a stable three-dimensional network of cellulose-based derivatives was possible due to the occurrence of the hydroxyl reactive groups in each polymeric unit and it is the result of physical or chemical crosslinking of polysaccharide chains [106, 202, 32, 214, 34]. Methods employed for physical crosslinking of cellulose-based materials are: freeze–thaw [249], self-assembling [175], instantaneous gelation [70, 161], reconstitution [230], inverse emulsion technique [63], and ionotropic gelation [164]. Chemical crosslinking can be obtained by chemical reaction [246], polymerization [114] or by radiation (gamma, microwave, ultraviolet) [162]. Some derivatives of the cellulose, used for the synthesis of cellulose-based hydrogels, were: esters (hydroxypropyl methylcellulose phthalate, hydroxypropyl methylcellulose acetate succinate), ethers (methylcellulose, ethylcellulose, hydroxypropyl cellulose, carboxymethyl cellulose), composites (interpenetrating polymer networks or polymeric blendings). Hydrogels based on cellulose derivatives are favorable adsorbents for various contaminants [6, 53, 194].

Quaternized cellulose, grafted with polyacrylic acid by free radical polymerization, was employed as adsorbent for Methylene blue removal [235]. Dye was adsorbed by quaternized polysaccharide due to electrostatic interactions between ammonium groups of cellulose-based gel and carboxylic groups of polyacrylic acid.

New magnetic superabsorbent hydrogel nanocomposites based on carboxymethyl cellulose were synthesized in one-pot reaction [94]. The magnetic iron oxide nanoparticles were incorporated into carboxymethyl cellulose grafted with poly(acrylic acid) via the addition of two magnetic hydrates, namely $FeCl_2 \cdot 4H_2O$ and $FeCl_3 \cdot 6H_2O$. The nanocomposites were employed for Crystal violet removal from aqueous solutions. Redlich–Peterson isotherm model fitted well with the experimental adsorption data.

Table 1 Removal performances of chitin/chitosan-based hydrogels for different dyes

Sorbent	Dye	Dye removal performance	References
Chitosan-g-poly(acrylamide-co-acrylic acid)	Methyl orange	90% removal efficiency	[121]
Chitosan-g-poly(acrylamide-co-sodium methacrylate)	Fuchsin	97.2% removal efficiency	[97]
Chitosan grafted with ethylenediamine/methyl acrylate	Congo red	1607 (mg/g)/ 1143 (mg/g)	[217]
Magnetic chitosan grafted with 2-acrylamido-2-methylpropane sulfonic acid	Methylene blue	1000 mg/L	[239]
Chitosan grafted with acrylamide (microwave-assisted/conventional method)	Acid blue 113	255.5 (mg/g)/ 151.7 (mg/g)	[50]
Biohybrid chitosan/carbon-clay	Methylene blue	86.08 mg/g	[147]
	Acid blue 29	132.04 mg/g	
Chitosan-modified β-cyclodextrin	C.I. Reactive blue 49	80% removal efficiency	[252]
	Reactive yellow 176, Reactive blue 14, Reactive black 5, Reactive red 141	–	
Magnetic chitosan nanocomposite	Methylene blue	20.408 mg/g	[174]
Chitosan/oil palm ash zeolite composite	Methylene blue	151.51 mg/g	[116]
	Acid blue 29	212.76 mg/g	
Nano-ZnO/chitosan composite	Reactive black 5	189.44 mg/g	[33]
Chitosan/Ag-hydroxyapatite nanocomposite	Rhodamine B	127.61 mg/g	[132]
Chitosan/montmorillonite intercalated composite	Reactive red 136	473 mg/g	[131]
Zr (IV) surface-immobilized chitosan/bentonite composite	Amido black 10B	418.4 mg/g	[250]
Chitin/graphene oxide hybrid gels	Remazol black	9.3×10^{-2} mmol/g	[80]
	Neutral red	57×10^{-2} mmol/g	
Chitin/reduced graphene oxide composite gels	Congo red	230.5 mg/g	[136]
Chitosan/polyacrylate/graphene oxide composite	Food yellow 3	296.5 ± 31.7 (mg/g)	[35]
	Methylene blue	280.3 ± 23.9 (mg/g)	

(continued)

Table 1 (continued)

Sorbent	Dye	Dye removal performance	References
β-chitosan/polyamine/graphene oxide hybrid gel	Reactive blue 221	56.1 mg/g (pH = 2) 37.2 mg/g (pH = 12)	[43]
Hybrid composite $Cu_3(btc)_2$ -immobilized chitosan/graphene oxide	Methylene blue	357.14 mg/g	[187]
Magnetic chitosan/graphene oxide-quaternary ammonium salt composite	Basic brown 4	650 mg/g	[54]
Magnetic chitosan/polypyrrole/graphene oxide nanohybrid	Ponceau 4R	85% removal efficiency (pH = 2)	[185]
β-cyclodextrin/chitosan functionalized graphene oxide hydrogel	Methylene blue	1134 mg/g	[139]

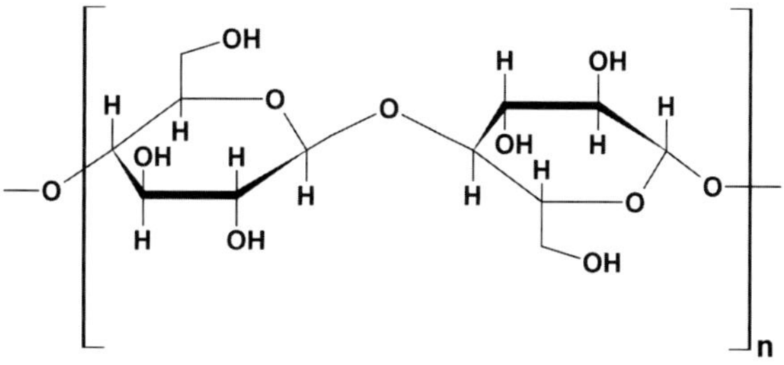

Fig. 3 Chemical structure of cellulose

New cellulose-based composites were obtained by mixing of hydroxypropyl cellulose powder with hydroxypropyl cellulose functionalized with molybdenum disulfide via esterification [37]. The derivatization of hydroxypropyl cellulose with nanosheets of molybdenum disulfide was done by treated MoS_2 with thioglycolic acid for the introduction of carboxyl reactive groups on its surface followed by the reaction with $SOCl_2$ for the conversion of MoS_2 –COOH to MoS_2–COCl. Finally, MoS_2–COCl nanosheets were grafted on hydroxypropyl cellulose under sonication. A high content of MoS_2-hydroxypropyl cellulose in the composite increased the sorption due to the existence of anionic groups and a large specific surface area of MoS_2-HPC. The composite was used for the catalyzation of Methylene blue degradation and it can be recycled upon contact to the sun light.

CdS-quantum dots were integrated in various molar ratios inside carboxymethyl cellulose-g-poly(acrylic acid-co-2-acrylamido-2-methylpropane sulfonic acid) hydrogel matrix and the resulted nanocomposite was employed for Rhodamine B elimination from aqueous solution [95]. Quantum dots are semiconductor nanoparticles having special optical and electronic properties [173]. Cadmium sulfide is known because of its availability, low price, thermal stability, low cytotoxicity [128]. The inclusion of CdS-quantum dots nanoparticles into polymeric network improved the thermal stability because the nanoparticles behaved as heat protective barricades. Desorption studies showed great regeneration capacity at various temperatures (25, 35, 45, 55 °C). Thus, after five cycles of adsorption–desorption, dye removal was between 95 and 75%.

Sepiolite is a fibrous magnesium hydrosilicate with a unique pore structure composed of an alteration of blocks and channels extending in the fiber direction. An organic–inorganic hybrid adsorbent based on cellulose and sepiolite was synthesized and employed for Malachite green dye elimination [105]. Hybrid was carried out by the adding of pretreated sepiolite over cellulose dissolved in NaOH/urea aqueous solution and the mixture was dropped into a diluted HCl–CaCl$_2$ solution. The resulted hybrid beads showed a high thermal resistance due to thermal insulation effect of sepiolite molecules.

Carboxymethyl cellulose-acrylamide-graphene oxide hydrogels, prepared by free-radical polymerization by varying carboxymethyl cellulose content, were used for Acid Blue-133 elimination from aqueous solution [227]. The swelling kinetic data showed that the gels have super Case II diffusion transport mechanism. Removal capacity varied between 47 and 99.97%, depending on graphene oxide percent in the hydrogels.

Cellulose modified magnetic nanoparticles were incorporated together with reduced graphene oxide into poly(ethylene glycol) dimethacrylate-based hydrogels using photo-polymerization [85]. Magnetic hydrogels were employed for Methylene blue removal. Cellulose having bound magnetic particles was prepared by co-precipitation reaction of Fe^{2+} and Fe^{3+} in NH$_4$OH and post-coating with the polysaccharide. The reduced graphene oxide was prepared from graphene oxide through hydrazine reduction. The resulted hybrid gels proved a high thermal stability. Magnetic reduced graphene oxide-charged hydrogel can be regenerated with the maintaining of the adsorption ability.

Langmuir isotherm and pseudo second-order model best followed the experimental adsorption data for most of cellulose-based gels. A part of the most recent dyes elimination studies by using cellulose-based gels are collected in Table 2.

3.3 Starch-Based Hydrogels

Starch is an abundant, losw cost, biocompatible and biodegradable biopolymer which occurs in plants (wheat, maize, potatoes, rice) as a reserve carbohydrate. Granules of starch consist of a mixture of amylose (20%) and amylopectin (80%). Amylose comprises of linear α-D-glucose units linked by α(1 → 4) glycosidic bonds while amylopectin consists of highly branched α-D-glucose units linked by α(1 → 4) or α(1 → 6) glycosidic bonds (Fig. 4) [229].

Native starch does not have adsorption ability but the problem can be resolved by chemical modification of the polysaccharide. General methods for obtaining starch derivatives are: grafting (free radical, living polymerization, ionic), crosslinking,

Table 2 Dyes elimination results of cellulose-based gels

Sorbent	Dye	Dye removal performance	References
Cellulose grafted with acrylamide	Methylene blue	734.816 mg/g	[235]
Chemi-mechanical pretreated cellulose-based superabsorbent gel	Methylene blue	3,003 mg/g	[135]
Carboxymethyl cellulose-based magnetic superabsorbent nanocomposites	Crystal violet	200 mg/L	[94]
Hydroxypropyl cellulose-based molybdenum disulfide composite	Methylene blue	6153 mg/g	[37]
Cellulose-g-poly(acrylic acid-co-acrylamide)	Acid blue 93 Methylene blue	85% removal for both dyes in single system; 60% removal for both dyes in dual system	[137]
Carboxymethyl cellulose/graphitic-carbon nitride/zinc oxide composite	Methyl violet	96.43 mg/g	[195]
Hydroxypropylmethyl cellulose-based nanocomposites	Crystal violet	76% removal efficiency (5 cycles)	[141]
Lignocellulose-g-poly(acrylic acid)/montmorillonite nanocomposites	Methylene blue	1,994.38 mg/g	[197]
Carboxymethyl cellulose-g-poly(acrylic acid) gel	Methyl Orange	84.2% removal ability	[248]
	Disperse Blue BLN	79.6% removal ability	
	Malachite green chloride	99.9% removal ability	
CdS quantum dots templated hydrogel nanocomposites	Rhodamine B	137 mg/g	[95]
Organic–inorganic hybrid beads from sepiolite and cellulose	Malachite green	314.47 mg/g	[105]
Carboxymethyl cellulose/acrylamide/graphene oxide hydrogels	Acid blue-133	185.45 mg/g	[227]
Polyvinyl alcohol/carboxymethyl cellulose/graphene oxide/bentonite hydrogels	Methylene blue	172.14 mg/g (30 °C)	[52]
Carboxymethyl cellulose/carboxylated graphene oxide composite microbeads	Methylene blue	180.32 mg/g	[65]
Magnetic reduced graphene oxide loaded hydrogels	Methylene blue	119 mg/g	[85]
Cellulose grafted with acrylamidomethylated-β-cyclodextrin	Methylene blue	15 mg/g	[77]
	Methyl orange	12 mg/g	

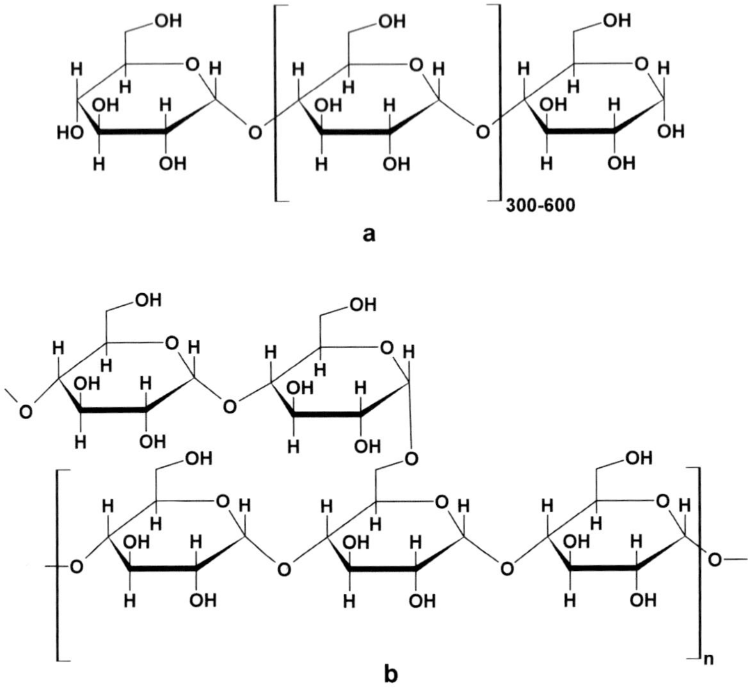

Fig. 4 Chemical structure of amylose **a** and amylopectine **b**

etherification, esterification and dual modification [38, 47]. Several derivatives of starch were used as sorbents in wastewater treatment.

Magnetic nanocomposite hydrogel beads were obtained by instantaneous gelation in boric acid of carboxymethyl starch-g-polyvinyl imidazole with a mixture composed by poly(vinyl alcohol) and Fe_3O_4, followed by crosslinking with glutaraldehyde [169]. The hydrogel beads were employed for Crystal violet and Congo red adsorption studies. Thermodynamic studies showed that chemisorption process was spontaneous and endothermic.

Composite hydrogel beads based on starch and humic acid were obtained by inverse suspension crosslinking method, using epichlorohydrin as crosslinker [38]. The composite was employed for Methylene blue adsorption studies. Dye was retained on the composite gel by $\pi-\pi$ interactions and ion exchanging. The regeneration capacity of the composite was high (95%), even after five cycles of adsorption–desorption.

Starch hydroxypropyl sulphate was obtained from 2-hydroxy-3-chloropropyl sulphate and starch in basic medium (NaOH), by using epichlorohydrin as crosslinking agent [90]. The polymer proved to be an effective adsorbent for Methylene blue removal due to good sorption results and its full regeneration capacity.

Porous nanocomposite hydrogel based on starch was prepared by free radical copolymerization of acrylamide and 2-acrylamido-2-methylpropane sulfonic acid onto the polysaccharide in the presence of $CaCO_3$ and graphene oxide [170]. The gel was employed for the elimination of Methylene blue from aqueous solution. Calcium carbonate particles acted as porogen agent and in the end of the synthesis these inorganic solid nanoparticles were removed by dissolving the gel in HCl solution. The adsorption capacity of Methylene blue increased by the augmentation of hydrogel porosity. After five cycles of adsorption–desorption, the removal efficiency of Methylene blue was 95.4%.

Starch-g-poly(acrylamide)/graphene oxide/hydroxyapatite nanocomposite, prepared by free radical copolymerization in the presence of graphene oxide nanosheets and different contents of hydroxyapatite, was used for Malachite green removal [96]. Thermogravimetric studies revealed that hydroxyapatite nanoparticles operate as thermal barriers, so their presence in nanocomposite structure increased the initial decomposition temperatures. Desorption efficiency increased with the decrease of the amount of hydroxyapatite occurred in the hydrogel structure. The introduction of hydroxyapatite in sorbent matrix contracts the free spaces of the network, and diminishes the porosity percent. Thus, after fifth cycles adsorption–desorption, 27, 19 and 14% of Malachite green were released from gel nanocomposite having 11, 25 and 30% porosity, respectively.

Langmuir isotherm and pseudo second order kinetic data best fitted with the experimental data for most of starch-based gels. Some of the latest dyes removal studies by using starch-based gels are summed up in Table 3.

3.4   Sodium Alginate-Based Hydrogels

Alginic acid is a natural polysaccharide derived from brown seaweeds but it can also be synthesized by microbial fermentation [84] (Fig. 5). It has applications in different areas, like: pharmaceutical industry [31, 42, 72], wastewater treatment [71, 103, 104], food industry [7, 186, 256], catalysts [58, 206]. Sodium alginate was the majority compound extracted from seaweed. It is a white or light yellow powder which is able to form an extremely viscous aqueous solution [93]. This property gives to sodium alginate the capacity to form gels, films or spinning fibers. Sodium alginate is composed of two alternating blocks 1-4-linked, namely β-D-mannuronic acid and α-L-guluronic acid which were arranged in a non-regular linear form [28]. This polysaccharide is biocompatible and non-toxic but its availability is confined due to its poor mechanical strength. For improving its mechanical properties, alginate was modified either by crosslinking (ionic or chemical) or via grafting polymerization. Some of the alginate-based hydrogels were further mixed with different (nano)fillers (carbon materials, oxides, organic matter) for obtaining composites which can selectively absorb antibiotics, dyes, heavy metal ions, or other contaminants.

Acrylamide/potassium 3-sulfopropyl methacrylate/sodium alginate/bentonite composite gels were synthesized by free radical solution polymerization [111].

Table 3 Dyes removal performances of starch-based hydrogels

Sorbent	Dye	Dye removal performance	References
Magnetic (carboxymethyl starch-g-polyvinyl imidazole/poly(vinyl alcohol)/ Fe$_3$O$_4$) nanocomposite hydrogel beads	Crystal violet	91.58 mg/g	[169]
	Congo red	83.66 mg/g	
Magnetic Fe$_3$O$_4$-based starch-g-poly (acrylic acid) nanocomposite hydrogel	Methylene violet	31.847 mg/g	[182]
Crosslinked amphoteric starch with carboxymethyl and quaternary ammonium groups	Acid light yellow 2G	227.39 mg/g	[240]
	Acid red G,	217.27 mg/g	
	Methyl green	133.33 mg/g	
	Methyl violet	333.33 mg/g	
Dithiocarbamate-modified starch gel	Acid orange 7	282 mg/g	[41]
	Acid orange 10	204 mg/g	
	Acid black 1	221 mg/g	
	Acid green 25	276 mg/g	
	Acid red 18	271 mg/g	
Chitosan/oxidized starch/silica hybrid membrane	Blue 71	75 mg/g	[89]
	Red 31	55 mg/g	
Composite hydrogel beads based on starch and humic acid	Methylene blue	111.10 mg/g	[38]
Starch-g-(acrylic acid-co-acrylamide)/ polydopamine hydrogel	Methylene blue	2276 mg/ g (pH 9)	[144]
Starch hydroxypropyl sulphate	Methylene blue	100% dye removal	[90]
Starch grafted with 2-acrylamido-propanesulphonic acid and dimethylaminoethyl methacrylate/benzyl chloride gel	Basic Violet 7	600 mg/g	[69]
Hydroxyethyl starch-g-poly(N,N-dimethyl acrylamide-co-acrylic acid) hydrogel	Malachite green	390 mg/g	[119]
Starch/poly(N,N-Diethylaminoethyl methacrylate) graft copolymer	Direct red 81	112 mg/g	[1]
Graphene oxide/hydrogel nanocomposites	Methylene blue	769.23 mg/g	[170]
Starch-g-poly(acrylamide)/graphene oxide/hydroxyapatite nanocomposite hydrogel	Malachite green	297 mg/g	[96]

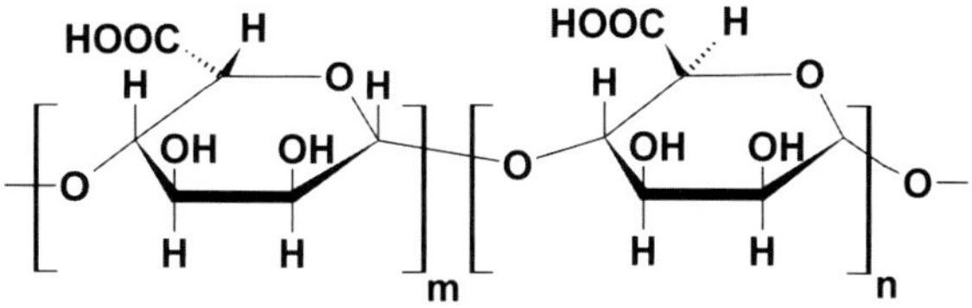

Fig. 5 Chemical structure of alginic acid

Ammonium persulfate/N,N,N',N'-tetramethylethylenediamine was the redox pair of initiators while poly(ethylene glycol) diacrylate was the crosslinker. The composites were employed to retain by hydrogen bonds cationic dye Lauths violet from aqueous solution. The values of Lauths violet sorption percent of the hydrogels were between 87.11 and 96.39%.

Interpenetrating polymeric hydrogels were prepared by mixing sodium alginate and acrylamide in various molar ratios under the irradiation with ^{60}Co-γ source [203]. The percentage of conversion of the synthesis was 100%. These hydrogels were employed for removal of some reactive dyes, namely Magenta, Safranine-O, Methylene Blue and Methyl Violet.

A composite copolymer based upon sodium alginate was prepared through *in-situ* controlled precipitation via the reaction between $AgNO_3$ and NaCl in the gel matrix of the polysaccharide included poly(acrylic acid-co-(hydroxyethyl)methacrylate) [25]. N,N'-methylene-bis-acryl amide was the crosslinker and potassium persulfate/sodium metabisulphite were the redox pair of initiators. The composite was utilized for the removal of Brilliant cresyl blue and the adsorption studies were done in fixed bed column. Dye was photocatalytic degraded in the presence of UV rays. Recycling of hydrogels was pretty good, thus, after five cycles of adsorption–desorption, the composite copolymer kept 94% of initial adsorption.

Hydrogel nanocomposites, synthesized by free radical graft copolymerization of acrylic acid on sodium alginate in the presence of TiO_2 nanoparticle, were used for Methyl violet dye removal [219]. TiO_2 increased dye sorption process due to the augmentation of both surface area and the pore volume of the nanocomposites.

Composite hydrogels, achieved by hydrothermal treatment of sodium alginate and graphene oxide solutions followed by ionically crosslinking of several metal ions (Cu^{2+}, Ca^{2+}, Ba^{2+}, Fe^{3+}), were utilized for elimination of some dyes (Methyl orange, Neutral red, Rhodamine B, Methylene blue and Congo red) [238]. Alginate/reduced graphene oxide composite hydrogel crosslinked by Fe^{3+} ions was more stable, having the highest compressive strength and lowest swelling ratio between the studied composite gels. Reusability studies showed for Fe^{3+}-composite hydrogel a high adsorption percentage (90% of its original adsorption capacity) after ten adsorption/desorption cycles.

The hydrogel composite based on sodium alginate crosslinked with acrylic acid/graphite was utilized in the elimination of Malachite green dye from aqueous solution [228]. Graphite has a tridimensional structure of graphene sheets held by van der Waals forces and it is often used due to its good reactivity, thermal stability and mechanical strength [24, 68]. Graphite is frequently used as nanofiller for improving the mechanical strength and sorption capacity of the polymers. The composite revealed 91% adsorption capacity after three adsorption–desorption cycles.

Langmuir isotherm and pseudo-second-order kinetic model obeyed with the experimental adsorption data for the most xanthan-based gels. A part of the more recent dyes elimination studies by using sodium alginate-based hydrogels are summarized in Table 4.

3.5   Xanthan-Based Hydrogels

Xanthan is a bacterial and anionic polysaccharide obtained from *Xanthomonas campestris* (Fig. 6). The main chain of the polysaccharide is composed of $(1{\rightarrow}4)$-linked β-D-glucopyranose units. The side chain contains two β-D- mannose and one β-D-glucuronic acid. In the end position, mannopyranose moieties are $(1 \rightarrow 4)$-linked to glucuronic acid and these are $(1 \rightarrow 2)$-bound to non-terminal α-D-mannopyranose. Xanthan and its derivatives have several applications, such as agriculture [76], petrol refining industry [113], wastewater treatment [2], biomedicine [17, 30], food industry [61].

Hydrogel nanocomposite, based on xanthan-g-poly(acrylic acid-co-acrylamide) with included Fe_3O_4 magnetic nanoparticles, was used for Malachite green dye removal from aqueous solution [149]. The composite was obtained by graft copolymerization of acrylic acid and acrylamide onto xanthan, followed by inclusion of magnetic nanoparticles Fe_3O_4 within the polymer network. The binding sites of the gel could be simply regenerated by using 0.1 M HCl solution with significance loss in adsorption capacity.

Nanocomposite containing xanthan gum, methionine and bentonite was used for the removal of Congo red dye [5]. L-methionine was employed for the modification of nanofiller bentonite. Desorption studies showed that the nanocomposite could be regenerated (65%) up to five sorption–desorption cycles by employing 0.1 M NaOH solution.

Xanthan-g-poly(acrylic acid)/oxidized multi-walled carbon nanotubes hydrogel nanocomposites were used for Methylene blue removal [145]. Graft copolymerization of acrylic acid on the xanthan was inducted by microwave radiations and various weight percents of oxidized multi-walled carbon nanotubes were included into the gels network during free radical polymerization. It is known that multi-walled carbon nanotubes act as good sorbents due to their great surface area, porous and layered structure [178]. A high dye removal capacity (96%) could be observed after four sorption–desorption cycles.

Xanthan-g-poly(acrylic acid)/reduced graphene oxide hydrogel composite was prepared by microwave-assisted-free radical graft copolymerization of acrylic acid onto xanthan and was further employed as sorbent for two dyes (Methylene blue and Methyl violet) removal [146]. Reduced graphene oxide solution was introduced during the graft copolymerization reaction and its role was to increase the adsorption capacity and thermal stability of the xanthan-based composite. Dye removal capacity

Table 4 Dyes elimination results of sodium alginate-based gels

Sorbent	Dye	Dye removal performance	References
Acrylamide/potassium 3-sulfopropyl methacrylate/sodium alginate/bentonite hybrid gels	Lauths violet	100% dye removal	[111]
Chitin/alginate magnetic nanogel beads	Methyl orange	107.5 mg/g	[129]
Sodium alginate/titania nanoparticles composite	Direct Red 80	163.934 mg/g	[143]
	Acid Green 25	151.515 mg/g	
(Alginate/acrylamide)-interpenetrating polymeric hydrogels	Magenta Safranine-O Methylene blue, Methyl violet	0.05 g hydrogel, 25 °C, 0.1 ionic strength, pH 7.0 (optimal conditions for dye elimination)	[203]
Sodium alginate-g-poly(sodium acrylate-co-styrene)/organo-illite/smectite clay	Methylene blue	1843.46 mg/g	[230]
Alginate-g-poly(acrylic acid-co-(hydroxyethyl)methacrylate)/AgCl composite	Brilliant cresyl blue	UV photocatalytic degradation of dye	[25]
Sodium alginate-g-acrylic acid/TiO_2 hydrogel nanocomposites	Methyl violet	1156.61 mg/g	[219]
Nano-iron oxide-loaded alginate microspheres	Malachite green	93.9% dye elimination	[204]
Zn–Al–Fe_3O_4 sodium alginate beads	Rhodamin B	97% dye elimination	[122]
Montmorillonite–alginate nanobiocomposite	Basic red 46	35 mg/g	[86]
Magnetic Fe_3O_4/activated charcoal/β-cyclodextrin/alginate polymer nanocomposite	Methylene blue	2,079 mg/g 99.53% dye removal (pH 6)	[241]
Glucose oxidase/laccase/$MnFe_2O_4$/calcium alginate nanocomposites	Indigo	44.03% dye removal (pH 5)	[200]
	Methylene blue	93.46% dye removal (pH 9)	
	Acid red 14	46.5% dye removal (pH 3)	
Alginate/reduced graphene oxide/Cu^{2+}, Ca^{2+}, Ba^{2+}, Fe^{3+} composite hydrogels	Methyl orange	23.8 mg/g	[238]
	Neutral red,	20.5 mg/g	
	Rhodamine B	18.4 mg/g	
	Methylene blue	23.8 mg/g	
	Congo red	8.3 mg/g	

(continued)

Table 4 (continued)

Sorbent	Dye	Dye removal performance	References
Alginate/graphene oxide hydrogel beads	Methylene blue	245 mg/g	[75]
	Rhodamine B	252 mg/g	
	Vat green 1	122 mg/g	
	Methyl orange	111 mg/g	
Sodium alginate/graphite hybrid hydrogel	Malachite green	628.93 mg/g	[228]

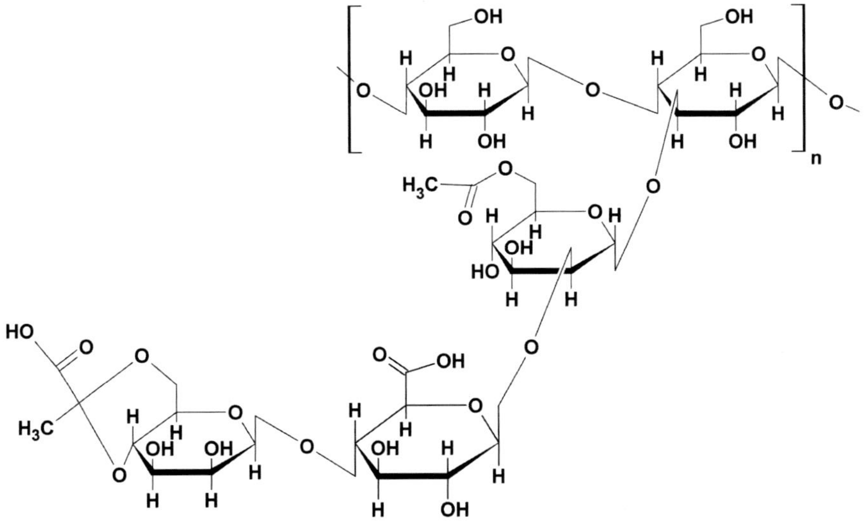

Fig. 6 Chemical structure of xanthan

remained high even after four cycles of sorption–desorption (96% for Methylene blue and 95% for Methyl violet).

Adsorption process followed the pseudo-second-order rate model and Langmuir adsorption isotherm for most of xanthan-based hydrogels. Some of the latest dyes removal studies by using xanthan-based gels are summarized in Table 5.

Table 5 Dyes removal performances of xanthan-based hydrogels

Sorbent	Dye	Dye removal performance	References
Xanthan-g-poly(acrylic acid-co-acrylamide)/Fe_3O_4 magnetic nanoparticles hydrogel nanocomposites	Malachite green	497.15 mg/g	[149]
Xanthan-g-poly(acrylic acid)/Fe_3O_4 magnetic nanoparticles hydrogel nanocomposites	Methyl violet	642.0 mg/g	[154]
Xanthan/methionine-bentonite hydrogel nanocomposite	Congo red	142.71 mg/g (30 °C) 278.84 mg/g (40 °C) 530.54 mg/g (50 °C)	[5]
Xanthan-g-poly(acrylic acid)/oxidized multi-walled carbon nanotubes hydrogel nanocomposite	Methylene blue	521.0 mg/g (30 °C)	[145]
Xanthan-g-poly(acrylic acid)/reduced graphene oxide hydrogel composite	Methylene blue	793.65 mg/g	[146]
	Methyl violet	1052.63 mg/g	
Poly(vinyl alcohol)-xanthan gum composite hydrogel	Methylene blue	80% dye removal (after five cycles)	[251]
Xanthan-psyllium-g-poly(acrylic acid-co-itaconic acid)	Auramine-O	95.63% dye removal	[36]

3.6 Miscellaneous Natural Polysaccharides-Based Hydrogels

3.6.1 Gum (Ghatti, Karaya, Guar) Polysaccharides-Based Hydrogels

Gum ghatti is an anionic natural polysaccharide containing alternating 4-O-substituted and 2-O-substituted α-D-mannopyranose units and also, 1 → 6 linked β-D-galactopyranose with side chains of l-arabinofuranose moieties (Fig. 7a). It is known that gum ghatti powder fortifies the immune system, reduces the cholesterol level, speeds up the healing of wounds, diminishes the side effects after chemotherapy, decelerates the ageing and also, it is employed as supplement for weight loss [108, 155]. Hydrogels and nanocomposites based on gum ghatti had several applications, one of them being the removal of different contaminants (dyes, heavy metal cations) from wastewater.

Gum ghatti–g-poly(acrylic acid-co-acrylamide)/Fe_3O_4 magnetic nanocomposites were synthesized and used as sorbent for Rhodamine B removal [150]. Acrylic acid and acrylamide were grafted onto gum polysaccharide by free radical copolymerization. The presence of Fe_3O_4 in nanocomposites structure enhanced the adsorption capacity due to the augmentation of pore volume and surface area. Also,

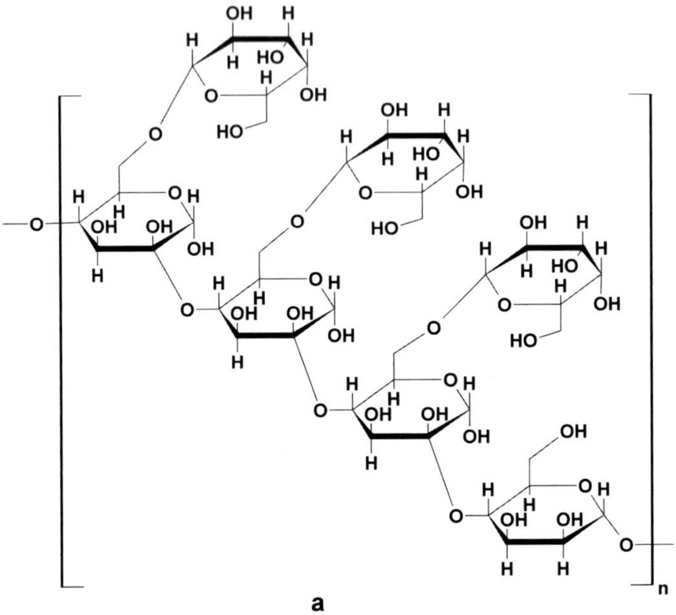

Fig. 7 Chemical structure of gum ghatti **a** and gum guar **b**

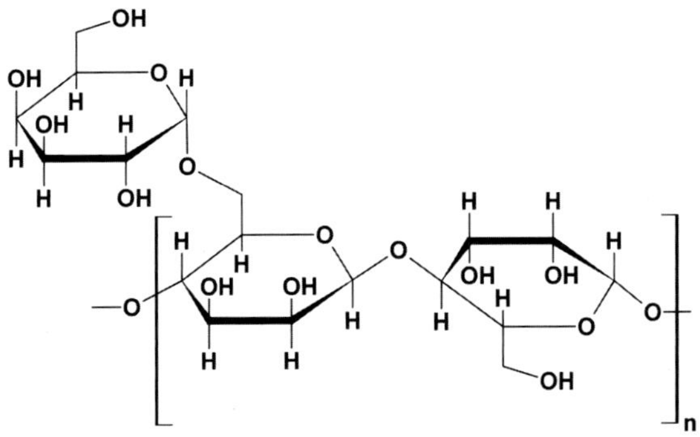

thermal stability increased after the incorporation of Fe_3O_4 in the polymeric network. Adsorption efficiency slightly decreased (12%) after five sorption–desorption cycles.

Gum karaya is an acetylated polysaccharide, having about 37% galacturonic acid. It is partially acetylated (8% acetyl groups by weight) [22]. This gum polysaccharide and its derivatives are used as adhesives, binding agents and sorbents due to their capacity to retain great quantities of water. Nanosilica-containing hydrogel nanocomposite of gum karaya grafted with acrylic acid and acrylamide was employed for Methylene blue elimination from aqueous solutions [153]. The regeneration capacity of the nanocomposite in acidic medium was 100% over three adsorption–desorption cycles.

Guar gum is a non-ionic polysaccharide, consisting of $(1 \rightarrow 4)$-linked β-D-mannopyranose units having as side chains $(1 \rightarrow 6)$-linked-α-D-galactopyranose (Fig. 7b). This gum polysaccharide acts as laxative and so, determines regular bowel movements, being used in the treatment of chronical diseases, like diverticulosis and colitis [21]. A hybrid nanocomposite based on guar gum grafted with acrylamide and having silica included in its structure was synthesized and used for the removal of Reactive blue 4 and Congo red dyes [163]. The reusability studies showed 83.46% of Reactive blue 4 and 82.09% of Congo red (pH = 10) desorption after four sorption–adsorption cycles.

3.6.2 Pullulan and Dextran-Based Hydrogels

Pullulan is a biocompatible and non-toxic polysaccharide. It consisting of maltotriose units bound by α(1–4) glycosidic bond, while consecutive maltotriose moieties are linked by α(1–6) glycosidic bonds (Fig. 8a). The polysaccharide and its derivatives

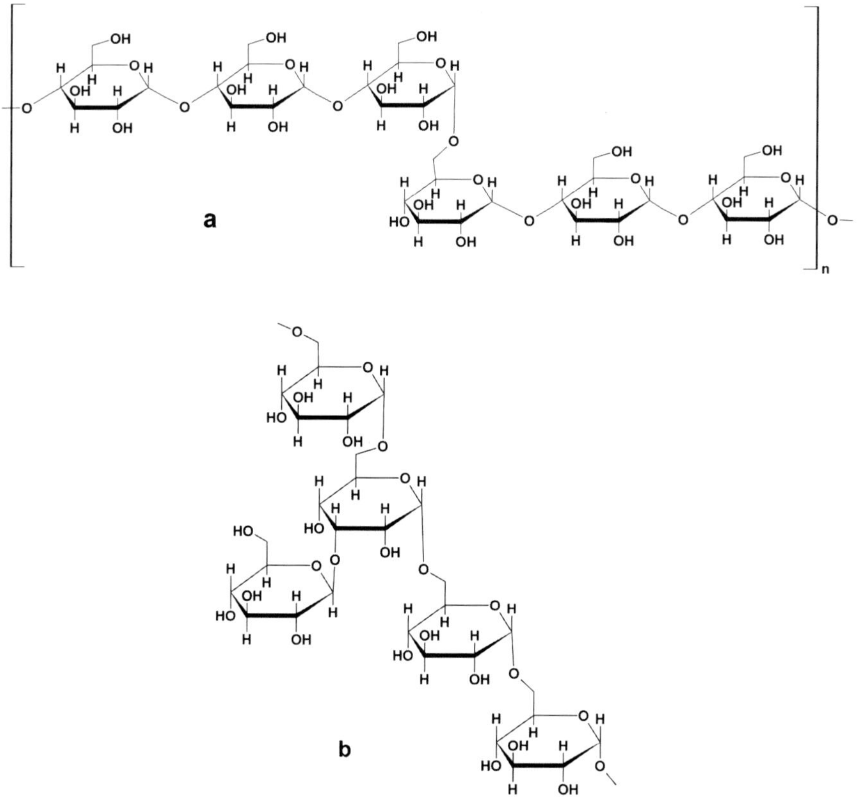

Fig. 8 Chemical structure of pullulan **a** and dextran **b**

have numerous applications, like biomedicine, pharmaceuticals, food industry, electronics. Some of its derivatives were employed as adsorbents of different pollutants in wastewater treatment.

Pullulan-graft-poly(3-acrylamidopropyl trimethylammonium chloride) microspheres were synthesized in two stages of reaction and were further used for several dyes (Azocarmine B, Acid Orange 7, Methyl Orange, Ponceau 6R, Congo Red) adsorption studies [44]. In the first step of the synthesis, 3-acrylamidopropyl trimethylammonium chloride was grafted onto the polysaccharide while in the second stage, the obtained graft polymer was crosslinked with epichlorohydrin.

Dextran is a biocompatible and non-toxic polysaccharide. Its main chain is composed of α-D-glucopyranose units which are linked by linear α-1,6 glycosidic bonds, with a reduced degree of α-1,3-linked side chains (Fig. 8b). Some of its applications included: biomedical, pharmaceutical, food and chemical industries. Certain dextran derivatives were employed as sorbents in wastewater detoxification.

An amphiphilic cationic dextran hydrogel, with two types of quaternary ammonium side-chains with different polarities and having the same molar ratio between hydrophilic and hydrophobic pendant groups, was prepared and tested as sorbent for several dyes (Methyl orange, Indigo Carmin, Orange II, Rose-bengal) [208]. The sorption capacity of the dextran-based hydrogel was higher, for the same dyes, compared to inorganic–polymer hybrids. A rapid and full regeneration of the hydrogel was obtained by using the consecutive addition of water, NaCl 0.5 M, and methanol.

For all above-mentioned gels based on miscellaneous natural polysaccharides, the sorption process obeyed the pseudo-second-order rate model and Langmuir adsorption isotherm while thermodynamic studies showed a spontaneous evolution. Table 6 condensed the latest dyes sorption studies by using miscellaneous polysaccharides-based hydrogels.

4   Synthetic Polymers-Based Hydrogels

Several (co)polymers hydrogels, based on: acrylic acid, (meth)acrylamide, vinylalcohol, vinylpyrrolidone, vinyl phosphonic acid, 3-(methacryloylamino)propyltrimethyl ammonium chloride, sodium p-styrene sulfonate, ethylenimine, 2-acylamido-2-methylpropanesulfonate and N-isopropylacrylamide, were synthesized and used for dyes elimination in wastewater treatment (Table 7).

Different fillers, like clays (kaolin, laponite, attapulgite, montmorillonite, bentonite), metals (Co, Cu) or metal oxides (Fe_3O_4, TiO_2), carbon-based materials (sulfonated graphene, graphene oxide) were incorporated in the hydrogel matrix, thus, increasing thermal and mechanical stability, chemical strength and also, the adsorption potential [165].

For the most sorption studies of synthetic polymers-based gels, Langmuir isotherm and pseudo-second-order kinetic were the models that fitted pretty well with the experimental data.

Table 6 Dyes elimination results of miscellaneous polysaccharides-based gels

Sorbent	Dye	Dye removal performance	References
Gum ghatti–g-poly(acrylic acid-co-acrylamide)/Fe$_3$O$_4$ nanocomposite	Rhodamine B	654.87 mg/L	[150]
Gum ghatti–g-poly(acrylic acid-co-methacrylamide) hydrogel	Methylene blue	694.44 mg/g	[152]
	Methyl violet	543.478 mg/g	
Gum ghatti–g-poly(acrylic acid)/Fe$_3$O$_4$ nanocomposite	Methylene blue	671.14 mg/g	[151]
Gum karaya-g-poly(acrylic acid-co-acrylamide) /nanosilica nanocomposite	Methylene blue	1,408.67 mg/g	[153]
Gum guar-g-poly(acrylamide)/silica hybrid nanocomposite	Reactive blue 4	579.01 mg/g	[163]
	Congo red	233.24 mg/g	
Pullulan-g-poly(3-acrylamidopropyl trimethylammonium chloride) microspheres	Azocarmine B	113.63 mg/g	[44]
	Acid Orange 7, Congo Red Methyl Orange, Ponceau 6R	–	
Pullulan-g-polyacrylamide porous hydrogel	Methylene blue	386.81 mg/g	[181]
	Reactive blue 2	273.24 mg/g	
Montmorillonite-included pullulan-based nanocomposite hydrogel	Crystal violet	80.00 mg/g	[213]
Polydopamine/montmorillonite-embedded pullulan hydrogels	Crystal violet	112.45 mg/g	[172]
Amphiphilic cationic dextran hydrogel	Methyl orange	705 mg/g	[208]
	Indigo carmin	732 mg/g	
	Orange II	652 mg/g	
	Rose-bengal	654 mg/g	
Amphiphilic cationic dextran hydrogels	Methyl orange	893 mg/g	[207]
	Rose-bengal	1718 mg/g	
Dextran-glycidyl methacrylate/acrylic acid hydrogel	Methylene blue	1994 mg/g	[245]
	Crystal violet	2390 mg/g	

5 The Influence of Different Parameters on Sorption Ability

5.1 Contact Time

Dye removal from aqueous solution showed a rapid increase at the beginning of adsorption, followed by a slowing of the process, and the reaching of an equilibrium state in the end. Initially, dyes molecules were rapidly attached to gel surface via

Table 7 Dyes removal performances of synthetic polymers-based hydrogels

Sorbent	Dye	Dye removal performance	References
Poly(acrylic acid)/kaolin composite hydrogel	Brilliant green	26.31 mg/g	[198]
Poly(acrylamide)/kaolin composite hydrogel	Crystal violet	23.80 mg/g	[199]
Poly(acrylamide)/laponite composite hydrogels	Crystal violet	44.33–95.56% dye removal	[133]
Poly(acrylamide-g-itaconic acid)/montmorillonite composite hydrogel	Brilliant Cresyl blue	458.1 mg/g	[110]
Poly(acrylamide co-acrylic acid)/attapulgite composite hydrogel	Methyl violet	1194 mg/g	[62]
Poly(acrylamide-co-maleic acid) montmorillonite nanocomposite hydrogel	Crystal violet	20.36 mg/g	[13]
Poly(vinyl alcohol)/laponite/Fe$_3$O$_4$ magnetic nanocomposite hydrogel	Methylene blue	251 mg/g 93% dye removal (4 cycles)	[142]
Poly(vinyl alcohol)/bentonite composite hydrogel	Methyl orange, Methyl red, Methylene blue	99% dye removal	[188]
Poly(vinyl alcohol)/sulfonated graphene gel	Methylene blue	27.7 mg/g	[130]
	Malachite green	22.5 mg/g	

(continued)

Table 7 (continued)

Sorbent	Dye	Dye removal performance	References
Poly(acrylamide-g-stearyl methacrylate)/graphene oxide composite hydrogel	Methylene blue	95% dye removal (6 cycles)	[49]
	Congo red	88% dye removal (6 cycles)	
Poly(acrylamide)/graphene oxide composite hydrogel	Methylene blue	292.84 mg/g	[243]
	Rhodamine 6G	288 mg/g	
Poly(ethylenimine)/graphene oxide composite hydrogel	Methylene blue	323.95 mg/g	[83]
	Rhodamine B	114.41 mg/g	
3-(methacryloylamino)propyl-trimethyl ammonium chloride and sodium p-styrene sulfonate-based gel/graphene oxide composite gel	Methylene blue	12.842 mg/g	[233]
	Rhodamine B	12.264 mg/g	
Poly(vinylpyrrolidone-co-acrylic acid)/graphene oxide composite hydrogel	Congo red	77.5193 mg/g	[16]
Poly(acrylic acid)/magnetic Co nanoparticles composite hydrogel	Methylene blue	836.5 mg/g	[10]
Poly(acrylic acid-co-acrylamide)/Co–Cu nanoparticles composite hydrogel	Malachite green	238.09 mg/g	[158]
Poly(acrylamide co-acrylic acid)/TiO$_2$ composite hydrogel	Methylene blue	4.86 mg/g (5% TiO$_2$)	[109]

(continued)

Table 7 (continued)

Sorbent	Dye	Dye removal performance	References
Poly(vinyl phosphonic acid)/Fe_3O_4 magnetic nanocomposite hydrogels	Methylene blue	13.79 mg/g	[192]
	Rhodamine 6G	15.47 mg/g	
Poly(2-acylamido-2-methylpropanesulfonate-co-N-isopropylacrylamide)/Fe_3O_4.Cu_2O.Fe_3O_4 magnetic nanocomposite hydrogels	Methylene blue	792.4 mg/g	[15]

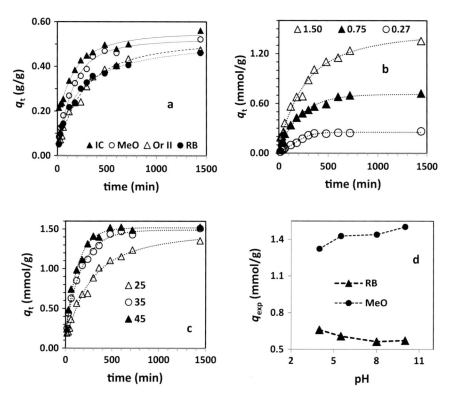

Fig. 9 Variation of dye adsorption against time, as a function of dye nature **a**, initial Orange II concentration (0.27, 0.75 and 1.5 mM) **b**, temperature (25, 35, 45 °C) (for Orange II) **c**, pH (for Rose bengal and Methyl orange) **d**. Initial dye concentration was 1.5 mM **a, c, d** and temperature 25 °C **a, b, d**. *Reprinted by permission from Springer Nature*, Stanciu and Nichifor [208]

surface mass transfer. Further, sorption was slowed down due to the progressive reduction of hydrogel external sites left unoccupied and also due to the slow diffusion of dye molecules into polymeric matrix to the internal sites. In the end, the available sorption sites became rarer and the equilibrium state is reached. Indigo-carmin (IC), Methyl orange (MO), Orange II (O II) and Rose-bengal (RB)) were employed as dyes for the study of the influence of contact time on removal performances by using dextran-based hydrogel as sorbent (Fig. 9a–c) [208].

5.2 Dye Nature

The adsorption efficiency was influenced by dye nature. So, for dyes having high molar masses there was a limitation of their amount which could be adsorbed by the hydrogel. The quantity of Rose bengal adsorbed by dextran-based hydrogel was the smallest compared with other dyes (Methyl orange, Indigo Carmine, Orange II) due to Rose bengal bulky structure and its high molar mass [208].

Also, the presence of two or more functional groups in chemical structure of dye led to the binding of more active sites of the sorbent, thus, decreasing the amount of adsorbed dye. An example in this regard was Indigo carmin dye which was adsorbed in a smaller amount in comparison with other dyes (Orange II, Methyl orange) due to the existence of two sulphonate groups in its chemical structure [208] (Fig. 9a).

5.3   Initial Dye Concentration

The increase of initial dye concentration determines an augmentation of dye amount which was adsorbed on the gel, due to the improvement of dye mass transfer between aqueous solution and solid sorbent (Fig. 9b). But, the percent of dye removal decreases with the augmentation of initial dye concentration as a result of the saturation of hydrogel' active sites.

5.4   Temperature

The increase of temperature can favor or not the adsorption process. Generally, the augmentation of the temperature determines a higher mobility of dye inside the adsorbent and/or a pore size enlargement. The increase of dye removal with the augmentation of temperature proves an endothermic process while its decrease with the increase of the temperature, shows an exothermic adsorption. The sorption of several dyes (Methyl orange, Indigo Carmin, Rose bengal and Orange II) on dextran gel, having two types of quaternary ammonium pendant groups with different polarities, was endothermic (Fig. 9c) [208].

5.5   pH

pH is a key factor in dye adsorption, if the main forces between sorbent and adsorbate are electrostatic. Studies regarding the adsorption of Methyl orange and Rose bengal dyes on dextran gel showed the influence of pH value on sorption process [208]. Thus, for a pH range between 4 and 10, the adsorption capability showed a reduced variation for both dyes, but the trend was different for each dye. The amount of Methyl orange adsorbed by the hydrogel thinly increased with pH augmentation due to the increase of gel ionization degree. The amount of Rose bengal retained by the hydrogel slightly decreased with pH increase because of the occurrence of electrostatic attraction between one dye molecule and two positively charged sites of the hydrogel, which reduced the number of dye molecules adsorbed by dextran-based gel (Fig. 9d).

6 Adsorption Kinetics

Adsorption kinetics, representing the variation of water uptake versus time at a permanent concentration, gives details about the rate of adsorption steps and the sorption mechanism. There are four stages in the adsorption process. The first stage consists in the migration of dye molecules from bulk liquid phase to the boundary layer. The second stage is the diffusion of the sorbate via the boundary layer to the external surface of the adsorbent (external diffusion). The third stage is composed of transport of dye molecules into the adsorbent pores (internal diffusion). The last stage consists in the interactions between dye molecules and the adsorbent active sites [222, 232]. Pseudo-first-order, pseudo-second-order and intraparticle diffusion models are among the most used kinetic models in sorption studies. The correlation coefficient (R^2) was utilized to determine the most appropriate equations.

6.1 Pseudo-First-Order Kinetic Model

Pseudo-first-order kinetic model, also named Lagergren model, takes into consideration that the mass transfer between solution and solid is the predominant mechanism in adsorption process [201]. The rate of sorption is directly proportional with the difference between the amount of adsorbed dye on hydrogel at equilibrium time and a fixed time. The linear form of pseudo-first-order kinetic equation is depicted in Eq. (4).

$$\log(q_{\exp} - q_t) = \log Q_1 - \frac{k_1 t}{2.303} \tag{4}$$

where t is the contact time (min), k_1 (min^{-1}) is the rate constant for the first-order kinetic model, q_{exp} (mmol/g) is the amount of adsorbed dye at equilibrium, q_t (mmol/g) is the dye quantity which was sorbed at moment t and Q_1 (mmol/g) is the calculated maximum adsorption value. Obtaining of a straight line by plotting $\log(q_{\exp} - q_t)$ against t shows the fitting of Lagergren model for adsorption kinetics. k_1 is found out from the slope of the graph and Q_1 from its intercept.

6.2 Pseudo-Second-Order Kinetic Model

Pseudo-second-order kinetic model, also called Ho and McKay rate equation model, is frequently related with chemisorption process, in which electrons were shared or exchanged between the adsorbent and the sorbate [201]. The linear form of pseudo-second-order kinetic equation is showed in Eq. (5).

$$\frac{t}{q_t} = \frac{1}{k_2 Q_2^2} + \frac{t}{Q_2} \tag{5}$$

where t is the contact time (min), k_2 (g/mmol min) is the rate constants for the second-order kinetic model. q_t (mmol/g) is the dye quantity which was sorbed at moment t while Q_2 (mmol/g) is the calculated maximum adsorption value. A straight line achieved by plotting t/q_t against t is the indication that Ho and McKay rate equation model is suitable to describe the adsorption kinetics. Q_2 is determined from the slope of the graph and k_2 from its intercept.

6.3 Intraparticle Diffusion Kinetic Model

The intraparticle diffusion model, designed by Weber and Morris, is used for the evaluation of adsorption diffusion mechanism [236]. The linear form of this kinetic model is revealed in Eq. (6).

$$q_t = k_{id} \sqrt{t} + C \tag{6}$$

k_{id} (mmol/g min$^{0.5}$) is the rate constant for the intraparticle diffusion model while C (mmol/g) is a parameter direct proportional with the boundary layer thickness. q_t (mmol/g) is the dye quantity which was sorbed at moment t and t(min) is the time. In case of Weber and Morris equation, the obtaining of a single straight line, who crossed the origin, proved that intraparticle diffusion is the only sorption mechanism. A multi-linear plot proved the existence of more stages in the adsorption process and each line corresponded to one step: (1) surface adsorption, (2) intraparticle diffusion and (3) adsorption near to equilibrium. Sometimes, because of the fast surface adsorption, the first two stages merged. That is the case of Methyl orange sorption on dextran-based hydrogels having quaternary ammonium groups as side-chains (Fig. 10).

Thus, the first line of the graph corresponded to intraparticle diffusion and it was characterized by $k_{id,1}$ constant, while the second line described the saturation process and it was defined by $k_{id,2}$ constant.

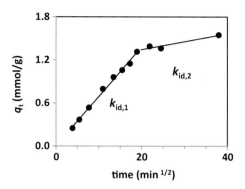

Fig. 10 Linear plot of Methyl orange sorption on dextran gel using intraparticle diffusion as kinetic model. ($C_i =$ 1.5 mM; t °C = 25 °C. *Reprinted by permission from Springer Nature,* Stanciu and Nichifor [208]

7 Adsorption Isotherms

Adsorption isotherms supply the relations between the concentration of sorbate retained on the solid phase and its concentration in solution at equilibrium state and indicate possible interactions between adsorbate and sorbent. Also, the adsorption efficiency of the sorbents can be evaluated and compared with the help of the isotherm models. Two-parameter model isotherms such as Langmuir, Freundlich and Dubinin–Raduskevich were often selected to describe the sorption characteristics at equilibrium state. A correlation coefficient, R^2, very close to 1 is an indication of a good fitting of the experimental data with respective to isotherm model.

7.1 Langmuir Model

Langmuir isotherm model supposes a monolayer homogeneous sorption on equivalent binding sites [40, 125]. The linear form of the binding isotherm model is depicted in Eq. (7).

$$\frac{C_{eq}}{q_{exp}} = \frac{C_{eq}}{Q_L} + \frac{1}{K_L Q_L} \tag{7}$$

where q_{exp} (mmol/g) is the experimental dye amount adsorbed at equilibrium, C_{eq} (mM) is concentration of sorbate at equilibrium, Q_L is maximum adsorbent capacity (mmol/g) and K_L (L/mmol) is Langmuir equilibrium constant. Another parameter of Langmuir equation is R_L, a non-dimensional separation factor stated by Eq. 8:

$$R_L = \frac{1}{1 + K_L C_i} \tag{8}$$

where C_i (mM) is the initial concentration of adsorbate. R_L values specify the sorption efficacy. Thus, adsorption can be unfavorable ($R_L > 1$), linear ($R_L = 1$), favorable ($0 < R_L < 1$) or irreversible ($R_L = 0$).

7.2   Freundlich Model

Freundlich isotherm considers a reversible and non-ideal sorption that happens on heterogeneous surfaces and consists in a multi-layer adsorption (Eq. 9) [40, 74].

$$\log q_{\exp} = \log K_F + \frac{1}{n_F} \log C_{eq} \qquad (9)$$

where $q_{\exp}$ (mmol/g) is the experimental dye amount adsorbed at equilibrium, C_{eq} (mM) is concentration of sorbate at equilibrium, K_F (mmol/g) is Freundlich equilibrium constant. n_F, the heterogeneity factor, is a measure of type of adsorption and heterogeneity of the adsorbate sites. If $0 < 1/n_F < 1$, the sorption process is favorable.

7.3   Dubinin–Radushkevich Model

Dubinin–Radushkevich isotherm takes into consideration that adsorption energy has a Gaussian distribution on the heterogeneous surfaces [51, 60]. Its linear form is showed in Eq. (10):

$$ln q_{\exp} = ln Q_{RD} - \beta \varepsilon^2 \qquad (10)$$

where $q_{\exp}$ (mmol/g) is the experimental dye amount adsorbed at equilibrium, Q_{DR} is maximum adsorbent capacity (mmol/g). β (mol^2/J^2) is a constant associated with the mean free energy per molecule of adsorbate for shifting from its location in the solution to the infinity and ε is Polanyi potential (Eq. 11)

$$\varepsilon = RT ln \left(1 + \frac{1}{C_{eq}} \right) \qquad (11)$$

where R is the universal gas constant (8.314 J/mol K) and T is the soluon temperature expressed in Kelvin scale (K).

Dubinin–Radushkevich model affords to find the mean free energy of adsorption for ligand molecules, E (kJ/mol). This parameter is essential to reveal the main forces of sorption (physical or chemical ones). When the energy of sorption is less than 8 kJ/mol, physical forces control the sorption, when E is in the range 8–16 kJ/mol, the adsorption mechanism is based on ion exchange and if E is greater than 16 kJ/mol, the sorption is derived by chemisorption.

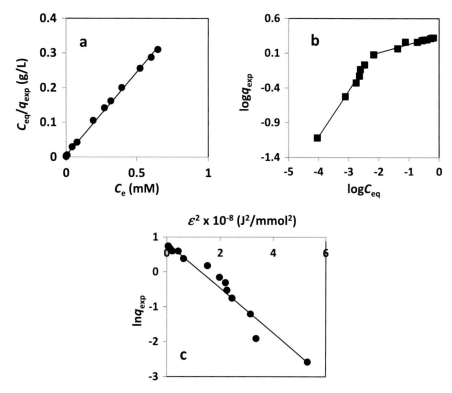

Fig. 11 Linear plots of Methyl orange adsorption data on dextran hydrogel in accordance with Langmuir **a**, Freundlich **b** and Dubinin–Radushkevich **c** isotherm models. *Reprinted by permission from Springer Nature,* Stanciu and Nichifor [208]

Langmuir and Dubinin–Radushkevich models depicted pretty well the experimental data for Methyl orange sorption on dextran-based cationic hydrogel ($R^2 >$ 0.96) [208]. Instead, the usability of Freundlich equation begins after Methyl orange was bound to 40% of gel cationic sites. So, the linear form of Freundlich isotherm gave two straight lines and Freundlich parameters were calculated from the breakpoint connecting the lines (Fig. 11).

8   Recycling of the Hydrogel-Based Sorbents

Regeneration of adsorbents is essential because it helps in reducing both the total cost of wastewater treatment and the generated trash. Another benefit of gels recycling studies is a better clarification of the adsorption mechanism. Desorption studies are the basis for the research of gels reusability. Thus, several adsorption–desorption cycles are resumed by using every time a fresh dye solution, the sorbent being washed and dried after each adsorption process. Recycling capacity of a sorbent is determined by the number of cycles for which the adsorption ability of the hydrogel is close to its initial value.

9 Conclusions

Several dyes, generated from agriculture and industry, pollute different water resources, producing a worldwide worry connected with their effect on the health of plants, animals and humans. Different methods were used for dyes removal from wastewater and adsorption is one of the most efficient, cheap and simple handling techniques. This chapter details dyes adsorption efficiency of the latest natural or synthetic (co)polymers-based hydrogels. Furthermore, factors that influence the sorption process together with kinetics and isotherm models employed in sorption studies are revealed.

References

1. Abdel-Halim ES (2013) Preparation of starch/poly(N, N-Diethylaminoethyl methacrylate) hydrogel and its use in dye removal from aqueous solutions. React Funct Polym 73(11):1531–1536. https://doi.org/10.1016/j.reactfunctpolym.2013.08.003
2. Abu Elella MH, Sabaa MW, Elhafeez EA, Mohamed RR (2019) Crystal violet dye removal using crosslinked grafted xanthan gum. Int J Biol Macromol 137:1086–1101. https://doi.org/10.1016/j.ijbiomac.2019.06.243
3. Adeyemo AA, Adeoye IO, Bello OS (2017) Adsorption of dyes using different types of clay: a review. Appl Water Sci 7:543–568. https://doi.org/10.1007/s13201-015-0322-y
4. Ahmad A, Mohd-Setapar SH, Chuong CS, Khatoon A, Wani WA, Kumar R, Rafatullah M (2015) Recent advances in new generation dye removal technologies: novel search for approaches to reprocess wastewater. RSC Adv 5:30801–30818. https://doi.org/10.1039/C4R A16959J
5. Ahmad R, Mirza A (2017) Green synthesis of xanthan gum/methionine-bentonite nanocomposite for sequestering toxic anionic dye. Surf Interfac 8:65–72. https://doi.org/10.1016/j.sur fin.2017.05.001
6. Akter M, Bhattacharjee M, Dhar AK, Rahman FBA, Haque S, Rashid TU, Kabir SMF (2021) Cellulose-based hydrogels for wastewater treatment: a concise review. Gels 7(1):30. https://doi.org/10.3390/gels7010030
7. Albrecht MA, Evans CW, Raston CL (2006) Green chemistry and the health implications of nanoparticles. Green Chem 8:417–432. https://doi.org/10.1039/B517131H
8. Alupei IC, Popa M, Hamcerencu M, Abadie MJM (2002) Superabsorbant hydrogels based on xanthan and poly (vinyl alcohol): 1. The study of the swelling properties. Eur Polym J 38:2313–2320. https://doi.org/10.1016/S0014-3057(02)00106-4
9. An Y-M, Liu T, Tian R, Liu S-X, Han Y-N, Wang Q-Q, Sheng W-J (2015) Synthesis of novel temperature responsive PEG-b-[PCL-g-P(MEO2MA-co-OEGMA)]-b-PEG (tBG) triblock-graft copolymers and preparation of tBG/graphene oxide composite hydrogels via click chemistry. React Funct Polym 94:1–8. https://doi.org/10.1016/j.reactfunctpolym.2015.05.011
10. Ansari TM, Ajmal M, Saeed S, Naeem H, Ahmad HB, Mahmood K, Farooqi ZH (2019) Synthesis and characterization of magnetic poly(acrylic acid) hydrogel fabricated with cobalt

nanoparticles for adsorption and catalytic applications. J Iran Chem Soc 16:2765–2776. https://doi.org/10.1007/s13738-019-01738-8

11. Aoki H, Al-Assaf S, Katayama T, Phillips GO (2007a) Characterization and properties of Acacia senegal (L.) Willd. var. senegal with enhanced properties (Acacia (sen) SUPER GUM^{TM}): part 2—mechanism of the maturation process. Food Hydrocoll 21:329–337. https://doi.org/10.1016/j.foodhyd.2006.04.002

12. Aoki H, Katayama T, Ogasawara T, Sasaki Y, Al-Assaf S, Phillips GO (2007b) Characterization and properties of Acacia senegal (L.) Willd. var. Senegal with enhanced properties (Acacia (sen) SUPER GUM^{TM}): part 5. Factors affecting the emulsification of Acacia senegal and Acacia (sen) super gum^{TM}. Food Hydrocoll 21:353–358. https://doi.org/10.1016/j.foodhyd.2006.04.014

13. Aref L, Navarchian AH, Dadkhah D (2017) Adsorption of crystal violet dye from aqueous solution by poly(acrylamide-co-maleic acid)/montmorillonite nanocomposite. J Polym Environ 25:628–639. https://doi.org/10.1007/s10924-016-0842-z

14. Athawale VD, Lele V (1998) Graft copolymerization onto starch. II. Grafting of acrylic acid and preparation of it's hydrogels. Carbohydr Polym 35:21–27. https://doi.org/10.1016/S0144-8617(97)00138-0

15. Atta AM, Al-Hussain SA, Al-Lohedan HA, Ezzat AO, Tawfeek AM, Al-Otabi T (2018) In situ preparation of magnetite/cuprous oxide/poly(AMPS/NIPAm) for removal of methylene blue from waste water. Polym Int 67:471–480. https://doi.org/10.1002/pi.5530

16. Atyaa AI, Radhy ND, Jasim LS (2019) Synthesis and characterization of graphene oxide/hydrogel composites and their applications to adsorptive removal congo red from aqueous solution. J Phys Conf Ser 1234:012095. https://doi.org/10.1088/1742-6596/1234/1/012095

17. Badwaik HR, Giri TK, Nakhate KT, Kashyap P, Tripathi DK (2013) Xanthan gum and its derivatives as a potential biopolymeric carrier for drug delivery system. Curr Drug Deliv 10:587–600. https://doi.org/10.2174/1567201811310050010

18. Bai H, Chen J, Wang Z, Wang L, Lamy E (2020) Simultaneous removal of organic dyes from aqueous solutions by renewable alginate hybridized with graphene oxide. J Chem Eng Data 65:4443–4451. https://doi.org/10.1021/acs.jced.0c00277

19. Banat IM, Nigam P, Singh D, Marchant R (1996) Microbial decolorization of textile-dye-containing effluents: a review. Bioresour Technol 58:217–227. https://doi.org/10.1016/S0960-8524(96)00113-7

20. Batukbhai GC, Augusto MRL, Cai W (2020) Functional biobased hydrogels for the removal of aqueous hazardous pollutants: current status, challenges, and future perspectives. J Mater Chem A 8:21585–21612. https://doi.org/10.1039/D0TA07028A

21. Behari K, Kumar R, Tripathi M, Pandey PK (2001) Graft copolymerization of methacrylamide onto guar gum using a potassium chromate/malonic acid redox pair. Macromol Chem Phys 202:1873–1877. https://doi.org/10.1002/1521-3935(20010601)202

22. BeMiller JN (2019) 16-gum arabic and other exudate gums. In: BeMiller JN (ed) Carbohydrate chemistry for food scientists (3rd edn). AACC International Press, pp 313–321. https://doi.org/10.1016/B978-0-12-812069-9.00016-9

23. Berger J, Reist M, Mayer JM, Felt O, Peppas NA, Gurny R (2004) Structure and interactions in covalently and ionically crosslinked chitosan hydrogels for biomedical applications. Eur J Pharm Biopharm 57:19–34. https://doi.org/10.1016/S0939-6411(03)00161-9

24. Bi Y, Wang H, Liu J, Wang M, Ge S, Zhang H, Zhang S (2018) Preparation and oxidation resistance of SiC-coated graphite powders via microwave-assisted molten salt synthesis. Surf Coat Tech 337:217–222. https://doi.org/10.1016/j.surfcoat.2018.01.017

25. Bhangi BK, Ray SK (2020) Nano silver chloride and alginate incorporated composite copolymer adsorbent for adsorption of a synthetic dye from water in a fixed bed column and its photocatalytic reduction. Int J Biol Macromol 144:801–812. https://doi.org/10.1016/j.ijbiomac.2019.09.070

26. Bhatnagar A, Sillanpää M (2009) Applications of chitin- and chitosan-derivatives for the detoxification of water and wastewater—a short review. Adv Colloid Interface Sci 152(1–2):26–38. https://doi.org/10.1016/j.cis.2009.09.003

27. Bornet A, Teissedre PL (2005) Applications and interest of chitin, chitosan and their derivatives in enology. J Int Sci Vigne et Vin 39:199–207. https://doi.org/10.20870/oeno-one.2005.39.4.890
28. Brownlee IA, Allen A, Pearson JP, Dettmar PW, Havler M, Atherton MR, Onsoyen E (2005) Alginate as a source of dietary fiber. Crit Rev Food Sci 45:497–510. https://doi.org/10.1080/10408390500285673
29. Buchholz F (1997) Absorbency and Super absorbency In: Buchholz FL, Graham AT (eds) Modern superabsorbent polymer technology. Wiley-VCH, New York. https://doi.org/10.1002/1097-0126(200011)
30. Cai X, Du X, Cui D, Wang X, Yang Z, Zhu G (2019) Improvement of stability of blueberry anthocyanins by carboxymethyl starch/xanthan gum combinations microencapsulation. Food Hydrocoll 91:238–245. https://doi.org/10.1016/j.foodhyd.2019.01.034
31. Castell DO, Dalton CB, Becker D, Sinclair J, Castell JA (1992) Alginic acid decreases postprandial upright gastroesophageal reflux-comparison with equal-strength antacid. Digest Dis Sci 37:589–593. https://doi.org/10.1007/BF01307584
32. Catalina M, Cot J, Balu AM, Serrano-Ruiz JC, Luque R (2012) Tailor-made biopolymers from leather waste valorisation. Green Chem 14:308–312. https://doi.org/10.1039/C2GC16330F
33. Çınar S, Kaynar ÜH, Aydemir T, Kaynar SÇ, Ayvacıklı M (2017) An efficient removal of RB5 from aqueous solution by adsorption onto nano-ZnO/Chitosan composite beads. Int J Biol Macromol 96:459–465. https://doi.org/10.1016/j.ijbiomac.2016.12.021
34. Chang C, Zhang L (2011) Cellulose-based hydrogels: present status and application prospects. Carbohydr Polym 84(1):40–53. https://doi.org/10.1016/j.carbpol.2010.12.023
35. Chang Z, Chen Y, Tang S, Yang J, Chen Y, Chen S, Li P, Yang Z (2020) Construction of chitosan/polyacrylate/graphene oxide composite physical hydrogel by semi-dissolution/acidification/sol-gel transition method and its simultaneous cationic and anionic dye adsorption properties. Carbohydr Polym 229:115431. https://doi.org/10.1016/j.carbpol.2019.115431
36. Chaudhary S, Sharma J, Kaith BS, Yadav S, Sharma AK, Goel A (2018) Gum xanthan-psyllium-cl-poly(acrylic acid-co-itaconic acid) based adsorbent for effective removal of cationic and anionic dyes: adsorption isotherms, kinetics and thermodynamic studies. Ecotoxicol Environ Saf 149:150–158. https://doi.org/10.1016/j.ecoenv.2017.11.030
37. Chen P, Liu X, Jin R, Nie W, Zhou Y (2017) Dye adsorption and photo-induced recycling of hydroxypropyl cellulose/molybdenum disulfide composite hydrogels. Carbohyd Polym 167:36–43. https://doi.org/10.1016/j.carbpol.2017.02.094
38. Chen Q, Yu H, Wang L, Abdin Z, Chen Y, Wang J, Zhou W, Yang X, Khan RU, Zhang H (2015) Recent progress in chemical modification of starch and its applications. RSC Adv 5:67459–67474. https://doi.org/10.1039/C5RA10849G
39. Chen R, Zhang Y, Shen L, Wang X, Chen J, Ma A, Jiang W (2015) Lead (II) and methylene blue removal using a fully biodegradable hydrogel based on starch immobilized humic acid. Chem Eng J 268:348–355. https://doi.org/10.1016/j.cej.2015.01.081
40. Chen X (2015) Modeling of experimental adsorption isotherm data. Information 6:14–22. https://doi.org/10.3390/info6010014
41. Cheng R, Xiang B, Li Y, Zhang M (2011) Application of dithiocarbamate-modified starch for dyes removal from aqueous solutions. J Hazard Mater 188(1–3):254–260. https://doi.org/10.1016/j.jhazmat.2011.01.104
42. Cheng Y, Yu S, Zhen X, Wang X, Wu W, Jiang X (2012) Alginic acid nanoparticles prepared through counterion complexation method as a drug delivery system. ACS Appl Mat Interfaces 4:5325–5332. https://doi.org/10.1021/am3012627
43. Chiu C-W, Wu M-T, Lin C-L, Li J-W, Huang C-Y, Soong Y-C, Lee JC-M, Lee Sanchez WA, Lin H-Y (2020) Adsorption performance for reactive blue 221 dye of β-chitosan/polyamine functionalized graphene oxide hybrid adsorbent with high acid-alkali resistance stability in different acid-alkaline environments. Nanomaterials 10(4):748. https://doi.org/10.3390/nano10040748

44. Constantin M, Asmarandei I, Harabagiu V, Ghimici L, Ascenzi P, Fundueanu G (2013) Removal of anionic dyes from aqueous solutions by an ion-exchanger based on pullulan microspheres. Carbohydr Polym 91(1):74–84. https://doi.org/10.1016/j.carbpol.2012.08.005

45. Crini G (2006) Non-conventional low-cost adsorbents for dye removal: a review. Bioresour Technol 97(9):1061–1085. https://doi.org/10.1016/j.biortech.2005.05.001

46. Crini G, Badot P-M (2008) Application of chitosan, a natural aminopolysaccharide, for dye removal from aqueous solutions by adsorption processes using batch studies: a review of recent literature. Prog Polym Sci 33(4):399–447. https://doi.org/10.1016/j.progpolymsci. 2007.11.001

47. Crini G, Badot P-M (2010) Starch-based biosorbents for dyes in textile wastewater treatment. Int J Environ Technol Manag 12(2–4):129–150. https://doi.org/10.1504/IJETM.2010.031524

48. Cruz A, García-Uriostegui L, Ortega A, Isoshima T, Burillo G (2017) Radiation grafting of N-vinyl caprolactamonto nano and macrogels of chitosan: synthesis and characterization. Carbohydr Polym 155:303–312. https://doi.org/10.1016/j.carbpol.2016.08.083

49. Cui W, Ji J, Cai FY, Li H, Rong R (2015) Robust, anti-fatigue, and self-healing graphene oxide/hydrophobic association composite hydrogels and their use as recyclable adsorbents for dye wastewater treatment. J Mater Chem A 3:17445–17458. https://doi.org/10.1039/C5T A04470G

50. da Silva RC, de Aguiar SB, da Cunha PLR, de Paula RCM, Feitosa JP (2020) Effect of microwave on the synthesis of polyacrylamide-g-chitosan gel for azo dye removal. React Funct Polym 148:104491. https://doi.org/10.1016/j.reactfunctpolym.2020.104491

51. Dabrowski A (2001) Adsorption-from theory to practice. Adv Colloid Interface Sci 93:135–224. https://doi.org/10.1016/S0001-8686(00)00082-8

52. Dai H, Huang Y, Huang H (2018) Eco-friendly polyvinyl alcohol/carboxymethyl cellulose hydrogels reinforced with graphene oxide and bentonite for enhanced adsorption of methylene blue. Carbohydr Polym 185:1–11. https://doi.org/10.1016/j.carbpol.2017.12.073

53. Dai L, Cheng T, Xi X, Nie S, Ke H, Liu Y, Tong S, Chen Z (2020) A versatile TOCN/CGG self-assembling hydrogel for integrated wastewater treatment. Cellulose 27:915–925. https://doi.org/10.1007/s10570-019-02834-x

54. de Figueiredo NT, Dalarme NB, da Silva PMM, Landers R, Picone CSF, Prediger P (2020) Novel magnetic chitosan/quaternary ammonium salt graphene oxide composite applied to dye removal. J Environ Chem Eng 8:103820. https://doi.org/10.1016/j.jece.2020.103820

55. de Jong SJ, De Smedt SC, Demeester J, van Nostrum CF, Kettenes-van den Bosch JJ, Hennink WE (2001) Biodegradable hydrogels based on stereocomplex formation between lactic acid oligomers grafted to dextran. J Control Release 72:47–56. https://doi.org/10.1016/S0168-365 9(01)00261-9

56. de Nooy AEJ, Masci G, Crescenzi V (1999) Versatile synthesis of polysaccharide hydrogels using the Passerini and Ugi multicomponent condensations. Macromolecules 32:1318–1320. https://doi.org/10.1021/ma9815455

57. de Nooy AEJ, Capitani D, Masci G, Crescenzi V (2000) Ionic polysaccharide hydrogels via the Passerini and Ugi multicomponent condensations: synthesis, behavior and solid-state NMR characterization. Biomacromol 1:259–267. https://doi.org/10.1021/bm005517h

58. Dekamin MG, Karimi Z, Latifidoost Z, Ilkhanizadeh S, Daemi H, Naimi-Jamal MR, Barikani M (2018) Alginic acid: a mild and renewable bifunctional heterogeneous biopolymeric organocatalyst for efficient and facile synthesis of polyhydroquinolines. Int J Biol Macromol 108:1273–1280. https://doi.org/10.1016/j.ijbiomac.2017.11.050

59. Derbyshire F, Jagtoyen M, Andrews R, Rao A, Martin-Gullon I, Grulke E (2001) Carbon materials in environmental applications. In: Radovic LR (ed) Chemistry and physics of carbon, vol 27. Marcel Dekker, New York, pp 1–66

60. Dubinin MM, Zaverina ED, Radushkevich LV (1947) Sorption and structure of active carbons. J Phys Chem 21:1351–1362

61. Ebrahiminezhad A, Moeeni F, Taghizadeh SM, Seifan M, Bautista C, Novin D, Ghasemi Y, Berenjian A (2019) Xanthan gum capped ZnO microstars as a promising dietary zinc supplementation. Foods 8:1–10. https://doi.org/10.3390/foods8030088

62. El-Zahhar AA, Awwad NS, El-Katori EE (2014) Removal of bromophenol blue dye from industrial waste water by synthesizing polymer-clay composite. J Mol Liq 199:454–461. https://doi.org/10.1016/j.molliq.2014.07.034

63. Elbedwehy AM, Atta AM (2020) Novel superadsorbent highly porous hydrogel based on arabic gum and acrylamide grafts for fast and efficient methylene blue removal. Polymers 12(2):338. https://doi.org/10.3390/polym12020338

64. Elsayed MM (2019) Hydrogel preparation technologies: relevance kinetics, thermodynamics and scaling up aspects. J Polym Environ 27:871–891. https://doi.org/10.1007/s10924-019-01376-4

65. Eltaweil AS, Elgarhy GS, El-Subruiti GM, Omer AM (2020) Novel carboxymethyl cellulose/carboxylated graphene oxide composite microbeads for efficient adsorption of cationic methylene blue dye. Int J Biol Macromol 154:307–318. https://doi.org/10.1016/j.ijbiomac.2020.03.122

66. Erickson IE, Kestle SR, Zellars KH, Dodge GR, Burdick JA, Mauck RL (2012) Improved cartilage repair via in vitro pre-maturation of MSC seeded hyaluronic acid hydrogels. Biomed Mater 7(2):024110. https://doi.org/10.1088/1748-6041/7/2/024110

67. Essawy HA, Ghazy MBM, El-Hai FA, Mohamed MF (2016) Superabsorbent hydrogels via graft polymerization of acrylic acid from chitosan-cellulose hybrid and their potential in controlled release of soil nutrients. Int J Biol Macromol 89:144–151. https://doi.org/10.1016/j.ijbiomac.2016.04.071

68. Evans JS, Guo T, Sun Y, Liu W, Peng L, Xu Z, Gao C, He S (2018) Shape-controlled of ten-nanometer-thick graphite and worm-like graphite by lithographic exfoliation. Carbon 135:248–252. https://doi.org/10.1016/j.carbon.2018.02.073

69. Farag AM, Sokker HH, Zayed EM, Eldien FAN, Alrahman Abd NM (2018) Removal of hazardous pollutants using bifunctional hydrogel obtained from modified starch by grafting copolymerization. Int J Biol Macromol 120(B):188–2199. https://doi.org/10.1016/j.ijbiomac.2018.06.171

70. Feng Y, Gong J-L, Zeng G-M, Niu Q-Y, Zhang H-Y, Niu C-G, Deng J-H, Yan M (2010) Adsorption of Cd (II) and Zn (II) from aqueous solutions using magnetic hydroxyapatite nanoparticles as adsorbents. Chem Eng J 162:487–494. https://doi.org/10.1016/j.cej.2010.05.049

71. Fernando IPS, Sanjeewa KKA, Kim SY, Lee JS, Jeon YJ (2018) Reduction of heavy metal (Pb^{2+}) biosorption in zebrafish model using alginic acid purified from Ecklonia cava and two of its synthetic derivatives. Int J Biol Macromol 106:330–337. https://doi.org/10.1016/j.ijbiomac.2017.08.027

72. Foocharoen C, Chunlertrith K, Mairiang P, Mahakkanukrauh A, Suwannaroj S, Namvijit S, Wantha O, Nanagara R (2017) Effectiveness of add-on therapy with domperidone vs alginic acid in proton pump inhibitor partial response gastroesophageal reflux disease in systemic sclerosis: randomized placebo-controlled trial. Rheumatology 56:214–222. https://doi.org/10.1093/rheumatology/kew216

73. Forgacs E, Cserhati T, Oros G (2004) Removal of synthetic dyes from wastewaters: a review. Environ Int 30:953–971. https://doi.org/10.1016/j.envint.2004.02.001

74. Freundlich HMF (1906) Over the adsorption in solution. J Phys Chem 57:385–471

75. Gan L, Li H, Chen L, Xu L, Liu J, Geng A, Mei C, Shang S (2018) Graphene oxide incorporated alginate hydrogel beads for the removal of various organic dyes and bisphenol A in water. Colloid Polym Sci 296:607–615. https://doi.org/10.1007/s00396-018-4281-3

76. García-Ochoa F, Santos VE, Casas JA, Gómez E (2000) Xanthan gum: production, recovery, and properties. Biotechnol Adv 18:549–579. https://doi.org/10.1016/S0734-9750(00)00050-1

77. Ghemati DJ, Aliouche DJ (2014) Study of the sorption of synthetic dyes from aqueous solution onto cellulosic modified polymer. J Water Chem Technol 36(6):265–272. https://doi.org/10.3103/S1063455X14060022

78. Giri Dev VR, Neelakandan R, Sudha S, Shamugasundram OL, Nadaraj RN (2005) Chitosan-a polymer with wider applications. Textile Mag 46:83–86. https://doi.org/10.3390/molecules25173981

79. Godiya CB, Ruotolo LAM, Cai W (2020) Functional biobased hydrogels for the removal of aqueous hazardous pollutants: current status, challenges, and future perspectives. J Mater Chem A 8:21585–21612. https://doi.org/10.1039/D0TA07028A
80. González JA, Villanueva ME, Piehl LL, Copello GJ (2015) Development of a chitin/graphene oxide hybrid composite for the removal of pollutant dyes: adsorption and desorption study. Chem Eng J 280:41–48. https://doi.org/10.1016/j.cej.2015.05.112
81. Grimaud E, Lecoq JC (1978) Comparison of gels used for molecular sieving of proteins by electron microscopy and pore parameters determinations. J Chromatogr 166:37–45. https://doi.org/10.1016/S0021-9673(00)92247-7
82. Guibal E (2005) Heterogeneous catalysis on chitosan-based materials: a review. Prog Polym Sci 30:71–109. https://doi.org/10.1016/j.progpolymsci.2004.12.001
83. Guo H, Jiao T, Zhang Q, Guo W, Peng Q, Yan X (2015) Preparation of graphene oxide-based hydrogels as efficient dye adsorbents for wastewater treatment. Nanoscale Res Lett 10(1):931. https://doi.org/10.1186/s11671-015-0931-2
84. Guo X, Wang Y, Qin Y, Shen P, Peng Q (2020) Structures, properties and application of alginic acid: a review. Int J Biol Macromol 162:618–628. https://doi.org/10.1016/j.ijbiomac.2020.06.180
85. Halouane F, Oz Y, Meziane D, Barras A, Juraszek J, Singh SK, Kurungot S, Shaw PK, Sanyal R, Boukherroub R, Sanyal A, Szunerits S (2017) Magnetic reduced graphene oxide loaded hydrogels: highly versatile and efficient adsorbents for dyes and selective Cr(VI) ions removal. J Colloid Interface Sci 507:360–369. https://doi.org/10.1016/j.jcis.2017.07.075
86. Hassani A, Soltani RDC, Karaca S, Khataee A (2015) Preparation of montmorillonite–alginate nanobiocomposite for adsorption of a textile dye in aqueous phase: Isotherm, kinetic and experimental design approaches. J Ind Eng Chem 21:1197–1207. https://doi.org/10.1016/j.jiec.2014.05.034
87. Hassan CM, Peppas NA (2000) Structure and applications of poly (vinyl alcohol) hydrogels produced by conventional crosslinking or by freezing/thawing methods. In: Biopolymers PVA hydrogels, anionic polymerisation, nanocomposites. Advances in polymer science, vol 153. Springer, Berlin, Heidelberg. https://doi.org/10.1007/3-540-46414-X_2
88. He H, Klinowski J, Forster M, Lerf A (1998) A new structural model for graphite oxide. Chem Phys Lett 287(1–2):53–56. https://doi.org/10.1016/S0009-2614(98)00144-4
89. He X, Du M, Li H, Zhou T (2016) Removal of direct dyes from aqueous solution by oxidized starch crosslinked chitosan/silica hybrid membrane. Int J Biol Macromol 82:174–181. https://doi.org/10.1016/j.ijbiomac.2015.11.005
90. Hebeish AA, Aly AA (2014) Synthesis, characterization and utilization of starch hydroxypropyl sulphate for cationic dye removal. Int J Org Chem 4:208–217. https://doi.org/10.4236/ijoc.2014.43024
91. Hennink WE, van Nostrum CF (2002) Novel crosslinking methods to design hydrogels. Adv Drug Deliv Rev 54:13–36. https://doi.org/10.1016/S0169-409X(01)00240-X
92. Hennink WE, De Jong SJ, Bos GW, Veldhuis TFJ, van Nostrum CF (2004) Biodegradable dextran hydrogels crosslinked by stereocomplex formation for the controlled release of pharmaceutical proteins. Int J Pharm 277:99–104. https://doi.org/10.1016/j.ijpharm.2003.02.002
93. Hirst EL, Jones JKN, Jones WO (1939) The structure of alginic acid—part I. J Chem Soc 1880–1885. https://doi.org/10.1039/JR9390001880
94. Hosseinzadeh H, Javadi A (2016) Fabrication and characterization of CMC-based magnetic superabsorbent hydrogel nanocomposites for crystal violet removal. Polym Adv Technol 27:1609–1616. https://doi.org/10.1002/pat.3838
95. Hosseinzadeh H, Bahador N (2017) Novel CdS quantum dots templated hydrogel nanocomposites: synthesis, characterization, swelling and dye adsorption properties. J Mol Liq 240:630–641. https://doi.org/10.1016/j.molliq.2017.05.129
96. Hosseinzadeh H, Ramin S (2018) Fabrication of starch-graft poly(acrylamide)/graphene oxide/hydroxyapatite nanocomposite hydrogel adsorbent for removal of malachite green dye from aqueous solution. Int J Biol Macromol 106:101–115. https://doi.org/10.1016/j.ijbiomac.2017.07.182

97. Ibrahim AG, Sayed AZ, Abd El-Wahab H, Sayah MM (2020) Synthesis of a hydrogel by grafting of acrylamide-co-sodium methacrylate onto chitosan for effective adsorption of fuchsin basic dye. Int J Biol Macromol 159:422–432. https://doi.org/10.1016/j.ijbiomac.2020.05.039

98. Irvine JC (1909) LXXII-A polarimetric method of identifying chitin. J Chem Soc Trans 95:564–570

99. Jayakumar R, Prabaharan M, Reis RL, Mano JF (2002) Graft copolymerized chitosan—present and status applications. Carbohydr Polym 62:142–158. https://doi.org/10.1016/j.carbpol.2005.07.017

100. Jayaramudu T, Raghavendra GM, Varaprasad K, Sadiku R, Raju KM (2013) Development of novel biodegradable Au nanocomposite hydrogels based on wheat: for inactivation of bacteria. Carbohydr Polym 92:2193–2200. https://doi.org/10.1016/j.carbpol.2012.12.006

101. Jayaramudu T, Li Y, Ko H-U, Shishir IR, Kim J (2016a) Poly(acrylic acid)-poly(vinyl alcohol) hydrogels for reconfigurable lens actuators. Int J Precis Eng Manuf Technol 3:375–379. https://doi.org/10.1007/s40684-016-0047-x

102. Jayaramudu T, Raghavendra GM, Varaprasad K, Raju KM, Sadiku ER, Kim J (2016b) 5-Fluorouracil encapsulated magnetic nanohydrogels for drug-delivery applications. J Appl Polym Sci 133:1–10. https://doi.org/10.1002/app.43921

103. Jeon C, Kwon YD, Park KH (2005) Removal of lead ions using phosphorylated alginic acid gel and beads. J Ind Eng Chem 11:643–647

104. Jeon C, Nah IW, Hwang KY (2007) Adsorption of heavy metals using magnetically modified alginic acid. Hydrometallurgy 86:140–146. https://doi.org/10.1016/j.hydromet.2006.11.010

105. Jiang X, Wang S, Ge L, Lin F, Lu Q, Wang T, Huang B, Lu B (2017) Development of organic-inorganic hybrid beads from sepiolite and cellulose for effective adsorption of malachite green. RSC Adv 7(62):38965–38972. https://doi.org/10.1039/C7RA06351B

106. Kabir SF, Sikdar PP, Haque B, Rahman Bhuiyan MA, Ali A, Islam MN (2018) Cellulose-based hydrogel materials: chemistry, properties and their prospective applications. Prog Biomater 7:153–174. https://doi.org/10.1007/s40204-018-0095-0

107. Kadhom M, Albayati N, Alalwan H, Al-Furaiji M (2020) Removal of dyes by agricultural waste. Sustain Chem Pharm 16:100259. https://doi.org/10.1016/j.scp.2020.100259

108. Kaith BS, Jindal R, Mittal H, Kumar K (2010) Temperature, pH and electric stimulus responsive hydrogels from Gum ghatti and polyacrylamide-synthesis, characterization and swelling studies. Chem Sin 1(2):44–45

109. Kangwansupamonkon W, Klaikaew N, Kiatkamjornwong S (2018) Green synthesis of titanium dioxide/acrylamide-based hydrogel composite, self degradation and environmental applications. Eur Polym J 107:118–131. https://doi.org/10.1016/j.eurpolymj.2018.08.004

110. Kaplan M, Kasgoz H (2011) Hydrogel nanocomposite sorbents for removal of basic dyes. Polym Bull 67:1153–1168. https://doi.org/10.1007/s00289-011-0444-9

111. Karadag E, Kasim ZD, Kundakci S, Uzum OB (2017) Acrylamide/potassium 3-sulfopropyl methacrylate/sodium alginate/bentonite hybrid hydrogels: synthesis, characterization and its application in lauths violet removal from aqueous solutions. Fibers Polym 18:9–21. https://doi.org/10.1007/s12221-017-5910-z

112. Kato Y, Onishi H, Machida Y (2004) N-succinyl-chitosan as a drug carrier: water-insoluble and water-soluble conjugates. Biomaterials 25:907–915. https://doi.org/10.1016/S0142-9612(03)00598-2

113. Katzbauer B (1998) Properties and applications of xanthan gum. Polym 59:81–84. https://doi.org/10.1016/S0141-3910(97)00180-8

114. Khan M, Lo IMC (2016) A holistic review of hydrogel applications in the adsorptive removal of aqueous pollutants: recent progress, challenges, and perspectives. Water Res 106:259–271. https://doi.org/10.1016/j.watres.2016.10.008

115. Khan ZU, Kausar A, Ullah H, Badshah A, Khan WU (2016) A review of graphene oxide, graphene buckypaper, and polymer/graphene composites: properties and fabrication techniques. J Plast Film Sheet 32(4):336–379. https://doi.org/10.1177/8756087915614612

116. Khanday WA, Asif M, Hameed BH (2017) Crosslinked beads of activated oil palm ash zeolite/chitosan composite as a bio-adsorbent for the removal of methylene blue and acid blue 29 dyes. Int J Biol Macromol 95:895–902. https://doi.org/10.1016/j.ijbiomac.2016.10.075

117. Khor E, Lim LY (2003) Implantable applications of chitin and chitosan. Biomaterials 24:2339–2349. https://doi.org/10.1016/S0142-9612(03)00026-7

118. Kishi R, Ichijo H, Hirasa O (1993) Thermo-responsive devices using poly (vinyl methyl ether) hydrogels. J Intell Mater Syst Struct 4:533–537. https://doi.org/10.1177/1045389X9300400413

119. Kolya H, Tripathy T (2013) Hydroxyethyl starch-g-Poly-(N, N-dimethylacrylamide-co-acrylic acid): an efficient dye removing agent. Eur Polym J 49(12):4265–4275. https://doi.org/10.1016/j.eurpolymj.2013.10.012

120. Krajewska B (2004) Applications of chitin- and chitosan-based materials for enzyme immobilizations: a review. Enzyme Microbial Technol 35:126–139. https://doi.org/10.1016/j.enzmictec.2003.12.013

121. Kumar D, Gihar S, Shrivash MK, Kumar P, Kundu PP (2020) A review on the synthesis of graft copolymers of chitosan and their potential applications. Int Biol Macromol 163:2097–2112. https://doi.org/10.1016/j.ijbiomac.2020.09.060

122. Kumar M, Vijayakumar G, Tamilarasan R (2019) Synthesis, characterization and experimental studies of nano Zn–Al–Fe3O4 blended alginate/Ca beads for the adsorption of rhodamin B. J Polym Environ 27(1):106–117. https://doi.org/10.1007/s10924-018-1318-0

123. Kyzas GZ, Bikiaris DN, Mitropoulos AC (2017) Chitosan adsorbents for dye removal: a review. Polym Int 66(12):1800–1811. https://doi.org/10.1002/pi.5467

124. Lamb GF (1916) The chemical constitution of chitin. Science 44:866–868

125. Langmuir I (1916) The constitution and fundamental properties of solids and liquids. Part I. Solods. J Am Chem Soc 38:2221–2295. https://doi.org/10.1021/ja02268a002

126. Lerf A, He H, Riedl T, Forster M, Klinowski J (1997) 13C and 1H MAS NMR studies of graphite oxide and its chemically modified derivatives. Solid State Ion 101–103(2):857–862. https://doi.org/10.1016/S0167-2738(97)00319-6

127. Lerf A, He H, Forster M, Klinowski J (1998) Structure of graphite oxide revisited. J Phys Chem B 102:4477–4482. https://doi.org/10.1021/jp9731821

128. Li F, Bi W, Kong T, Wang C, Li Z, Huang X (2009) Effect of sulfur sources on the crystal structure, morphology and luminescence of CdS nanocrystals prepared by a solvothermal method. J Alloys Compd 479(1–2):707–710. https://doi.org/10.1016/j.jallcom.2009.01.034

129. Li GX, Du YM, Tao YZ, Deng HB, Luo XG, Yang JH (2010) Iron(II) crosslinked chitin based gel beads: preparation, magnetic property and adsorption of methyl orange. Carbohydr Polym 82:706–713. https://doi.org/10.1016/j.carbpol.2010.05.040

130. Li H, Fan J, Shi Z, Lian M, Tian M, Yin J (2015) Preparation and characterization of sulfonated graphene-enhanced poly (vinyl alcohol) composite hydrogel and its application as dye absorbent. Polymer 60:96–106. https://doi.org/10.1016/j.polymer.2014.12.069

131. Li J, Cai J, Zhong L, Cheng H, Wang H, Ma Q (2019) Adsorption of reactive red 136 onto chitosan/montmorillonite intercalated composite from aqueous solution. Appl Clay Sci 167:9–22. https://doi.org/10.1016/j.clay.2018.10.003

132. Li L, Iqbal J, Zhu Y, Zhang P, Chen W, Bhatnagar A, Du Y (2018) Chitosan/Ag-hydroxyapatite nanocomposite beads as a potential adsorbent for the efficient removal of toxic aquatic pollutants. Int J Biol Macromol 120(B):1752–1759. https://doi.org/10.1016/j.ijbiomac.2018.09.190

133. Li P, Siddaramaiah KNH, Heo S-B, Lee J-H (2008) Novel PAAm/Laponite clay nanocomposite hydrogels with improved cationic dye adsorption behavior. Composit B: Eng 39(5):756–763. https://doi.org/10.1016/j.compositesb.2007.11.003

134. Li X, Zhang G, Bai X, Sun X, Wang X, Wang E, Dai H (2008) Highly conducting graphene sheets and Langmuir-Blodgett films. Nature Nanotech 3:538–542. https://doi.org/10.1038/nnano.2008.210

135. Lin Y, Fang GG, Deng YJ, Shen KZ, Wu T, Li M (2018) Highly effective removal of methylene blue using a chemi-mechanical pretreated cellulose-based superabsorbent hdrogel. BioResources 13:8709–8722. https://doi.org/10.15376/biores.13.4.8709-8722

136. Liu C, Liu H, Tang K, Zhang K, Zou Z, Gao X (2020) High-strength chitin based hydrogels reinforced by tannic acid functionalized graphene for congo red adsorption. J Polym Environ 28:984–994. https://doi.org/10.1007/s10924-020-01663-5

137. Liu L, Gao ZY, Su XP, Chen X, Jiang L, Yao JM (2015) Adsorption removal of dyes from single and binary solutions using a cellulose-based bioadsorbent. Acs Sustain Chem Eng 3:432–442. https://doi.org/10.1021/sc500848m

138. Liu Y, Vrana NE, Cahill PA, McGuinness GB (2009) Physically crosslinked composite hydrogels of PVA with natural macromolecules: structure, mechanical properties, and endothelial cell compatibility. J Biomed Mater Res B Appl Biomater 90(2):492–502. https://doi.org/10.1002/jbm.b.31310

139. Liu Y, Huang S, Zhao X, Zhang Y (2018) Fabrication of three-dimensional porous β-cyclodextrin/chitosan functionalized graphene oxide hydrogel for methylene blue removal from aqueous solution. Colloid Surf A: Physicochem Eng Asp 539:1–10. https://doi.org/10.1016/j.colsurfa.2017.11.066

140. Macquarrie DJ, Hardy JJE (2005) Applications of functionalised chitosane in catalysis. Ind Eng Chem Res 44:8499–8520. https://doi.org/10.1021/ie050007v

141. Mahdavinia GR, Hasanpour J, Rahmani Z, Karami S, Etemadi H (2013) Nanocomposite hydrogel from grafting of acrylamide onto HPMC using sodium montmorillonite nanoclay and removal of crystal violet dye. Cellulose 20:2591–2604. https://doi.org/10.1007/s10570-013-0004-6

142. Mahdavinia GR, Soleymani M, Sabzi M, Azimi H, Atlasi Z (2017) Novel magnetic polyvinyl alcohol/laponite RD nanocomposite hydrogels for efficient removal of methylene blue. J Environ Chem Eng 5(3):2617–2630. https://doi.org/10.1016/j.jece.2017.05.017

143. Mahmoodi NM, Hayati B, Arami M, Bahrami H (2011) Preparation, characterization and dye adsorption properties of biocompatible composite (alginate/titania nanoparticle). Desalination 275:93–101. https://doi.org/10.1016/j.desal.2011.02.034

144. Mahmoodi-Babolan N, Nematollahzadeh A, Heydari A, Merikhy A (2019) Bioinspired catecholamine/starch composites as superadsorbent for the environmental remediation. Int J Biol Macromol 125:690–699. https://doi.org/10.1016/j.ijbiomac.2018.12.032

145. Makhado E, Pandey S, Nomngongo PN, Ramontja J (2018a) Preparation and characterization of xanthan gum-cl-poly(acrylic acid)/o-MWCNTs hydrogel nanocomposite as highly effective re-usable adsorbent for removal of methylene blue from aqueous solutions. J Colloid Interface Sci 513:700–714. https://doi.org/10.1016/j.jcis.2017.11.060

146. Makhado E, Pandey S, Ramontja J (2018b) Microwave assisted synthesis of xanthan gum-cl-poly (acrylic acid) based-reduced graphene oxide hydrogel composite for adsorption of methylene blue and methyl violet fromaqueous solution. Int J Biol Macromol 119:255–269. https://doi.org/10.1016/j.ijbiomac.2018.07.104

147. Marrakchi F, Hameed B, Hummadi E (2020) Mesoporous biohybrid epichlorohydrin crosslinked chitosan/carbon–clay adsorbent for effective cationic and anionic dyes adsorption. Int J Biol Macromol 163:1079–1086. https://doi.org/10.1016/j.ijbiomac.2020.07.032

148. Masruchin N, Park B-D, Causin V (2015) Influence of sonication treatment on supramolecular cellulose microfibril-based hydrogels induced by ionic interaction. J Ind Eng Chem 29:265–272. https://doi.org/10.1016/j.jiec.2015.03.034

149. Mittal H, Parashar V, Mishra SB, Mishra AK (2014a) Fe_3O_4 MNPs and gum xanthan based hydrogels nanocomposites for the efficient capture of malachite green from aqueous solution. Chem Eng J 255:471–482. https://doi.org/10.1016/j.cej.2014.04.098

150. Mittal H, Mishra SB (2014b) Gum ghatti and Fe_3O_4 magnetic nanoparticles based nanocomposites for the effective adsorption of rhodamine B. Carbohydr Polym 101:1255–1264. https://doi.org/10.1016/j.carbpol.2013.09.045

151. Mittal H, Ballav N, Mishra SB (2014c) Gum ghatti and Fe_3O_4 magnetic nanoparticles based nanocomposites for the effective adsorption of methylene blue from aqueous solution. J Ind Eng Chem 20(4):2184–2192. https://doi.org/10.1016/j.jiec.2013.09.049

152. Mittal H, Maity A, Ray SS (2015a) Effective removal of cationic dyes from aqueous solution using gum ghatti-based biodegradable hydrogel. Int J Biol Macromol 79:8–20. https://doi.org/10.1016/j.ijbiomac.2015.04.045

153. Mittal H, Maity A, Ray SS (2015b) Synthesis of co-polymer-grafted gum karaya and silica hybrid organic–inorganic hydrogel nanocomposite for the highly effective removal of methylene blue. Chem Eng J 279:166–179. https://doi.org/10.1016/j.cej.2015.05.002
154. Mittal H, Kumar V, Saruchi RSS (2016a) Adsorption of methyl violet from aqueous solution using gum xanthan/Fe_3O_4 based nanocomposite hydrogel. Int J Biol Macromol 89:1–11. https://doi.org/10.1016/j.ijbiomac.2016.04.050
155. Mittal H, Ray SS, Okamoto M (2016b) Recent progress on the design and applications of polysaccharide-based graft copolymer hydrogels as adsorbents for wastewater purification. Macromol Mater Eng 301:496–522. https://doi.org/10.1002/mame.201500399
156. Muzzarelli RAA (1973) Natural chelating polymers: alginic acid, chitin, and chitosan. Pergamon Press, Oxford, UK
157. Muzzarelli RAA, Muzzarelli C (2005) Chitosan chemistry: relevance to the biomedical sciences. In: Heinze T (ed) Polysaccharides I. Advances in polymer science, vol 186. Springer, Berlin, Heidelberg, pp 151–209. https://doi.org/10.1007/b136820
158. Naseer F, Ajmal M, Bibi F, Farooqi ZH, Siddiq M (2018) Copper and cobalt nanoparticles containing poly(acrylic acid-co-acrylamide) hydrogel composites for rapid reduction of 4-nitrophenol and fast removal of malachite green from aqueous medium. Polym Compos 39:3187–3198. https://doi.org/10.1002/pc.24329
159. Novoselov K, Geim A, Morozov S, Jiang D, Katsnelson MI, Grigorieva IV, Dubonos SV, Firsov AA (2005) Two-dimensional gas of massless Dirac fermions in graphene. Nature 438:197–200. https://doi.org/10.1038/nature04233
160. Omidian H, Zohuriaan-Mehr MJ, Kabiri K, Shah K (2004) Polymer chemistry attractiveness: synthesis and swelling studies of gluttonous hydrogels in the advanced academic laboratory. J Polym Mater 21:281–291
161. Ozay O, Ekici S, Baran Y, Aktas N, Sahiner N (2009) Removal of toxic metal ions with magnetic hydrogels. Water Res 43:4403–4411. https://doi.org/10.1016/j.watres.2009.06.058
162. Pakdel PM, Peighambardoust SJ (2018) A review on acrylic based hydrogels and their applications in wastewater treatment. J Environ Manag 217:123–143. https://doi.org/10.1016/j.jenvman.2018.03.076
163. Pal S, Patra AS, Ghorai S, Sarkar AK, Mahato V, Sarkar S, Singh RP (2015) Efficient and rapid adsorption characteristics of templating modified guar gum and silica nanocomposite toward removal of toxic reactive blue and Congo red dyes. Bioresour Technol Sep 191:291–9. https://doi.org/10.1016/j.biortech.2015.04.099
164. Pedroso-Santana S, Fleitas-Salazar N (2020) Ionotropic gelation method in the synthesis of nanoparticles/microparticles for biomedical purposes. Polym Int 69:443–447. https://doi.org/10.1002/pi.5970
165. Pereira AGB, Rodrigues FHA, Paulino AT, Martins AF, Fajardo AR (2021) Recent advances on composite hydrogels designed for the remediation of dye-contaminated water and wastewater: a review. J Clean Prod 284:124703. https://doi.org/10.1016/j.jclepro.2020.124703
166. Pereira L, Alves M (2012) Dyes-environmental impact and remediation. In: Malik A, Grohmann E (eds) Environmental protection strategies for sustainable development. Strategies for sustainability. Springer, Dordrecht, pp 111–162. https://doi.org/10.1007/978-94-007-1591-2_4
167. Plieva FM, Karlsson M, Aguilar MR, Gomez D, Mikhalovsky S, Galaev IY, Mattiasson B (2006) Pore structure of macroporous monolithic cryogels prepared from poly(vinyl alcohol). J Appl Polym Sci 100:1057–1066. https://doi.org/10.1002/app.23200
168. Pooresmaeil M, Namazi H (2020) Application of polysaccharide-based hydrogels for water treatments. In: Yu C (ed) Hydrogels based on natural polymers. Elsevier, pp 411–455. https://doi.org/10.1016/B978-0-12-816421-1.00014-8
169. Pour ZS, Ghaemy M (2015) Removal of dyes and heavy metal ions from water by magnetic hydrogel beads based on poly(vinyl alcohol)/carboxymethyl starch-g-poly(vinyl imidazole). RSC Adv 5:64106–64118. https://doi.org/10.1039/C5RA08025H

170. Pourjavadi A, Nazari M, Kabiri B, Hosseini SH, Bennett C (2016) Preparation of porous graphene oxide/hydrogel nanocomposites and their ability for efficient adsorption of methylene blue. RSC Adv 6:10430–10437. https://doi.org/10.1039/C5RA21629J

171. Prabaharan M, Mano JF (2005) Chitosan-based particles as controlled drug delivery systems. Drug Deliv 12:41–57. https://doi.org/10.1080/10717540590889781

172. Qi X, Zeng Q, Tong X, Su T, Xie L, Yuan K, Xu J, Shen J (2021) Polydopamine/montmorillonite-embedded pullulan hydrogels as efficient adsorbents for removing crystal violet. J Hazard Mat 402:123359. https://doi.org/10.1016/j.jhazmat.2020.123359

173. Qian J, Chen J, Ruan S, Shen S, He Q, Jiang X, Zhu J, Gao H (2014) Preparation and biological evaluation of photoluminescent carbonaceous nanospheres. J Colloid Interface Sci 429:77–82. https://doi.org/10.1016/j.jcis.2014.05.016

174. Rahmi I, Irfan M (2019) Methylene blue removal from water using H_2SO_4 crosslinked magnetic chitosan nanocomposite beads. Microchem J 144:397–402. https://doi.org/10.1016/j.microc.2018.09.032

175. Rajbhandary A, Nilsson BL (2016) Self-assembling hydrogels. In GELS Handbook: fundamentals, properties and applications volume 1: fundamentals of hydrogels. World Scientific Tuck Link, Singapore, pp 219–250

176. Ravi Kumar MNV (2000) A review of chitin and chitosan applications. React Funct Polym 46:1–27. https://doi.org/10.1016/S1381-5148(00)00038-9

177. Ravi Kumar MNV, Muzzarelli RAA, Muzzarelli C, Sashiwa H, Domb AJ (2004) Chitosan chemistry and pharmaceutical perspectives. Chem Rev 104:6017–6084. https://doi.org/10.1021/cr030441b

178. Ren X, Chen C, Nagatsu M, Wang X (2011) Carbon nanotubes as adsorbents in environmental pollution management: a review. Chem Eng J 170(2–3):395–410. https://doi.org/10.1016/j.cej.2010.08.045

179. Rinaudo M (2006) Chitin and chitosan: properties and applications. Prog Polym Sci 31:603–632. https://doi.org/10.1016/j.progpolymsci.2006.06.001

180. Robinson T, McMullan G, Marchant R, Nigam P (2001) Remediation of dyes in textile effluent: a critical review on current treatment technologies with a proposed alternative. Bioresour Technol 77:247–255. https://doi.org/10.1016/S0960-8524(00)00080-8

181. Saber-Samandari S, Gulcan HO, Saber-Samandari S, Gazi M (2014) Efficient removal of anionic and cationic dyes from an aqueous solution using pullulan-graft-polyacrylamide porous hydrogel. Water Air Soil Pollut 225:2177. https://doi.org/10.1007/s11270-014-2177-5

182. Saberi A, Alipour E, Sadeghi M (2019) Superabsorbent magnetic Fe_3O_4-based starch-poly (acrylic acid) nanocomposite hydrogel for efficient removal of dyes and heavy metal ions from water. J Polym Res 26:271. https://doi.org/10.1007/s10965-019-1917-z

183. Sadat Ebrahimi M-M, Voss Y, Schönherr H (2015) Rapid detection of escherichia coli via enzymatically triggered reactions in self-reporting chitosan hydrogels. ACS Appl Mater Interfaces 7:20190–20199. https://doi.org/10.1021/acsami.5b05746

184. Said HM, Alla SGA, El-Naggar AWM (2004) Synthesis and characterization of novel gels based on carboxymethyl cellulose/acrylic acid prepared by electron beam irradiation. React Funct Polym 61:397–404. https://doi.org/10.1016/j.reactfunctpolym.2004.07.002

185. Salahuddin NA, EL-Daly HA, Sharkawy RGE, Nasr BT (2020) Nano-hybrid based on polypyrrole/chitosan/graphene oxide magnetite decoration for dual function in water remediation and its application to form fashionable colored product. Adv Powder Technol 31:1587–1596. https://doi.org/10.1016/j.apt.2020.01.030

186. Saltz A, Kandalam U (2016) Mesenchymal stem cells and alginate microcarriers for craniofacial bone tissue engineering: a review. J Biomed Mater Res A 104:1276–1284. https://doi.org/10.1002/jbm.a.35647

187. Samuel MS, Suman S, Venkateshkannan SE, Mathimani T, Pugazhendhi A (2020) Immobilization of $Cu_3(btc)_2$ on graphene oxide-chitosan hybrid composite for the adsorption and photocatalytic degradation of methylene blue. J Photochem Photobiol B 204:111809. https://doi.org/10.1016/j.jphotobiol.2020.111809

188. Sanchez LM, Alvarez VA, Ollier RP (2019) Acid-treated bentonite as filler in the development of novel composite PVA hydrogels. J Appl Polym Sci 136:47663. https://doi.org/10.1002/app. 47663

189. Sashiwa H, Aiba S (2004) Chemically modified chitin and chitosan as biomaterials. Prog Polym Sci 29:887–908. https://doi.org/10.1016/j.progpolymsci.2004.04.001

190. Schulze J, Hendrikx S, Schulz-Siegmund M, Aigner A (2016) Microparticulate poly(vinyl alcohol) hydrogel formulations for embedding and controlled release of polyethylenimine (PEI)-based nanoparticles. Acta Biomater 45:210–222. https://doi.org/10.1016/j.actbio.2016. 08.056

191. Senel S, McClure SJ (2004) Potential applications of chitosan in veterinary medicine. Adv Drug Delivery Rev 56:1467–1480. https://doi.org/10.1016/j.addr.2004.02.007

192. Sengel SB, Sahiner N (2016) Poly(vinyl phosphonic acid) nanogels with tailored properties and their use for biomedical and environmental applications. Eur Polym J 75:264–275. https:// doi.org/10.1016/j.eurpolymj.2016.01.007

193. Shalla AH, Bhat MA, Yaseen Z (2018) Hydrogels for removal of recalcitrant organic dyes: a conceptual overview. J Environ Chem Eng 6(5):5938–5949. https://doi.org/10.1016/j.jece. 2018.08.063

194. Sharma G, Kumar A, Naushad M, García-Peñas A, Al-Muhtaseb AH, Ghfar AA, Sharma V, Ahamad T, Stadler FJ (2018) Fabrication and characterization of gum arabic-cl-poly(acrylamide) nanohydrogel for effective adsorption of crystal violet dye. Carbohydr Polym 202:444–453. https://doi.org/10.1016/j.carbpol.2018.09.004

195. Sharma G, Kumar A, Sharma S, Naushad M, Ghfar AA, Al-Muhtaseb AH, Ahamad T, Sharma N, Stadler FJ (2020) Carboxymethyl cellulose structured nano-adsorbent for removal of methyl violet from aqueous solution: Isotherm and kinetic analyses. Cellulose 27:3677–3691. https://doi.org/10.1007/s10570-020-02989-y

196. Sharma S, Tiwari S (2020) A review on biomacromolecular hydrogel classification and its applications. Int J Biol Macromol 162:737–747. https://doi.org/10.1016/j.ijbiomac.2020. 06.110

197. Shi Y, Xue Z, Wang X, Wang L, Wang A (2013) Removal of methylene blue from aqueous solution by sorption on lignocellulose-g-poly(acrylic acid)/montmorillonite three-dimensional crosslinked polymeric network hydrogels. Polym Bull 70:1163–1179. https://doi.org/10. 1007/s00289-012-0898-4

198. Shirsath SR, Patil AP, Patil R, Naik JB, Gogate PR, Sonawane SH (2013) Removal of brilliant green from wastewater using conventional and ultrasonically prepared poly(acrylic acid) hydrogel loaded with kaolin clay: a comparative study. Ultrason Sonochem 20(3):914–923. https://doi.org/10.1016/j.ultsonch.2012.11.010

199. Shirsath SR, Patil AP, Bhanvase BA, Sonawane SH (2015) Ultrasonically prepared poly(acrylamide)-kaolin composite hydrogel for removal of crystal violet dye from wastewater. J Environ Chem Eng 3(2):1152–1162. https://doi.org/10.1016/j.jece.2015.04.016

200. Shojaat R, Saadatjoo N, Karimi A, Aber S (2016) Simultaneous adsorption-degradation of organic dyes using MnFe$_2$O$_4$/calcium alginate nano-composites coupled with GOx and laccase. J Environ Chem Eng 4(2):1722–1730. https://doi.org/10.1016/j.jece.2016.02.029

201. Simonin J-P (2016) On the comparison of pseudo-first order and pseudo-second order rate laws in the modeling of adsorption kinetics. Chem Eng J 300:254–263. https://doi.org/10. 1016/j.cej.2016.04.079

202. Sinha V, Chakma S (2019) Advances in the preparation of hydrogel for wastewater treatment: a concise review. J Environ Chem Eng 7(5):103295. https://doi.org/10.1016/j.jece.2019.103295

203. Solpan D, Torun M, Gueven G (2008) The usability of (sodium alginate/acrylamide) semi-interpenetrating polymer networks on removal of some textile dyes. J Appl Polym Sci 108:3787–3795. https://doi.org/10.1002/app.27945

204. Soni A, Tiwari A, Bajpai AK (2014) Removal of malachite green from aqueous solution using nano-iron oxide-loaded alginate microspheres: batch and column studies. Res Chem Intermed 40(3):913–930. https://doi.org/10.1007/s11164-012-1011-1

205. Sperinde JJ, Griffith LG (1997) Synthesis and characterization of enzymatically-crosslinked poly(ethylene glycol) hydrogels. Macromolecules 30:5255–5264. https://doi.org/10.1021/ma970345a

206. Srivastava A, Yadav A, Samanta S (2015) Biopolymeric alginic acid: an efficient recyclable green catalyst for the friedel-crafts reaction of indoles with isoquinoline-1,3,4-triones in water. Tetrahedron Lett 56:6003–6007. https://doi.org/10.1016/j.tetlet.2015.09.041

207. Stanciu MC, Nichifor M (2018) Influence of dextran hydrogel characteristics on adsorption capacity for anionic dyes. Carbohyd Polym 199:75–83. https://doi.org/10.1016/j.carbpol.2018.07.011

208. Stanciu MC, Nichifor M (2019) Adsorption of anionic dyes on a cationic amphiphilic dextran hydrogel: equilibrium, kinetic, and thermodynamic studies. Colloid Polym Sci 297:45–57. https://doi.org/10.1007/s00396-018-4439-z

209. Stanciu MC, Nichifor M, Ailiesei GL (2021) Bile salts adsorption on dextran-based hydrogels. Int J Biol Macromol 190:270–283. https://doi.org/10.1016/j.ijbiomac.2021.08.205

210. Stankovich S, Dikin DA, Piner RD, Kohlhaas KA, Kleinhammes A, Jia Y, Wu Y, Nguyen ST, Ruoff RS (2007) Synthesis of graphene-based nanosheets via chemical reduction of exfoliated graphite oxide. Carbon 45(7):1558–1565. https://doi.org/10.1016/j.carbon.2007.02.034

211. Stoller MD, Park S, Zhu Y, An J, Ruoff RS (2008) Graphene based ultracapacitors. Nano Lett 8:3498–3502. https://doi.org/10.1021/nl802558y

212. Struszczyk MH (2002) Chitin and chitosan-part I. Properties and production. Polimery 47:316–325. https://doi.org/10.14314/polimery.2002.316

213. Su T, Wu L, Pan X, Zhang C, Shi M, Gao R, Qi X, Dong W (2019) Pullulan-derived nanocomposite hydrogels for wastewater remediation: synthesis and characterization. J Colloid Interface Sci 542:253–262. https://doi.org/10.1016/j.jcis.2019.02.025

214. Sun X, Peng B, Ji Y, Chen J, Li D (2009) Chitosan(chitin)/cellulose composite biosorbents prepared using ionic liquid for heavy metal ions adsorption. Aiche J 55:2062–2069. https://doi.org/10.1002/aic.11797

215. Suzuki M, Hirasa O (1993) An approach to artificial muscle using polymer gels formed by micro-phase separation. In: Dušek K (ed) Responsive gels: volume transitions II. Advances in polymer science, vol 110. Springer, Berlin, Heidelberg. https://doi.org/10.1007/BFb0021135

216. Szabó T, Berkesi O, Dékány I (2005) DRIFT study of deuterium-exchanged graphite oxide. Carbon 43(15):3186–3189. https://doi.org/10.1016/j.carbon.2005.07.013

217. Tahira I, Aslam Z, Abbas A, Monim-ul-Mehboob M, Ali S, Asghar A (2019) Adsorptive removal of acidic dye onto grafted chitosan: a plausible grafting and adsorption mechanism. Int J Biol Macromol 136:1209–1218. https://doi.org/10.1016/j.ijbiomac.2019.06.173

218. Takigami M, Amada H, Nagasawa N, Yagi T, Kasahara T, Takigami S, Tamada M (2007) Preparation and properties of CMC gel. Trans Res Soc Japan 32(3):713–716. https://doi.org/10.14723/tmrsj.32.713

219. Thakur S, Arotiba O (2018) Synthesis, characterization and adsorption studies of an acrylic acid-grafted sodium alginate-based TiO_2 hydrogel nanocomposite. Adsorp Sci Technol 36(1–2):458–477. https://doi.org/10.1177/0263617417700636

220. Tsuji H, Horii F, Nakagawa M, Ikada Y, Odani H, Kitamaru R (1992) Stereocomplex formation between enantiomeric poly(lactic acid)s. 7. Phase structure of the stereocomplex crystallized from a dilute acetonitrile solution as studied by high resolution solid-state carbon-13 NMR spectroscopy. Macromolecules 25:4114–4118. https://doi.org/10.1021/ma00042a011

221. Tsuji H (2005) Poly (lactide) stereocomplexes: formation, structure, properties, degradation, and applications. Macromol Biosci 5:569–597. https://doi.org/10.1002/mabi.200500062

222. Unuabonah EI, Omorogie MO, Oladoja NA (2019) 5-Modeling in adsorption: fundamentals and applications. In: Kyzas GZ, Mitropoulos AC (eds) Micro and nano technologies, composite nanoadsorbents. Elsevier, pp 85–118. https://doi.org/10.1016/B978-0-12-814132-8.00005-8

223. Van Tran V, Park D, Lee YC (2018) Hydrogel applications for adsorption of contaminants in water and wastewater treatment. Environ Sci Pollut Res 25:24569–24599. https://doi.org/10.1007/s11356-018-2605-y

224. Varaprasad K, Mohan YM, Ravindra S, Reddy NN, Vimala K, Monika K, Sreedhar B, Raju KM (2010) Hydrogel–silver nanoparticle composites: a new generation of antimicrobials. J Appl Polym Sci 115:1199–1207. https://doi.org/10.1002/app.31249

225. Varaprasad K, Sadiku R (2015) Development of microbial protective kolliphor-based nanocomposite hydrogels. J Appl Polym Sci 132(46):42781–42787. https://doi.org/10.1002/app.42781

226. Varaprasad K, Raghavendra GM, Jayaramudu T, Yallapu MM, Sadiku R (2017a) A mini review on hydrogels classification and recent developments in miscellaneous applications. Mater Sci Eng C Mater Biol Appl 79:958–971. https://doi.org/10.1016/j.msec.2017.05.096

227. Varaprasad K, Jayaramudu T, Sadiku ER (2017b) Removal of dye by carboxymethyl cellulose, acrylamide and graphene oxide via a free radical polymerization process. Carbohyd Polym 164:186–194. https://doi.org/10.1016/j.carbpol.2017.01.094

228. Verma A, Thakur S, Mamba G, Prateek GRK, Thakur P, Thakur VK (2020) Graphite modified sodium alginate hydrogel composite for efficient removal of malachite green dye. Int J Biol Macromol 148:1130–1139. https://doi.org/10.1016/j.ijbiomac.2020.01.142

229. Visakh PM, Mathew AP, Oksman K, Thomas S (2012) Starch-based bionanocomposites: processing and properties. Polysaccharide building blocks: a sustainable approach to the development of renewable biomaterials. In: Habibi Y, Lucia LA (eds) Polysaccharide building blocks: a sustainable approach to the development of renewable biomaterials. Wiley, Inc, pp 287–306. https://doi.org/10.1002/9781118229484.ch11

230. Wang J, Wei L, Ma Y, Li K, Li M, Yu Y, Wang L, Qiu H (2013) Collagen/cellulose hydrogel beads reconstituted from ionic liquid solution for Cu(II) adsorption. Carbohyd Polym 98:736–743. https://doi.org/10.1016/j.carbpol.2013.06.001

231. Wang J, Wei J (2016) Hydrogel brushes grafted from stainless steel via surface-initiated atom transfer radical polymerization for marine antifouling. Appl Surf Sci 382:202–216. https://doi.org/10.1016/j.apsusc.2016.03.223

232. Wang J, Guo X (2020) Adsorption kinetic models: physical meanings, applications, and solving methods. J Hazard Mat 390:122156. https://doi.org/10.1016/j.jhazmat.2020.122156

233. Wang R, Yu Q, He Y, Bai J, Jiao T, Zhang L, Bai Z, Zhou J, Peng Q (2019) Self-assembled polyelectrolyte-based composite hydrogels with enhanced stretchable and adsorption performances. J Mol Liq 294:111576. https://doi.org/10.1016/j.molliq.2019.111576

234. Wang Y, Wang W, Wang A (2013) Efficient adsorption of methylene blue on an alginate-based nanocomposite hydrogel enhanced by organo-illite/smectite clay. Chem Eng J 228:132–139. https://doi.org/10.1016/j.cej.2013.04.090

235. Wang Y, Zhang C, Zhao L, Meng G, Wu J, Liu Z (2017) Cellulose-based porous adsorbents with high capacity for methylene blue adsorption from aqueous solutions. Fibers Polym 18:891–899. https://doi.org/10.1007/s12221-017-6956-7

236. Weber W, Morris J (1963) Kinetics of adsorption on carbon from solution. J Sanit Eng Div Am Soc Civ Eng 89:31–60

237. Wei Q, Xu M, Liao C, Wu Q, Liu M, Zhang Y, Wu C, Cheng L, Wang Q (2016) Printable hybrid hydrogel by dual enzymatic polymerization with superactivity. Chem Sci 7:2748–2752. https://doi.org/10.1039/C5SC02234G

238. Xiao D, He M, Liu Y, Xiong L, Zhang Q, Wei L, Li L, Yu X (2020) Strong alginate/reduced graphene oxide composite hydrogels with enhanced dye adsorption performance. Polym Bull 77:6609–6623. https://doi.org/10.1007/s00289-020-03105-7

239. Xu B, Zheng H, Wang Y, An Y, Luo K, Zhao C, Xiang W (2018) Poly (2-acrylamido-2-methylpropane sulfonic acid) grafted magnetic chitosan microspheres: preparation, characterization and dye adsorption. Int J Biol Macromol 112:648–655. https://doi.org/10.1016/j.ijbiomac.2018.02.024

240. Xu S, Wang J, Wu R, Wang J, Li H (2006) Adsorption behaviors of acid and basic dyes on crosslinked amphoteric starch. Chem Eng J 117:161–167. https://doi.org/10.1016/j.cej.2005.12.012

241. Yadav S, Asthana A, Chakraborty R, Jain B, Singh AK, Carabineiro SAC, Hasan Susan MAB (2020) Cationic dye removal using novel magnetic/activated charcoal/β-cyclodextrin/alginate polymer nanocomposite. Nanomaterials 10(1):170. https://doi.org/10.3390/nano10010170

242. Yagub MT, Sen TK, Afroze S, Ang HM (2014) Dye and its removal from aqueous solution by adsorption: a review. Adv Colloid Interface Sci 209:172–184. https://doi.org/10.1016/j.cis.2014.04.002

243. Yang Y, Song S, Zhao Z (2017) Graphene oxide (GO)/polyacrylamide (PAM) composite hydrogels as efficient cationic dye adsorbents. Colloid Surf A: Physicochem Eng Asp 513:315–324. https://doi.org/10.1016/j.colsurfa.2016.10.060

244. Ye X, Li X, Shen Y, Chang G, Yang J, Gu Z (2017) Self-healing pH-sensitive cytosine- and guanosine-modified hyaluronic acid hydrogels via hydrogen bonding. Polymer (Guildf) 108:348–360. https://doi.org/10.1016/j.polymer.2016.11.063

245. Yuan Z, Wang J, Wang Y, Liu Q, Zhong Y, Wang Y, Li L, Lincoln SF, Guo X (2019) Preparation of a poly(acrylic acid) based hydrogel with fast adsorption rate and high adsorption capacity for the removal of cationic dyes. RSC Adv 9(37):21075–21085. https://doi.org/10.1039/c9ra03077h

246. Zainal SH, Mohd NH, Suhaili N, Anuar FH, Lazim AM, Othaman R (2021) Preparation of cellulose-based hydrogel: a review. J Mater Res Technol 10:935–952. https://doi.org/10.1016/j.jmrt.2020.12.012

247. Zhai M, Yoshii F, Kume T, Hashim K (2002) Syntheses of PVA/starch grafted hydrogels by irradiation. Carbohydr Polym 50:295–303. https://doi.org/10.1016/S0144-8617(02)00031-0

248. Zhang G, Yi L, Deng H, Sun P (2014) Dyes adsorption using a synthetic carboxymethyl cellulose-acrylic acid adsorbent. J Environ Sci 26:1203–1211. https://doi.org/10.1016/S1001-0742(13)60513-6

249. Zhang H, Zhang F, Wu J (2013) Physically crosslinked hydrogels from polysaccharides prepared by freeze–thaw technique. React Funct Polym 73:923–928. https://doi.org/10.1016/j.reactfunctpolym.2012.12.014

250. Zhang L, Hu P, Wang J, Huang R (2016) Adsorption of amido black 10B from aqueous solutions onto Zr(IV) surface-immobilized crosslinked chitosan/bentonite composite. Appl Surf Sci 369:558–566. https://doi.org/10.1016/j.apsusc.2016.01.217

251. Zhang Q, Hu XM, Wu MY, Wang MM, Zhao YY, Li TT (2019) Synthesis and performance characterization of poly(vinyl alcohol)-xanthan gum composite hydrogel. React Funct Polym 136:34–43. https://doi.org/10.1016/j.reactfunctpolym.2019.01.002

252. Zhao J, Zou Z, Ren R, Sui X, Mao Z, Xu H, Zhong Y, Zhang L, Wang B (2018) Chitosan adsorbent reinforced with citric acid modified β-cyclodextrin for highly efficient removal of dyes from reactive dyeing effluents. Eur Polym J 108:212–218. https://doi.org/10.1016/j.eurpolymj.2018.08.044

253. Zhao L, Mitomo H, Zhai M, Yoshii F, Nagasawa N, Kume T (2003) Synthesis of antibacterial PVA/CM-chitosan blend hydrogels with electron beam irradiation. Carbohydr Polym 53:439–446. https://doi.org/10.1016/S0144-8617(03)00103-6

254. Zhao QS, Ji QX, Xing K, Li XY, Liu CS, Chen XG (2009) Preparation and characteristics of novel porous hydrogel films based on chitosan and glycerophosphate. Carbohydr Polym 76:410–416. https://doi.org/10.1016/j.carbpol.2008.11.020

255. Zhou Y, Lu J, Zhou Y, Liu Y (2019) Recent advances for dyes removal using novel adsorbents: a review. Environ Pollut 252:352–365. https://doi.org/10.1016/j.envpol.2019.05.072

256. Zhu BW, Yao Z (2020) Marine oligosaccharides originated from seaweeds: source, preparation, structure, physiological activity and applications. Crit Rev Food Sci Nutr 61(1):60–74. https://doi.org/10.1080/10408398.2020.1716207

Bacterial Extracellular Polymeric Substances for Degradation of Textile Dyes

Ghulam Mustafa, Muhammad Tariq Zahid, Sidra Ihsan, Itrash Zia, Syed Zaghum Abbas, and Mohd Rafatullah

1 Introduction

Synthetic chemicals are the major cause of Environmental pollution in present era. These chemicals have adverse effects of the health of all type of life and now need to be monitored [72]. The most precious compound on earth is water. In recent eras, the worth of water has been worsening because of anthropogenetic activities, unplanned urbanization, population growth, rapid industrialization, and amateurish employment of the natural water assets and result in water pollution. Implausible extent of dyes pour forth frequently from several industries as textile, pharmaceuticals, paper-making, cosmetics, food, tannery, and dyes producing industries. Water contaminated with dyes expelled into fresh water channels without pre-treatment is the major threat to the human health and environment as well because most of the dyes content remains unaffected in the water for long time [6]. The industrial effluent produced during different manufacturing processes have adverse effect for aquaculture and other organisms due to high toxicity of toxins and different types of pollutants fabrics, leather, plastics, pulp, ink, soap, paper, and palm oil present in wastewater. The short-term exposure to these toxins may cause nervous disorders but long-term exposure may cause generalized hypoxia, thyroid dysfunction, and

G. Mustafa · M. T. Zahid · S. Ihsan · I. Zia
Department of Zoology, Government College University Lahore, Lahore 54000, Pakistan

S. Z. Abbas
School of Environment and Safety Engineering, Biofuels Institute, Jiangsu University, 301 Xuefu Road, Zhenjiang 212013, Jiangsu Province, China

M. Rafatullah (✉)
Division of Environmental Technology, School of Industrial Technology, Universiti Sains Malaysia, 11800 Penang, Malaysia
e-mail: mrafatullah@usm.my

weight loss. So the treatment of toxic water is mandatory before its final elimination in water channels for the protection of biodiversity and pure water reserves for future generation [28]. Natural and the synthetic dyes have been universally used to shade substances, including textile products, paper leather, plastics cosmetics, food, and pharmaceutical products. Synthetic dyes are in fabric and textile industry than natural dyes, which unsatisfactorily fulfill the demands of textile industries so that more used in food industry [52]. In industries there are about 10,000 different types of dyes used. Synthetic dyes are produced more than 8×10^5 tons annually worldwide and about 10–15% is released into fresh water [22, 68]. Synthetic dyes are non-biodegradable, and their persistence in waterways causes pollution, which has become a serious concern around the world [34]. Synthetic dyes are carcinogenic, mutagenic, and poisonous in nature. As a result, dyes should be properly degraded prior to wastewater disposal [16].

The dyes have high tectorial values so that researchers showed great interest for the treatment of wastewater. The physical and chemical properties of water significantly changed even the presence of dyes less than 1 ppm in water. The traditional wastewater treatment technology depends mainly on physical, chemical, and biological methods which contribute effectively to improve the quality of wastewater as biochemical oxygen demand (BOD), chemical oxygen demand (COD), turbidity and total suspension solids (TSS). Regrettably, these processes are unsatisfactory to confiscate the dyes from the effluent. Coagulation/flocculation is a prospective substitute, and an exceedingly competent technique for eliminating dyes from wastewaters. As a result, there is a pressing need to find low-cost substrates to replace expensive traditional media substrates and, as a result, lower bioflocculant manufacturing costs [75]. The biological treatment of azo dye wastewater has been thoroughly investigated, and various bacteria have been identified, e.g., *Acinetobacter sp.* [39], *Shewanella oneidensis* [39], *Pseudomonas sp.* [61] *and Bacillus sp.* [37], *Klebsiella sp.* [51] have been well-known from different sources. However, an anaerobic condition is dynamic to reduce azo dyes in the prokaryote kingdom, as the aerobic environment decreases the accessible electron donors for the breakdown of azo bond of dyes [55]. Bioremediation is the most effective technology for the treatment of environmental pollution. This technique is eco-friendly and cost-effective because living organisms such as bacteria, fungi, and algae involved in the treatment of pollutants in air, soil, and water [72]. The microbes masses present in biological wastewater treatment system forms biofilm, sludge flocs, and granules that result in the formation of mixture of highly complex with high molecular weight polymers known as extracellular polymeric substances (EPS). EPS have a significant impact on the physicochemical characteristics of microbial masses, including their structure, settling properties, surface charge, dewatering properties, adsorption ability, and flocculation. EPS have the ability to form and protect aggregate from dewatering by developing vast net-like structure by complex interaction for storage of plenty of water. EPS is the best source of carbon and energy for the cells in extreme conditions [70].

In last decade prompted development of economic and industries consequences increase in the pollutants, inorganic and organic increasing significantly in water channels. Aerobic and anaerobic advance biological wastewater treatment technologies were dominantly in wastewater treatment plants. Microbial communities play a promising role in removing pollutants from water. Even though they can remove exogenous compounds which are detrimental to the sludge by biodegradation and biosorption. The presence of exogenous compounds such as heavy metals and toxic organic substances stimulates the live cells to produce EPS for their own protection from the toxicity of these substances which reveals that exogenous compound and EPS have interaction. Recently, many researchers focus on the use of bacterial EPS in environmental applications. On the other hand, evidence of EPS's impact on environmental applications is sparse, and no accompanying report has been published to yet that highlights the characteristics of EPS and their significance to environmental applications. Because they retain both hydrophobic and hydrophilic sections within their structures, EPS have considerable binding capacities toward heavy metal ions and organic contaminants [29]. In addition, microbial strains manufacture more EPS to protect themselves from unfavorable environments, such as harmful chemicals. As a result, EPS could play a significant role in the reduction or detoxification of organic and inorganic contaminants in wastewater and/or polluted soil. Many researches have looked at the composition and structure of EPS from natural strains and active sludges when it comes to EPS characterization [8]. EPS are considered as an operative adsorbent and broadly used for wastewater treatment due to their copious binding sites and functional groups [42]. Metabolization/transformation, adsorption is the basic mechanism involved in the bioremediation of textile dyes by bacterial cell mass. The bacterial consortiums exist in the form of biofilm that has potential to decolorize and metabolize the textile dyes while the intracellular processes can perform degradation and biosorption of dyes. Biofilm is a mixture of bacterial cells surrounded by exopolysaccharides and the natural environment for biofilm is same as bacterial cells. Biofilms are made up of a diverse microbial population of cells that grow on surfaces encased in exopolysaccharides. As a result, biofilm is the dominant microbial lifestyle in most natural habitats. The EPS generated by different bacterial strains is primarily used to protect bacteria against desiccation and predation, as well as to aid adherence to surfaces. EPS is formed in two forms: capsule, which is strongly connected to the bacterium's surface, and slime, which is only loosely attached to the bacterium's surface. EPS are also considered as the most immediate interfacial boundary between the bulk aqueous phase and the bacterial cells. EPS generally consist of a wide variety of macromolecular compounds including acidic polysaccharides and proteins, as well as lipids. The goal of this study was to see if a biofilm made up of selected indigenous and foreign bacteria could be used to bioremediate color from textile effluent. This work will provide useful data for future EPS research.

2 Extracellular Polymeric Substances

Secretion, excretion, cell lysis, and sorption in bacteria produce EPS, which are particularly unusual macromolecules. Polysaccharides, DNA/nucleic acids, lipids, protein, uronic acids, humic-like compounds, and other micro-molecules make up EPS chemically [2, 8, 13, 14, 46]. Due to the presence of functional groups (e.g., hydroxyl, carboxylic, sulfhydryl, phosphate groups, etc.) in the EPS, it forms a protective coating surrounding the bacteria against the harsh external environment containing heavy metals and highly poisonous chemical compounds [5, 7, 21]. Polysaccharides have frequently been supposed most copious components of EPS in primary biofilm research [11]. That is why the term "EPS" has been used as abbreviation for "extracellular polysaccharides" or "exopolysaccharides." However, proteins and nucleic acids [54, 60], in addition to amphiphilic compounds with phospholipids, have also been seen in significant amounts in EPS preparations from pure cultures of bacteria, sewer biofilms, activated sludge, and trickling filter biofilms. Furthermore, some researchers pronounced humic substances as components of EPS grounds of soil and water biofilms [63]. EPS have attracted much concern and ubiquitous impact on the biosorption, sludge performance, sludge settle ability, sludge bioflocculability, surface charge characteristic, oxidation–reduction property, hydrophilicity/hydrophobicity and biodegradability for environmental remediation [21, 46, 71]. According to the nature of their interaction with cells or the procedure used to remove EPS from microbial cells, EPS are classified as slime (S-EPS), capsular (CEPS), tightly bound (TB-EPS), and loosely bound (LB-EPS).

Extracellular polymeric substances (EPS) are polymers produced by a variety of microorganism strains. Polysaccharides, proteins, and DNA make up the majority of them. Environmental cues are principally responsible for the development of these slimes. Because their biosynthesis is so costly, they should provide a benefit to the producer microbe [19, 20]. These substances are reflected as sustainable, eco-friendly, and cost-effective as compared to auxiliary, the prevailing chemical compounds [49]. EPS exist both in internal and external environment of microbial cells. They are glycocalyx or slime that are found on or near the microbial cell surface, and speed up and facilitate microbial adhesion to the substratum. Bacterial secretions, shedding materials from the cell surface, cell lysates, and hydrolysates, and the adsorption of organic molecules from the surrounding environment are all examples of EPS [63]. Microorganisms secrete EPS, a complex mixture of biomolecules (proteins, polysaccharides, nucleic acids, lipids, and other macromolecules) that hold microbial aggregation together [70]. Proteins and exopolysaccharides, which constitute 40–95% of EPS, are the most important components of macromolecules. "Organic polymers of bacterial origin are usually responsible for the binding of cells and other particulate matter together in biofilm systems (cohesion) and bonding to the matrix (adhesion)," according to the definition. The abbreviation EPS is used for "Extracellular Polymeric Substances," "exo-polysaccharides," "extracellular polysaccharides," and "exo-polymers" [53, 70]. Nielsen and Jahn [53] suggested that sometimes extracellular polymeric substances are not anchored with murein layer or

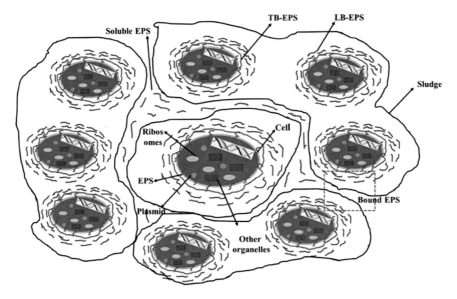

Fig. 1 The structural composition diagram of EPS (Adapted from [67])

external membrane of the cell. Moreover, selected organic matters from effluent can also be adhering to EPS [41]. In the study of microbial aggregates, EPS linked with microbial metabolism may be more useful and can change their physical and chemical properties. Outside of the cells, EPS is split into two types: bound EPS (loosely bound polymers, sheaths, capsular polymers, and condensed gels) and soluble EPS (soluble macromolecules, slimes, and colloids) [38]. EPS can be easily separated by centrifugation: soluble EPS in supernatant and bound EPS in the pellet of cells. The origin of both types is still unknown. There is a weak interaction of soluble EPS with cell as compared to bound EPS and has crucial effects on the microbial activities and structure of sludge surface. However, the soluble EPS has limited study. The structure of the bound EPS is generally described by a double-layer model (Fig. 1). The inner layer is made up of tightly bound EPS (TB-EPS) with a defined form and a strong bond to the cell surface. The outer layer is made up of loosely bound EPS (LB-EPS), a dispersible mucus layer with no defined form. Because the fraction of LB-EPS in microbial aggregates is always lower than that of TB-EPS, it may have an effect on microbial aggregate properties [62].

EPS are macromolecules that have features such as precipitation, adsorption, flocculation, and dehydration. They are made from activated sludge by microbial metabolite excretion, shredding of cell surface components, cell lysis, and adsorption of organic or inorganic compounds from the external environment to the cell wall surface [67]. EPS play an important role in biological wastewater treatment. To begin with, they build a dense protective barrier around cells to protect them from environmental toxins (such as heavy metals, PAHs, phenols, antibiotics, and so on). Second, they stick to activated sludge and form a spatial network to absorb and degrade contaminants, third, in the absence of nutrition, they offer energy and carbon for vital life functions [47].

3 Bacterial Extracellular Polymeric Substances

Microbial cells have EPS as a key component (bacteria, algae, fungi, and yeast). EPS is often made up of sugar polymers containing charged functional groups such as phosphate, amine, acetate, carbonyl, and hydroxyl. They convert them into natural ligand sources, providing binding sites for other charged molecules as well as sequestering or immobilizing them through the formation of numerous complexes. Bacterial extracellular polymeric substances (EPS) secreted by bacterial strains may also act as outstanding adsorbents. They give slimy appearance to bacterial masses. Bacteria are generally present in an aggregated form where they are surrounded by EPS [60]. Bacterial EPS are mainly composed of curli (a proteinaceous surface appendages), lipids, anocellulose-rich polysaccharides, and nucleic acids [45, 50]. Bacteria can produce extracellular DNA during growth. But it is still unknown whether DNA is actively secreted or passively released due to rise in cell envelope permeability [23, 43]. The spontaneous release of integral cellular components such as lipopolysaccharides (LPS) from Gram-negative bacteria's outer membrane is another route for extracellular polymer release [3, 4, 15]. EPS can be found outside archaeobacteria's cell wall (pseudomurein layer), Gram-positive eubacteria's cell wall (peptidoglycan layer), and Gram-negative eubacteria's outer membrane. Because polysaccharides were recognized as common elements in bacterial EPS, the term "glycocalyx" was coined to describe polysaccharide-containing structures of bacterial origin [11, 58]. EPS-containing structures such as capsules, slimes, and sheaths are referred to as "glycocalyx." "Integral capsules" are unique structures with distinct shapes that have a firm link with the cell surface. Non-covalent interactions bind capsular polymers to the cell surface, although they can also be covalently linked to phospholipid at the cell surface. Sheaths are present in filamentous bacteria (for example, certain cyanobacteria or bacteria of the genera Sphaerotilus and Leptothrix). S-layers are also considered as glycocalyx components [11]. These are noncovalent interactions between monomolecular proteins or glycoprotein complexes and underlying integral cell surface components. In Bacteria and Archaea, S-layers are one of the most common cell surface structures. MFX (EPS from *Klebsiella* sp. J1) has been used in previous study for adsorbing tetracycline and other pollutants. An attempt was made for efficient degradation of textile dyes, Congo red, methyl orange, and food coloring agents (yellow and green) with EPS-stabilized AgNPs. AgNPs and bacterial degradation were shown to be inferior. These EPS-stabilized AgNPs can be employed as a low-cost and environmentally benign method for degrading hazardous colors, with applications in textiles and food additives [1].

4 Characteristics of Bacterial Extracellular Polymeric Substances

The EPS in the bacterial clumps can adsorb metals and organics in many places, such as the aromatic and aliphatic in protein and the hydrophobic area in carbohydrates. This reveals the latent role of bacterial EPS in heavy metal and dyes adsorption with bacterial cells and transportation in the environment [26]. The carboxyl group, phosphoric acid group, sulfhydryl group, phenol group, and hydroxyl group are all functional groups in EPS that can form complexes with heavy metals [25, 32]. EPS is thought to have a very high binding power potential based on the projected number of carboxyl and hydroxyl groups [24]. Heavy metals and dyes have the ability to combine with proteins, nucleic acids, and carbohydrates in the EPS [73]. The adsorption isotherm obeyed the Langmuir isotherm or Freundlich isotherm equations, and the binding ability and power of the bonds existent between heavy metals/dyes and EPS were recognized to be high [48]. Similarly, soluble EPS has a greater dye adsorptive capacity than bound EPS from a slurry [9]. Various mechanisms involved for adsorption of metals and dye removal are presented in literature. EPS can also adsorb other pollutants, such as textile dyes [63], reactive brilliant red X-3B, phenanthrene [40] benzene [64], and humic acids [17]. EPS contain specific hydrophobic regions for the attachment of dyes and metals [18]. Späth et al. [65] described that EPS can absorb 60% or more of benzene, m-xylene, and toluene, and only a small portion of these contaminants were adsorbed by bacterial cells. Because EPS particles are negatively charged, they can easily bond with cationic pollutants via an electrostatic contact [17]. Furthermore, unlike humic acid, the proteins have a strong binding ability and proficiency because loosed EPS contain higher percentage of proteins than bounded EPS, due to higher binding capacity [56].

EPS are the source of carbon and energy for the bacterial aggregates. Generally, carbohydrates and proteins are the major components of EPS, and enzymes that breakdown these polymers are common in biological wastewater treatment reactors. These bacteria can utilize EPS emitted by other bacteria for metabolic activities in activated sludge [74]. However, [38] discussed that microorganisms are unable to degrade certain components of EPS. Wang et al. [69], explained a portion of bacterial EPS from aerobic sludge was biodegradable, and it was revealed that the bacterial EPS in the periphery of aerobic granular sludge could not be biodegraded, but those in the core layer could. Park and Novak [57], demonstrated how the biodegradability of bacterial EPS collected using various procedures differed. For example, EPS extracted using a cationic exchange resin methodology is aerobic biodegradable, whereas EPS extracted using a sulfide method is anaerobic biodegradable. In times of nutrient scarcity, the small molecular compounds created by EPS breakdown can be utilized as carbon and energy sources for cell development. Sludge flocs can be deflocculated as a result of EPS breakdown. The nondegradable part of EPS may mix with reactor waste, lowering the effluent's prominence.

EPS in microbial cell aggregates have a polar group and several functional groups (e.g., hydroxyl carboxyl, sulfhydryl, phenolic, and phosphoric groups) (e.g., aliphatic in proteins aromatics and hydrophobic parts in the carbohydrates). The development of hydrophobic regions in the EPS would be advantageous for organic contaminant adsorption (Spath et al., 1998). The presence of hydrophobic and hydrophilic groups specifies that EPS are the amphoteric compounds. The composition of EPS is based on the relative ratio of two groups. Jorand et al. [31], to separate the hydrophobic and hydrophilic EPS fractions by using XAD resin, found that nearly 7% were the hydrophobic EPS and mostly included proteins, while the hydrophilic portion chiefly comprised of carbohydrates. The analysis of amino acid and monosaccharide contents in EPS revealed that about 25% of the amino acids were negatively charged and about 24% were hydrophobic after hydrolysis. The hydrophobicity or hydrophilicity of the EPS is likely to expressively impact the hydrophobicity of the microbial cell aggregates and their development in the bioreactors [41]. It also determines the significance of the EPS as the sorption places for organic contaminants [18].

5 Chemical Composition of Extracellular Polymeric Substances

Chemical composition of EPS varies in different microbes. The general composition of EPS is given in the Fig. 2. The quantity and chemical composition of EPS depend upon the kind of microorganisms, age of the biofilms, and ecological actors during biofilm formation. It has been discovered, for example, that EPS productivity is high during the early phases of biofilm development [67]. Under difficult conditions, EPS production is generally significantly increased. For example, [30] discovered that the carbohydrate-to-protein ratios for acidophilic microbial biofilms were much greater than previously reported ratios. To some extent, EPS production can reflect the physiological status of biofilms.

5.1 Carbohydrates

Carbohydrates are the most important component of EPS. Neutral carbohydrates (mainly hexose, seldom pentose) and uronic acids make up the microbial secretions (galacturonic, glucuronic, and mannuronic acids). The nature of EPS macromolecules (anionic, cationic, neutral) can be determined by the presence of these components. Exopolysaccharides are polymers with molecular weights ranging from 500 to 2000 kDa. Exopolysaccharides are long, linear, or branching substances found in bacteria [20]. They have been termed as "adhesive polymers" due to their adhesive and cohesive interactions and play an important role in stabilizing the structural

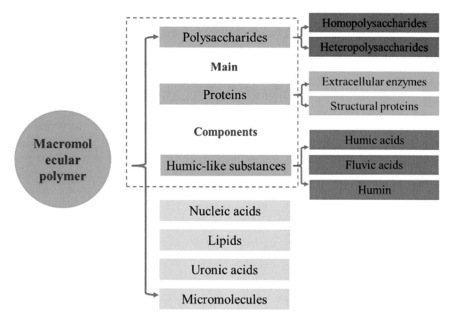

Fig. 2 The chemical composition of EPS (Adapted from [67])

integrity of biofilms. They are of two types, either homo-polysaccharides or hetero-polysaccharides. Homo-polysaccharides, in general, are neutral and made up of only one monosaccharide, such as L-fructose or D-glucose. Homo-polysaccharides are divided into three categories. The first group consists of Leuconostoc mesenteroides-produced a-D-glucans. The second group consists of b-D-glucan produced by the bacteria such as *Streptococcus* sp. and *Pediococcus* sp., and the third group consists of fructans which are produced by *Streptococcus salivarius*. Homo-polysaccharides include Dextran, Cellulose, and Curdlan, to name a few. Lactic acid bacteria produce hetero-polysaccharides, which are made up of repeating units of monosaccharides like D-glucose, L-fructose, and D-galactose. The bonding nature between monosaccharides units and the branching of the chains determine the physical properties of microbial hetero-polysaccharides. Most hetero-polysaccharides are also succinate, pyruvate, and formate process substitutes. Alginate is a well-studied hetero-polysaccharide that plays a role in the synthesis of EPS in pathogens like *Azotobacter* and *Pseudomonas aeruginosa* [20].

5.2 Protein

EPS contain proteins such as enzymatic proteins and structural proteins. Numerous enzymes have been detected in EPS matrix, many of which are involved in the EPS degradation. Enzymatic proteins play an important role in metabolism and act as an efficient external digestive system [20]. EPS can be successfully degraded by enzymes released by bacteria during hunger. These enzymes can attack EPS from the same bacteria or from different species [20]. *Bacillus subtilis* makes peptide antibiotics that are resistant to protease hydrolysis. The formation of polysaccharides matrix network is aided by EPS structural proteins, which serve as a link between the extracellular and bacterial surfaces [20]. For example, lectin-like protein is found in the matrix of activated sludge flocs that aids in the formation of flocs via bacterial aggregation [57]. Sugar and protein are covalently linked to produce glycoproteins. They aid in signaling and intercellular communications structural integrity and stability. In prokaryotes, there are two forms of glycosylation: N-glycosylation and O-glycosylation. Membrane-associated, surface-layered, cellular, and secretory glycoproteins are the four types of glycoproteins found in bacterial cells. Glycoproteins secreted by gram-positive bacteria have been discovered (e.g., *Bacillus macerans*, *Bacillus amyloliquefaciens*, *Cellulomonas fimi*, and *Clostridium acetobutylicum*) and gram-negative bacteria (e.g., *Mycobacterium tuberculosis*, *Flavobacterium meningosepticum*, and *Mycobacterium bovis*). Microbial adhesion to various solid surfaces is regulated by structural proteins. Electrostatic forces between charged proteins in EPS, according to [35], could contribute cohesive stability to the biofilm matrix.

5.3   Extracellular DNA

Extracellular DNA has been discovered in EPS from a variety of sources (e.g., soils, wastewater sludge, and marine sediments) (E-DNA). It is a key component of Staphylococcus aureus' EPS matrix, but just a small component of *Staphylococcus epidermidis'* EPS [20]. E-DNA is the part of the EPS matrix of *Bacillus cereus* [44]. The export of competence-signaling peptides to the medium is connected to E-DNA in the EPS matrix of *Streptococcus mutans*, which aids horizontal gene transfer.

5.4   Lipids and Surfactant

The polysaccharides-linked methyl and acetyl groups were found to be responsible for the hydrophobic properties of EPS. Lipids and lipid derivatives are also present in the EPS matrix. Adhesion is aided by lipopolysaccharides (e.g., attachment of Thiobacillus *ferrooxidans* to pyrite surfaces) [20]. Biosurfactants come in a variety of forms (viscosin, surfactin, and emulsion). They are vital for bacterial adhesion and dissociation from oil droplets and can have antibacterial and antifungal properties. Rhamnolipid (glycolipid) found in the EPS matrix of *Pseudomonas aeruginosa* is an example of biosurfactants that aid colonization and resource competition [44].

5.5 Humic Substances

Although humic substances are not secreted by microbes, however, they are the essential part of EPS. EPS matrix adsorbed these substances from the natural environment (e.g., soils, wastewater sludge). They can influence the EPS properties, for instance, biodegradability and adsorption ability [70]. These chemicals make up around 30–80% of total dissolved organic matter in natural aquatic systems. Humin, humic acids, and fulvic acids are the three components that make up humin. The majority of humic compounds are composed of humic acids. The alkali-soluble humus fragments are represented by humic acids and fulvic acids, whereas the insoluble remainder is represented by humin.

6 Mechanism of Degradation

The extracellular mechanism of degradation of dyes existed by nonspecific cytoplasmic enzymes present in the extracellular substances such as reductive enzymes (azoreductase, laccase, and peroxidases) through chemical reduction by transferring the electrons from reduced flavin group to the molecule of dye. Keck et al. [36], elaborated those redox mediators help to transport azo dye molecules from the membrane of bacterial cells, during the metabolic activity these mediators are quinone dependent. It was confirmed by many scientists from University of Stuttgart, Germany that cytoplasmic extracellular bacterial azoreductase involved in the in vivo degradation of azo dyes. Stolz et al. [66], described that *Sphingomonas xenophaga*, a novel gram-negative bacterium, has the ability to "consume foreign compound." They had explored that NADPH-dependent flavin oxidoreductases can be cloned and overexpressed in *S. xenophaga* and *E. coli* strains. The azoreductase activity confirmed by extracts shows that they have improved their ability to degrade the dyes. Russ et al. [59], also explored the evidence that aerobic azoreductases have higher flavin-dependent reductase activities. Mustafa et al. [51], explored a novel strain *Klebsiella pneumoniae* involved in the reduction of azo disperse blue 284 dye and it was confirmed that azoreductase gene sequence is present in its genomic DNA. The modification in the color, toxicity, and high inorganic load of textile industrial wastewater is due to dyes. When this toxic, highly inorganic load containing water enters into the natural sources of water, they not only destroy the quality of water but also destroy the marine life present there. During the last few decades, textile industrial wastewaters have been treated with a potentially accepted technology known as biosorption. The bacteria have high biosorption capacity due to the presence of hetero-polysaccharide and lipids EPS in the cell wall that consist up of diverse amount of functional groups for example, hydroxyl, amino, phosphate, and carboxyl, etc., that is mainly responsible for attraction between the cell wall and the dye. Today several studies have shown the applicability of EPS as a biosorbent material. There are several species of bacteria like *Exiguobacterium, Staphylococcus Bacillus, Klebsiella*, and *Pseudomonas* that have the ability to use EPS in the removal of textile dye, while using EPS as a biosorbent element the efficiency in decolorization was 99.9%. EPS give 93% efficiency in degradation, while working with basic fuchsine and 28% efficiency with chromium (VI) of 280 mg/L. The efficiency of the decolorization or biosorbents depends on a lot of factors like EPS concentration, temperature, contact time, pH, and concentration of dye, ionic strength, microorganism, and structure of dye. EPS produced by *Bacillus species* were studied by [27] for removal of a basic brown dye at altered pH and temperature. They studied that with the increase in temperature the rate of dye adsorption increase, whereas with the reduction in the rate of temperature the adsorption capacity also reduces. The pH 5 was count as the maximum pH at which dye adsorption occurs effectively. The EPS formed by *Klebsiella mobilis* in the dairy wastewater were competent in flocculating aqueous solutions containing disperse dyes and have degradation efficiency of about 91% for disperse violet dye [69]. In case of hazardous dyes, the EPS produced by *Proteus mirabilis* have proved effective in degradation of basic blue 54 in batch systems from aqueous solution. The rate of uptake observed by EPS is 2.005 g/g which was observed while using basic blue 54. According to [12], the adsorption of dye occurs due to the existence of a huge number of binding sites of high molecular weight present in EPS matrix along with stronger van der Waals forces, which attaches with the molecules of dye. As a consequence of this EPS matrix was entirely rooted in adsorbed molecules of dye. The consortiums of containing bacteria hold high dye degradation efficiency on comparison to the single bacterium. Costa et al. [10] studied that the EPS formed by a consortium of *Pseudomonas* sp. and *Staphylococcus* sp. had capability of printing dyeing and treating wastewater. Kamath Miyar et al. [33] had studied that the EPS obtained from the backwashed mud exposed high decolorization efficiency of 82% in fast blue aqueous and malachite green solutions. Overall, the EPS exposed the probability to become a good biosorbent for degradation and improvement of dye from diverse wastewaters and solutions. A large number of researchers are determined on developing strategies to employ EPS for biosorption. The information about the biosorption using EPS is expected to rise in near future as it is a striking cost-efficient treatment choice.

7   Summary and Future Perspectives

Most bacteria that dwell in contaminated environments can develop biofilms to help them survive. Because of their complex EPS matrix structure, intercellular gene transfer, quorum sensing, and chemotaxis features, they have been shown to be feasible candidates for bioremediation of a variety of harmful and resistant substances. Numerous studies have found that bacterial biofilms made up of single or multiple species in various environments can be successfully used in in-situ and ex-situ bioremediation procedures. This chapter aims to offer a comprehensive knowledge of the creation, growth, and many uses of biofilms in the bioremediation of contaminated settings, in light of the rising interest in biofilm research across the world. More study into successful gene transfer, the creation of genetic engineering biofilms, and the correct examination of bioremediation of aquatic and terrestrial systems should result in a long-term green solution for pollution management. Bacterial fauna that dwell in a contaminated environment have the potential to create a biofilm in order to survive. Because of their sophisticated organized EPS matrix, intercellular gene transfer, quorum sensing, and chemotaxis capabilities, they have proven to be a potential competitor for the bioremediation of various harmful recalcitrant substances. Several studies have found that bacterial biofilms including single or multiple species from various habitats can be successfully exploited for both in-situ and ex-situ bioremediation techniques. This chapter should have offered a solid grasp of biofilm formation, development, and many aspects of biofilm application for bioremediation of a contaminated environment, given the growing interest in biofilm research around the world. More study into effective gene transfer for producing a genetically engineered biofilm, as well as adequate bioremediation for both aquatic and terrestrial systems, could lead to a long-term green pollution control solution.

References

1. Akanya S, Krishna AR, Chockalingam J, Balakrishnan L, Janarthanan NT, Ramachandran P, Palanisami SD, Dhanaraj SS, Joseph S, Jayalekshmi SK (2020) Extraction of exopolysaccharide from enterococcus faecium grdaa and the application of its silver nanoparticles in degradation of azo dyes. J Adv Sci Res 11:152–157
2. Bourven I, Costa G, Guibaud G (2012) Qualitative characterization of the protein fraction of exopolymeric substances (EPS) extracted with EDTA from sludge. Biores Technol 104:486–496
3. Cadieux J, Kuzio J, Milazzo F, Kropinski A (1983) Spontaneous release of lipopolysaccharide by Pseudomonas aeruginosa. J Bacteriol 155:817–825
4. Cania B, Vestergaard G, Krauss M, Fliessbach A, Schloter M, Schulz S (2019) A long-term field experiment demonstrates the influence of tillage on the bacterial potential to produce soil structure-stabilizing agents such as exopolysaccharides and lipopolysaccharides. Environ Microbiome 14:1–14
5. Cao B, Zhang W, Du Y, Wang R, Usher SP, Scales PJ, Wang D (2018) Compartmentalization of extracellular polymeric substances (EPS) solubilization and cake microstructure in relation to wastewater sludge dewatering behavior assisted by horizontal electric field: effect of operating conditions. Water Res 130:363–375
6. Carmen Z, Daniela S (2012) Textile organic dyes-characteristics, polluting effects and separation/elimination procedures from industrial effluents-a critical overview (IntechOpen). https://doi.org/10.5772/32373
7. Caudan C, Filali A, Spérandio M, Girbal-Neuhauser E (2014) Multiple EPS interactions involved in the cohesion and structure of aerobic granules. Chemosphere 117:262–270
8. Cescutti P, Toffanin R, Pollesello P, Sutherland IW (1999) Structural determination of the acidic exopolysaccharide produced by a pseudomonas sp. strain 1.15. Carbohyd Res 315:159–168
9. Comte S, Guibaud G, Baudu M (2006) Relations between extraction protocols for activated sludge extracellular polymeric substances (EPS) and EPS complexation properties: part I. comparison of the efficiency of eight EPS extraction methods. Enzyme Microb Technol 38:237–245

10. Costa OYA, Raaijmakers JM, Kuramae EE (2018) Microbial extracellular polymeric substances: ecological function and impact on soil aggregation. Front Microbiol 9:1636. https://doi.org/10.3389/fmicb.2018.01636
11. Costerton J, Irvin R, Cheng K (1981) The bacterial glycocalyx in nature and disease. Ann Rev Microbiol 35:299–324
12. Decho AW, Gutierrez T (2017) Microbial extracellular polymeric substances (EPSs) in ocean systems. Front Microbiol 8:922. https://doi.org/10.3389/fmicb.2017.00922
13. Ding Y, Tian Y, Li Z, Zuo W, Zhang J (2015) A comprehensive study into fouling properties of extracellular polymeric substance (EPS) extracted from bulk sludge and cake sludge in a mesophilic anaerobic membrane bioreactor. Biores Technol 192:105–114
14. Drakou E-M, Amorim CL, Castro PM, Panagiotou F, Vyrides I (2018) Wastewater valorization by pure bacterial cultures to extracellular polymeric substances (EPS) with high emulsifying potential and flocculation activities. Waste Biomass Valorization 9:2557–2564
15. Drulis-Kawa Z (2020) Editorial bacterial surface glycans as the virulence agent and the target for predators, therapy, and the immune system. Front Microbiol 11:3211
16. Duman O, Tunç S, Polat TG, Bozoğlan BK (2016) Synthesis of magnetic oxidized multiwalled carbon nanotube-κ-carrageenan-Fe_3O_4 nanocomposite adsorbent and its application in cationic Methylene Blue dye adsorption. Carbohyd Polym 147:79–88
17. Esparza-Soto M, Westerhoff P (2003) Biosorption of humic and fulvic acids to live activated sludge biomass. Water Res 37:2301–2310
18. Fleming ZL, Monks PS, Rickard AR, Heard DE, Bloss WJ, Seakins PW, Still T, Sommariva R, Pilling MJ, Morgan R (2006) Peroxy radical chemistry and the control of ozone photochemistry at Mace Head, Ireland during the summer of 2002. Atmos Chem Phys 6:2193–2214
19. Flemming H-C, Wingender J (2001) Relevance of microbial extracellular polymeric substances (EPSs)-part I: structural and ecological aspects. Water Sci Technol 43:1–8
20. Flemming H-C, Wingender J (2010) The biofilm matrix. Nat Rev Microbiol 8:623–633
21. Geyik AG, Kılıç B, Çeçen F (2016) Extracellular polymeric substances (EPS) and surface properties of activated sludges: effect of organic carbon sources. Environ Sci Pollut Res 23:1653–1663
22. Góes MM, Keller M, Oliveira VM, Villalobos LDG, Moraes JCG, Carvalho GM (2016) Polyurethane foams synthesized from cellulose-based wastes: kinetics studies of dye adsorption. Ind Crops Prod 85:149–158
23. Gotoh Y, Maruo T, Tanaka K, Ohashi S, Yoshida K-I, Suzuki T (2021) Loss of the intrinsic plasmid-encoded eps genes in Lactococcus lactis subsp. cremoris FC grown at elevated temperature abolishes exopolysaccharide biosynthesis. Food Sci Technol Res 27:241–248
24. Guibaud G, van Hullebusch E, Bordas F (2006) Lead and cadmium biosorption by extracellular polymeric substances (EPS) extracted from activated sludges: pH-sorption edge tests and mathematical equilibrium modelling. Chemosphere 64:1955–1962
25. Ha J, Gélabert A, Spormann AM, Brown GE Jr (2010) Role of extracellular polymeric substances in metal ion complexation on Shewanella oneidensis: batch uptake, thermodynamic modeling, ATR-FTIR, and EXAFS study. Geochim Cosmochim Acta 74:1–15
26. Hu Y-Q, Wei W, Gao M, Zhou Y, Wang G-X, Zhang Y (2019) Effect of pure oxygen aeration on extracellular polymeric substances (EPS) of activated sludge treating saline wastewater. Process Saf Environ Prot 123:344–350
27. Inbaraj BS, Chiu CP, Ho GH, Yang J, Chen BH (2008) Effects of temperature and pH on adsorption of basic brown 1 by the bacterial biopolymer poly(γ-glutamic acid). Bioresour Technol 99(5):1026–1035
28. Ishak SA, Murshed MF, Md Akil H, Ismail N, Md Rasib SZ, Al-Gheethi AAS (2020) The application of modified natural polymers in toxicant dye compounds wastewater: a review. Water 12:2032
29. Jia C, Li X, Zhang L, Francis D, Tai P, Gong Z, Liu W (2017) Extracellular polymeric substances from a fungus are more effective than those from a bacterium in polycyclic aromatic hydrocarbon biodegradation. Water Air Soil Pollut 228:195

30. Jiao Y, Cody GD, Harding AK, Wilmes P, Schrenk M, Wheeler KE, Banfield JF, Thelen MP (2010) Characterization of extracellular polymeric substances from acidophilic microbial biofilms. Appl Environ Microbiol 76(9):2916–2922. https://doi.org/10.1128/AEM.02289-09

31. Jorand F, Boué-Bigne F, Block JC, Urbain V (1998) Hydrophobic/hydrophilic properties of activated sludge exopolymeric substances. Water Sci Technol 37:307–315

32. Joshi PM, Juwarkar AA (2009) In vivo studies to elucidate the role of extracellular polymeric substances from Azotobacter in immobilization of heavy metals. Environ Sci Technol 43:5884–5889

33. Kamath MH, Pai A, Goveas LC (2021) Adsorption of malachite green by extracellular polymeric substance of Lysinibacillus sp. SS1: kinetics and isotherms. Heliyon 7(6):e07169. https://doi.org/10.1016/j.heliyon.2021.e07169

34. Kanmani P, Aravind J, Kamaraj M, Sureshbabu P, Karthikeyan S (2017) Environmental applications of chitosan and cellulosic biopolymers: a comprehensive outlook. Biores Technol 242:295–303

35. Karunakaran E, Biggs CA (2010) Mechanisms of Bacillus cereus biofilm formation: an investigation of the physicochemical characteristics of cell surfaces and extracellular proteins. Appl Microbiol Biotechnol 89(4):1161–1175. https://doi.org/10.1007/s00253-010-2919-2

36. Keck A, Klein J, Kudlich M, Stolz A, Knackmuss H-J, Mattes R (1997) Reduction of azo dyes by redox mediators originating in the naphthalenesulfonic acid degradation pathway of Sphingomonas sp. strain BN6. Appl Environ Microbiol 63:3684–3690

37. Kurade MB, Murugesan K, Selvam A, Yu S-M, Wong JW (2014) Ferric biogenic flocculant produced by Acidithiobacillus ferrooxidans enable rapid dewaterability of municipal sewage sludge: a comparison with commercial cationic polymer. Int Biodeterior Biodegradation 96:105–111

38. Laspidou CS, Rittmann BE (2002) A unified theory for extracellular polymeric substances, soluble microbial products, and active and inert biomass. Water Res 36:2711–2720

39. Li R, Ning X-A, Sun J, Wang Y, Liang J, Lin M, Zhang Y (2015) Decolorization and biodegradation of the Congo red by Acinetobacter baumannii YNWH 226 and its polymer production's flocculation and dewatering potential. Biores Technol 194:233–239

40. Liu S, Feng X, Gu F, Li X, Wang Y (2017) Sequential reduction/oxidation of azo dyes in a three-dimensional biofilm electrode reactor. Chemosphere 186:287–294

41. Liu Y, Fang HH (2003) Influences of extracellular polymeric substances (EPS) on flocculation, settling, and dewatering of activated sludge

42. Liu Y, Lam M, Fang H (2001) Adsorption of heavy metals by EPS of activated sludge. Water Sci Technol 43:59–66

43. Lorenz MG, Wackernagel W (1994) Bacterial gene transfer by natural genetic transformation in the environment. Microbiol Rev 58:563–602

44. Marvasi M, Visscher PT, Martinez LC (2010) Exopolymeric substances (EPS) from Bacillus subtilis : polymers and genes encoding their synthesis. FEMS Microbiol Lett 313(1):1–9

45. McDougald D, Rice SA, Barraud N, Steinberg PD, Kjelleberg S (2012) Should we stay or should we go: mechanisms and ecological consequences for biofilm dispersal. Nat Rev Microbiol 10:39–50

46. Miao L, Zhang Q, Wang S, Li B, Wang Z, Zhang S, Zhang M, Peng Y (2018) Characterization of EPS compositions and microbial community in an Anammox SBBR system treating landfill leachate. Biores Technol 249:108–116

47. Mohapatra RK, Behera SS, Patra JK, Thatoi H, Parhi PK (2020) Potential application of bacterial biofilm for bioremediation of toxic heavy metals and dye-contaminated environments. In: New and future developments in microbial biotechnology and bioengineering: microbial biofilms (Elsevier), pp 267–281

48. Moon DH, Dermatas D (2006) An evaluation of lead leachability from stabilized/solidified soils under modified semi-dynamic leaching conditions. Eng Geol 85:67–74

49. More TT, Yadav JSS, Yan S, Tyagi RD, Surampalli RY (2014) Extracellular polymeric substances of bacteria and their potential environmental applications. J Environ Manage 144:1–25

50. Mosharaf M, Tanvir M, Haque M, Haque M, Khan M, Molla AH, Alam MZ, Islam M, Talukder M (2018) Metal-adapted bacteria isolated from wastewaters produce biofilms by expressing proteinaceous curli fimbriae and cellulose nanofibers. Front Microbiol 9:1334

51. Mustafa G, Zahid MT, Ali S, Abbas SZ, Rafatullah M (2021) Biodegradation and discoloration of disperse blue-284 textile dye by Klebsiella pneumoniae GM-04 bacterial isolate. J King Saud Univ Sci 101442

52. Ngulube T, Gumbo JR, Masindi V, Maity A (2017) An update on synthetic dyes adsorption onto clay based minerals: a state-of-art review. J Environ Manage 191:35–57

53. Nielsen PH, Jahn A (1999) Extraction of EPS. In: Microbial extracellular polymeric substances (Springer), pp 49–72

54. Nielsen PH, Jahn A, Palmgren R (1997) Conceptual model for production and composition of exopolymers in biofilms. Water Sci Technol 36:11–19

55. Ning X-A, Yang C, Wang Y, Yang Z, Wang J, Li R (2014) Decolorization and biodegradation of the azo dye Congo red by an isolated Acinetobacter baumannii YNWH 226. Biotechnol Bioprocess Eng 19:687–695

56. Pan X, Liu J, Zhang D, Chen X, Li L, Song W, Yang J (2010) A comparison of five extraction methods for extracellular polymeric substances (EPS) from biofilm by using three dimensional excitation-emission matrix (3DEEM) fluorescence spectroscopy. Water SA 36:111–116

57. Park C, Novak JT (2007) Characterization of activated sludge exocellular polymers using several cation-associated extraction methods. Water Res 41:1679–1688

58. Rembe JD, Huelsboemer L, Plattfaut I, Besser M, Stuermer EK (2020) Antimicrobial hypochlorous wound irrigation solutions demonstrate lower anti-biofilm efficacy against bacterial biofilm in a complex in-vitro human plasma biofilm model (hpBIOM) than common wound antimicrobials. Front microbiol 11:564513

59. Russ R, Rau JR, Stolz A (2000) The function of cytoplasmic flavin reductases in the reduction of azo dyes by bacteria. Appl Environ Microbiol 66:1429–1434

60. Salama Y, Chennaoui M, Sylla A, Mountadar M, Rihani M, Assobhei O (2016) Characterization, structure, and function of extracellular polymeric substances (EPS) of microbial biofilm in biological wastewater treatment systems: a review. Desalin Water Treat 57:16220–16237

61. Sarayu K, Sandhya S (2010) Aerobic biodegradation pathway for Remazol Orange by Pseudomonas aeruginosa. Appl Biochem Biotechnol 160:1241–1253

62. Sheng GP, Yu HQ, Li XY (2006) Stability of sludge flocs under shear conditions: roles of extracellular polymeric substances (EPS). Biotechnol Bioeng 93:1095–1102

63. Sheng G-P, Yu H-Q, Li X-Y (2010) Extracellular polymeric substances (EPS) of microbial aggregates in biological wastewater treatment systems: a review. Biotechnol Adv 28:882–894

64. Spaeth R, Wuertz S (2000) Extraction and quantification of extracellular polymeric substances from wastewater. Biofilms: investigative methods and applications. Technomic Publishers, Lancaster, pp 51–68

65. Späth R, Flemming H-C, Wuertz S (1998) Sorption properties of biofilms. Water Sci Technol 37:207–210

66. Stolz A, Schmidt-Maag C, Denner E, Busse H, Egli T, Kämpfer P (2000) Description of Sphingomonas xenophaga sp. Nov. for strains BN6T and N, N which degrade xenobiotic aromatic compounds. Int J Syst Evol Microbiol 50:35–41

67. Tian X, Shen Z, Han Z, Zhou Y (2019) The effect of extracellular polymeric substances on exogenous highly toxic compounds in biological wastewater treatment: an overview. Bioresour Technol Rep 5:28–42

68. Wang L, Li J (2013) Adsorption of CI reactive red 228 dye from aqueous solution by modified cellulose from flax shive: kinetics, equilibrium, and thermodynamics. Ind Crops Prod 42:153–158

69. Wang Z-W, Liu Y, Tay J-H (2007) Biodegradability of extracellular polymeric substances produced by aerobic granules. Appl Microbiol Biotechnol 74:462–466

70. Wingender J, Neu TR, Flemming H-C (1999) What are bacterial extracellular polymeric substances? In: Microbial extracellular polymeric substances (Springer), pp 1–19

71. Xu Y, Li Z, Su K, Fan T, Cao L (2018) Mussel-inspired modification of PPS membrane to separate and remove the dyes from the wastewater. Chem Eng J 341:371–382
72. Yadav M, Yadav H (2015) Applications of ligninolytic enzymes to pollutants, wastewater, dyes, soil, coal, paper and polymers. Environ Chem Lett 13:309–318
73. Zhang W, Cao B, Wang D, Ma T, Xia H, Yu D (2016) Influence of wastewater sludge treatment using combined peroxyacetic acid oxidation and inorganic coagulants re-flocculation on characteristics of extracellular polymeric substances (EPS). Water Res 88:728–739
74. Zhang X, Bishop PL (2003) Biodegradability of biofilm extracellular polymeric substances. Chemosphere 50:63–69
75. Zhao G, Ma F, Wei L, Chua H (2012) Using rice straw fermentation liquor to produce bioflocculants during an anaerobic dry fermentation process. Biores Technol 113:83–88

Polymer-Derived Electrospun Ceramic Nanofibers Adsorbents for Textile Wastewater Treatment

Abhipsa Mahapatra, Manamohan Tripathy, and G. Hota

1 Introduction

Contamination of water due to inadequate disposal of industrial and mining waste is a serious threat to the biological ecosystem [20, 24, 29]. Over the years, this problem has been exacerbated globally due to increasing wastewater and its continual release into water bodies. The threat of insufficiency of water across the world is unable to reach the minimum daily requirement [31]. Major countries of Asia and Africa will be facing acute drinkable water shortages soon. According to recent statistics, surface water pollution has increased 20 times since the industrial revolution [25]. From the reported literature it is found that about half the world's population will be vulnerable to water shortages by 2025 [28]. The major cause of such a rise could be attributed directly to various anthropogenic activities such as agriculture, animal husbandry, urbanization, industrialization, etc. [19]. Most of the contaminants that affect water quality are organic pollutants (dyes, pesticides, pharmaceutical, fertilizers, etc.), inorganic (heavy metal ions, metal oxides, salts, etc.), pathogens, or particulates [1, 4, 5, 8, 9, 47]. Among them, organic pollutants have been demanded focus due to their several utilization in industries and day to day life, which, when discharged to aquatic environment or earth surface, has a longer persistence and negative impact on the biological ecosystem [7, 12].

Dyeing wastewater is one of the most hazardous wastewaters that is associated with water pollution because of its ability to change the quality of water even in minute concentrations. Dyeing wastewater usually results from many industries such as textile, cosmetic, tannery, photographs, food, and plastic industries. Dyeing or coloring products are used in industries, such as plastic, paper, leather, and textile. Among all, the textile industrial segment in India is spearheading at an alarming

A. Mahapatra · M. Tripathy · G. Hota (✉)
Department of Chemistry, NIT Rourkela, Rourkela, Odisha 769008, India
e-mail: garud@nitrkl.ac.in

© The Author(s), under exclusive license to Springer Nature Singapore Pte Ltd. 2022   193
A. Khadir and S. S. Muthu (eds.), *Polymer Technology in Dye-containing Wastewater*,
Sustainable Textiles: Production, Processing, Manufacturing & Chemistry,
https://doi.org/10.1007/978-981-19-0886-6_8

rate [21]. The textile industry is also known for its longest most complicated industrial chains in the manufacturing industry. Although the main issue inside the textile industry is the extensive water consumption, a significant amount of colored wastewater is generated, during the coloration process, which greatly endangers human and aquatic life [38]. The waste dyes can hinder light penetration and intervene in the photosynthetic process that is crucial for living beings [18]. It can also increase biochemical (BOD) and chemical (COD) oxygen demand in water bodies. The dyes mostly discharged from industries as pollutants have very complex composition, high concentration, high color, and contains a variety of toxic properties like carcinogenic, teratogenic, and mutagenic which are detrimental for all forms of living organisms [26]. In general, dyes can be defined as organic compounds that contain two key components: chromophores (responsible for the color of dye) and auxochromes (responsible for color intensity). As reported by the United States "Color Index," commodity dyes have touched around tens of thousands. Each year approximately 60,000 tons of dyes are released into the environment, out of which 80% are azo dyes [31]. In this context, various treatment methods (chemical, physical, biological, or hybrid) have been effective in removing color and the process of adsorption, in particular, has gained much prominence [6].

Adsorption is the process of separating and detoxifying toxic pollutants, compared with conventional methods it possess superior applicability because of the ease of operation, eco-friendly nature, and high efficiency [45, 46]. The adsorbate can be released and recycled easily. Inspite of elimination from the environment, the process of adsorption transfers pollutants from one medium to another. Figure 1 depicts how the adsorption process takes place on the adsorbent surface. In general, the adsorbate adsorbs on the adsorbent surface through chemisorption or physisorption as per the properties of the adsorbent and adsorbate. Therefore, it is believed that adsorption is a physicochemical process [33].

Adsorption efficiency depends not only on chemical properties but also shares a close relationship with the structure and morphologies of the adsorbents. Several nanostructures have been used as adsorbents but lately, reports on metal oxide nanostructure and their hybrid materials for dye removal are on the rise significantly. Few exceptional properties of metal oxide or ceramic nanostructure which

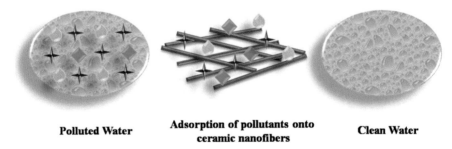

Polluted Water   **Adsorption of pollutants onto ceramic nanofibers**   **Clean Water**

Fig. 1 Diagrammatic illustration of adsorptive removal of contaminants by porous ceramic nanofibers

Table 1 Literature review of ceramic nanofibers and nanocomposite in the field of removal of dyes in wastewater treatment

Sl. No	Ceramic nanofiber/nanocomposite	Polymer in nanofiber	Dye studied	References
1	ZnO–CuO	–	Methylene blue	[41]
2	γ-Al$_2$O$_3$	–	Congo red, Methyl blue	[39]
3	α-Fe$_2$O$_3$	PVA	Methyl orange	[17]
4	TiO$_2$-SiO$_2$	PEO	Methylene blue	[52]
5	ZnO/SnO$_2$	PVP	Methyl orange, Methylene blue, Congo red, eosin red	[10]
6	CeO$_2$	PVP	Methyl orange	[53]
7	CuO-ZnO	PVA	Congo red	[34]
8	Alumina	PVP	Methyl orange	[27]

PVA polyvinyl alcohol, *PEO* polyethylene oxide, *PVP* polyvinyl pyrrolidone

make them potent in dye removal are small size, high surface area to volume ratio, greater porosity, free surface active sites, and their surface energies [13]. As previously reported, the temperature has a significant influence on dye removal efficiency of nanofibrous adsorbent which implies the endothermic nature of the adsorption process [22]. In basic conditions and high temperature, 85% highest adsorption capacity is found out nevertheless of the adsorbent composition. For industrial wastewater, dyeing processes the environmental conditions are more basic and high temperature.

Based on the previous literature review, in this chapter, we have focused on electrospun ceramic nanofibers (NFs) and also its nanocomposite for adsorption studies of dye removal. As per literature ceramic, NFs cannot be fabricated without the precursor solution, which is nothing but the polymer. Polymers help in binding the nanofibers with each other. It improves the mechanical strength and stability of the fiber membrane. Further, the obtained nanocomposite has great potential to assist with the difficulty of dye pollution in wastewater treatment. Table 1 briefly enlists the applications of various polymer-derived ceramic NFs and nanocomposites in dye wastewater treatment.

2 Outline of the Electrospinning Process

Electrospinning (E-spinning) is a simple and unique process that is specific for the fabrication of the perfect morphology of nanofiber. The origin of E-spinning could be traced way back to 1930. However, at the end of the twentieth century, noticeable advancements occur for the instrumentation and working parameters [43]. This technique is highly versatile, where the surface properties, fiber orientation, and morphology are controlled by controlling the solution properties. The fundamental of fiber formation is "**electrostatic attraction/interactions**" of charges. The instrument consists of three fragments: firstly, a high voltage DC power supply; secondly, a combination of syringe pump & with a syringe having a metallic needle, and thirdly, a metallic collector (Fig. 2). The solution should be at such a density that the self-repulsion of charges objects the liquid to stretch into a fiber in an electric field [44].

The basic principle of E-spinning is that the polymer drops into the needle tip controls surface tension to form a jet spray and then the solvent is evaporated to form ultrafine fibers. Several critical parameters are to be managed during the process of electrospinning as given in Table 2. During the process, if the voltage reaches the critical voltage, a "Taylor cone" is formed. On the further rise in voltage to form a jet, it forms nanofibers. The very high voltage causes a reduction in fiber collection efficiency. The production of porous, smooth, and defect-free nanofibrous membranes is severely influenced by operating conditions and solution parameters. The formation

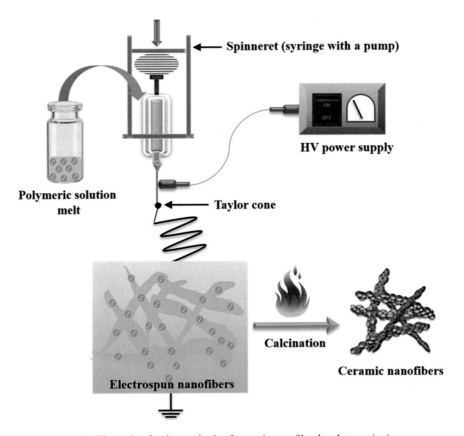

Fig. 2 Schematic illustration for the synthesis of ceramic nanofiber by electrospinning process

Table 2 Experimental components for electrospinning process

Solution parameters	Process parameters	Ambient parameters
Concentration	Electrostatic potential	Temperature
Viscosity	Electric field strength	Humidity
Surface tension	Electrostatic field shape	Local atmosphere flow
Conductivity	Working distance	Atmospheric composition
Dielectric constant	Feed rate	Pressure
Solvent volatility	Orifice diameter	

of fibers is also governed by the repulsive electrostatic force of attraction between the needle tip and the collector. The end architecture of the fiber is governed by electrostatic attraction in the charged fluid jet which results in jet instabilities, subsequently, the Taylor cone at needle tip will be formed [2].

The geometry of the collector is equally important. It can influence mat morphology, such as fiber alignment and pattern. Collectors are also of different types depending on the type of fiber production. The collector can be a stationary or rotating-type disk. Generally, on static collectors randomly oriented web fibers are collected and aligned fibers are collected on spinning substrates [42, 51]. Usually, an aluminum foil is wrapped on the collector surface which acts as the conducting material for the collection of the fibers. The type of E-spin instrument is based on the needle type in which the selection depends on instrumental parameters such as needle tip, solution volume, gauge diameter, and modified syringe. In industries, due to large-scale production needleless type of E-spinning is generally employed.

Apart from this due to progress in research, nowadays researchers also focused to develop a more convenient electrospinning setup for the fabrication of nanofibers matrix which possesses greater surface area and enhanced working performance. For the fabrication of high surface area hollow nanofibers, a setup was developed which contains dual syringe spinneret, known as coaxial spinneret, and the equipped setup is known as coaxial electrospinning setup (Fig. 3). This instrumental setup have two coaxial capillaries by which simultaneously two different polymeric solutions can be ejected to form a hybrid jet. This advancement helps for the fabrication of highly applicable nanofibers like nanotubes, core–shell fibers, and hollow nanofibers, etc. From the recent report, it is found that the formation of tubular hollow nanofibers by a single instrumental setup using sol–gel chemistry and electrospinning is mostly driven by an electro-hydrodynamic force, due to which a coaxial jet is formed. A high voltage DC power supply is applied to a pair of concentric needles that are used to inject two immiscible polymeric precursors solution which leads to the formation of a two-component liquid cone that elongates into coaxial liquid jets and forms hollow nanofibers [11, 48].

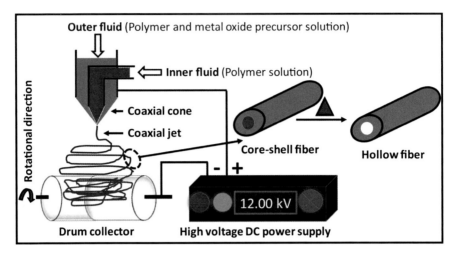

Fig. 3 Schematic representation of coaxial electrospinning setup

2.1 Nanofiber Generation by Electrospinning Process

E-spinning is the most efficient technique for the fabrication of continuous nanofibers. Nanofibers are included in the class of one-dimensional nanomaterial with a diameter range between 1 and 100 nm. The process of E-spinning is a typical arrangement of both electrospraying and conventional solutions of dry spinning. Due to their exceptional attributes such as very less diameter with outstanding pore connectivity, greater porosity, and high aspect ratio, nanofibers find immense potential in myriad applications. Apart from this, nanofibers possess enhanced thermal and mechanical behavior, and their surface can be readily functionalized, which results in adverse applicability [2]. In the E-spinning process, fiber formation is driven by repulsive electrostatic forces, where charged fluid jet dictates the end architecture of the fibers.

2.2 Electrospun Ceramic Nanofiber

In general, pristine ceramics are not spinnable and as per the reported literature, it is found that it required extremely high temperature for its formation. Therefore, it is always necessary that polymeric materials must be blended with the ceramic melt to synthesize suitable ceramic nanofibers [17]. Ceramic NFs were first formulated in the year 2002 by the combination of two fabrication methods viz. electrospinning and sol–gel. It is one of the most adaptable and industrially feasible methods, which can be adopted to fabricate ceramic fibers with a diameter range from nanometer to micrometer. Roughly ceramic NFs are of two types i.e., oxide nanofibers (Fe_2O_3,

Al_2O_3, TiO_2, SnO_2, CeO_2) and non-oxide nanofibers (carbides, nitrides, borides, silicates).

Fabrication of ceramic NFs is a distinctive method that comprises three major steps as shown in Fig. 4: a stable colloidal suspension (sol) is prepared by the sol–gel method by using a polymer and a solvent; then, by using electrospinning process fabrication of ceramic nanofiber is done; and lastly production of nanofibers matrix by selective elimination of the organic constituent by solvent extraction or calcinations [36, 49].

In the E-spinning process, a polymeric material plays a vital role. The polymer supplies a matrix for a ceramic nanofiber that is formed when the fibers are electro-spun with dissolved nanoparticle or metal oxide within the polymer solution. As most of the nanoparticles are already incorporated into the fibers, the fabricated nanofibers will be having a smooth surface. Moreover, the nanoparticles can be deposited on the surface of polymeric electrospun fibers [37]. Ceramic nanofibers have outstanding physico-chemical characteristics, which is proved to be good for water treatment applications. The electrostatic attraction, hydrophobic interactions, and hydrogen

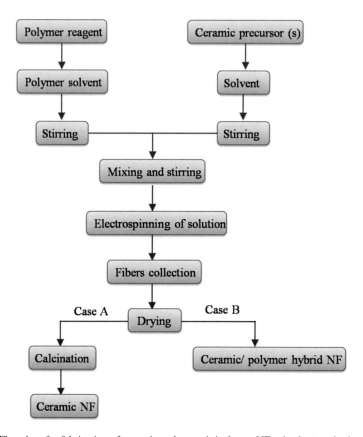

Fig. 4 Flowchart for fabrication of ceramic and ceramic/polymer NFs via electrospinning method

bonding between the nanofibers surface and dye molecule were found to dominate the adsorption process.

3 Adsorptive Application of Composite Nanofiber for Wastewater Treatment

Literature galore with numerous methods directed toward the remediation of organic contaminants from water based on their properties. As compared to various water treatment processes, it is always important to understand the different levels of contaminant elimination of polluted water. Several technologies viz. reverse osmosis, bioremediation, oxidation, electrolysis, etc., are best but these are expensive and also limit their utilization for adverse applications. Among all, adsorption is a highly effective and adaptable technique for water treatment applications [3]. In the last three decades, numerous advanced types of nanomaterials have been designed and fabricated in various shapes and sizes. For water treatment purposes metal oxide nanostructures materials possess greater concern due to their high surface area, semiconducting properties, and easy synthesis method. Also, electrospun composite nanofibers are better in adsorbing, and degrade dyes as compared to solo absorbents and promote convenient regeneration [40].

3.1 Cerium Oxide NFs as Adsorbent

As reported by [53], porous CeO_2 NFs with a diameter of 100–140 nm were fabricated by electrospinning process. Formation of CeO_2 NFs was carried out by using $Ce(NO_3)_3.6H_2O$/PVP reactant solution, and subsequent calcinations (550 °C). The CeO_2 porous nanofibers formation is mostly ascribed to the gas volatilization and the diffusion of the molecules. Based on the experimental analysis, a probable mechanism for the porous CeO_2 nanofibers formation is presented in Fig. 5. In the initial step (a), the solvents from the fibers' surface evaporate and the precursors of Ce decompose to form CeO_2. The $Ce(NO_3)_3$ present at the core of the fiber matrix is difficult to decay because of the thermal discrepancy, which causes a concentration gradient between $Ce(NO_3)_3$ and CeO_2 particles bidirectionally in the fiber cross-section. Therefore, in the second step (b) CeO_2 particles diffuse from the surface to the core and $Ce(NO_3)_3$ from the core to the surface. The production and diffusion of gases and the occurrence of the above activity are adequate for producing spaces between the nanocrystals [30]. Further increase in temperature the $Ce(NO_3)_3$ decompose completely. Apart from this, the undecomposed PVP inhibits the combination of CeO_2 and the folding of the nanostructure effectively (step (c)). Finally, after the complete decomposition of PVP, the CeO_2 porous nanofibers are achieved (step (d)).

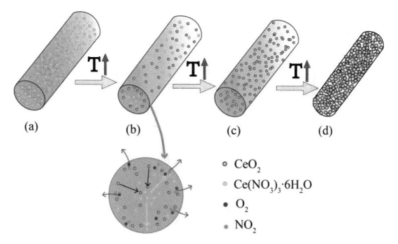

(a)　　(b)　　(c)　　(d)

○ CeO$_2$

○ Ce(NO$_3$)$_3$·6H$_2$O

● O$_2$

● NO$_2$

Fig. 5 Formation mechanism of the porous CeO$_2$ nanofibers

After successful synthesis, the materials were characterized suitably and applied as an efficient adsorbent toward the adsorption of methyl orange (MO). The adsorption efficiency for methyl orange removal was up to 50% within 5 min. The experimental data is best-fitted to the pseudo-second-order kinetics model. The porous CeO$_2$ NFs with greater specific surface area possess enhanced applicability toward water treatment [53].

3.2　Iron Oxide NFs as Adsorbent

One-dimensional magnetic nanofibers have novel properties. Iron-based NFs have greater magnetic properties, leading to unprecedented sorption capacity, and show a significant revolution in water treatment application. The shapes and sizes show substantial effects on the magnetic properties of iron oxide nanoparticles because of the variations in magnetic anisotropy. Studies by Ghasemi et al. show that magnetic α-Fe$_2$O$_3$ nanofibers were fabricated by E-spinning of PVA and Fe(NO$_3$)$_3$.9H$_2$O and subsequent calcinations [17]. The SEM image (Fig. 6) of the obtained nanofibers indicates that in calcination, the nanofibers bend and the surface becomes uneven, and the unbroken fibers show rupture axially, which is typically due to the decomposition of PVP during the calcination. After calcination, the fiber diameter decreases up to 50–90 nm.

These NFs are effective for the degradation and decolorization of the toxic methyl orange dye in aqueous solutions. The efficiency of decolorization is more than 99% in 10 min. Similarly, [16], have fabricated hollow α-Fe$_2$O$_3$ NFs made of rice-like nanorods via hydrothermal reaction on PVA nanofiber template followed by calcination [16]. Moreover, it shows efficient adsorption for methyl orange in water.

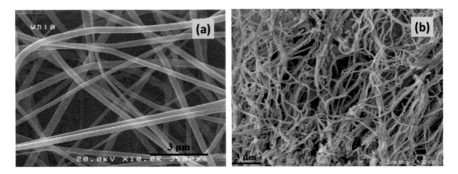

Fig. 6 SEM image of **a** PVA/Fe(NO$_3$)$_3$.9H$_2$O nanofibers and **b** α-Fe$_2$O$_3$ fibers

3.3 Titanium Dioxide Nanofiber as Adsorbent

TiO$_2$ NFs need no introduction in the world of wastewater remediation. TiO$_2$ NFs have excellent photoactivity, combined with high chemical stability and low cost. Several techniques have been adopted for the fabrication of pristine TiO$_2$ NFs. The NFs are considered good adsorbents for the removal of dye viz. Congo red, methyl orange, and methyl blue dye achieving rejections amount of 94%, 52%, and 92%, respectively. Marinho et al., 2020, have recently reviewed electrospun TiO$_2$ NFs and their applications in water and wastewater treatment [35].

3.4 Alumina NFs as Adsorbent

Alumina NF is an important ceramic material with numerous properties and applications. For the synthesis of alumina NFs, a combination of sol–gel and electrospinning methods is adopted. A series of phase transitions is seen during the process of calcination such as boehmite-γ-alumina-α-alumina from 500 to 1200 °C. Kim et al. [27] have also studied the adsorption property of methyl orange of the α- and γ-alumina both [27]. Similarly, [39], has fabricated porous hollow γ-alumina NFs by E-spinning and explored their potential application in environmental remediation [39]. According to the study, the prepared hollow and porous γ-alumina NFs were used as adsorbent to remediate most toxic dyes namely Congo red, methyl blue, and acid fuchsin from aqueous solution. The fabricated nanofibers possess enhanced removal efficacy of >90%.

3.5 Ceramic-Composite NFs/nanocomposite

Nanocomposites are studied as the result of binding two different components in nanoscale with a unique physical and chemical nature. Combining different materials not only incorporates their properties but also enhances the limited feature of a single species [50]. The composite nanoparticle shows significant applications. In this study, the CuO/PET nanocomposite is used for the removal of methylene blue, which was found to be 99% in just 30 min. The nanocomposite shows a rapid degradation rate for methylene blue. Electrospun reduced graphene oxide/poly(acrylonitrile-co-maleic acid) hybrid nanofibers were fabricated using the electrospinning method by Du et al. [14]. The nanocomposite is used to absorb and degrade malachite green and leucomalachite green from the water system, which shows 90.6% and 93.7% adsorption, respectively. It also exhibited good reusability for multiple cycles of operation for adsorption.

Nanocomposite with fine uniform domains is challenging because it is needed to control both the microstructural length scales as well as the elemental distributions. According to [32], mixed iron oxide-alumina nanocomposite can be applied as an adsorbent for decolorization of Congo red [32]. Three phases of mixed nanocomposite, i.e., as-prepared, 500 and 1000 °C have been compared and studied for removal. Among them, gamma-phase shows 100% removal of Congo red. Figure 7 depicts the UV–vis absorption spectra of Congo red solution with all three phases. Similarly, Fig. 8 shows the as-prepared sample rod-shaped with a size range of 100–300 nm.

In the last few years, researchers develop various composites ceramic nanofibers by electrospinning techniques. Kim et al. demonstrate the fabrication of multifunctional composite polyurethane (PU) membrane from a sol–gel system containing TiO_2 and fly ash (FA) nanoparticles (NPs). The material was well characterized and applied toward water treatment applications [23]. Esfandiyari et al. adopted a sol–gel

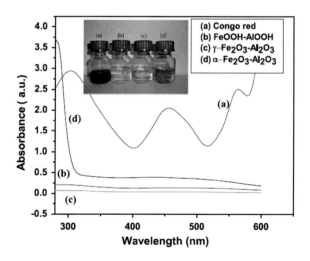

Fig. 7 UV–vis spectra of fresh Congo red solution and after being treated by various adsorbents [32]

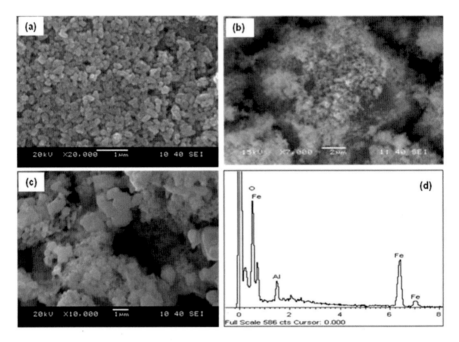

Fig. 8 Representation of the SEM image of **a** nanocomposite of mixed FeOOH and AlOOH, **b** mixed nanocomposite of γ-Fe_2O_3 and γ-Al_2O_3 calcined at 500 °C, **c** mixed α-Fe_2O_3 and α-Al_2O_3 calcined at 1000 °C, **d** EDAX spectra of mixed iron oxide and alumina nanocomposites

method as a favorable approach for the fabrication of multi-walled carbon nanotube-carbon ceramic composite (MWCNT-CCC) [15]. The synthesized materials were characterized and applied toward the adsorptive elimination of Basic Yellow 28 dye from the water system. The experimental outcomes indicate that about 85% of 15.0 ppm dye was adsorbed by using 1.5 g/L of adsorbent within 80 min of the contact time. After all, from the experimental data, it is found that pseudo-second-order kinetic and Freundlich isotherm is the best-fitted model. Furthermore, the thermodynamics investigation indicates the endothermic and spontaneous nature of the adsorption process. The overall study specifies that MWCNT-CCC was a favorable adsorbent toward the adsorption of basic dye from contaminated water.

4 Conclusion and Future Prospects

Here, we have focused on ceramic nanofibers synthesized by the electrospinning method and their applications in wastewater treatment. Primarily due to many interesting and controllable properties and functionalities, nanofibers are gaining fast ground for favorable future use. The interest in nanofiber-related research in the last decades regarding desalination and wastewater management shows us evidence of its adverse applicability in both academia and industry. Despite highly remarkable properties and potential for numerous applications, there are several challenges concerning the production of ceramic nanofibers. Among all, one of the important challenges is the upscaling potential of nanofiber fabrication. The electrospinning process is very costly as discussed before but the technique is more viable, facile, environmental-friendly, specific with ease of operation. The future of electrospun membrane is much more precise and efficient. They possess a greater specific surface area and enhance porosity. The flexibility of the surface makes them more functional and have stability by themselves which makes them potent for adsorption, photocatalysis, filtration, and other application. We have addressed the process of adsorption in the removal of dye from wastewater in this chapter. The overall aim of the research in the remediation of wastewater is to serve mankind globally with clean drinking water. The development of the novel ceramic nanofiber technique by electrospinning method for adsorptive removal of toxic dye from the contaminated water results in a new horizon toward achieving clean water.

Acknowledgements The authors would like to acknowledge NIT Rourkela, Odisha, India, for providing the research facility and funding to carry out this work.

References

1. Abbas Z, Ali S, Rizwan M et al (2018) A critical review of mechanisms involved in the adsorption of organic and inorganic contaminants through biochar. Arab J Geosci 11:1–23. https://doi.org/10.1007/s12517-018-3790-1
2. Ahmed FE, Lalia BS, Hashaikeh R (2015) A review on electrospinning for membrane fabrication: challenges and applications. Desalination 356:15–30. https://doi.org/10.1016/j.desal.2014.09.033
3. Ali AA, El-Sayed SR, Shama SA et al (2020) Fabrication and characterization of cerium oxide nanoparticles for the removal of naphthol green b dye. Desalin Water Treat 204:124–135. https://doi.org/10.5004/dwt.2020.26245
4. Ali I (2012) New generation adsorbents for water treatment. Chem Rev 112:5073–5091. https://doi.org/10.1021/cr300133d
5. Ambaye TG, Vaccari M, van Hullebusch ED et al (2020) Mechanisms and adsorption capacities of biochar for the removal of organic and inorganic pollutants from industrial wastewater. Int J Environ Sci Technol. https://doi.org/10.1007/s13762-020-03060-w
6. Ardila-Leal LD, Poutou-Piñales RA, Pedroza-Rodríguez AM, Quevedo-Hidalgo BE (2021) A brief history of colour, the environmental impact of synthetic dyes and removal by using laccases. Molecules 26:3813. https://doi.org/10.3390/molecules26133813
7. Awad AM, Shaikh SMR, Jalab R et al (2019) Adsorption of organic pollutants by natural and modified clays: a comprehensive review. Sep Purif Technol 228:115719. https://doi.org/10.1016/j.seppur.2019.115719
8. Bazan-Wozniak A, Pietrzak R (2020) Adsorption of organic and inorganic pollutants on activated bio-carbons prepared by chemical activation of residues of supercritical extraction of raw plants. Chem Eng J 393:124785. https://doi.org/10.1016/j.cej.2020.124785
9. Cao Y, Li X (2014) Adsorption of graphene for the removal of inorganic pollutants in water purification: a review. Adsorption 20:713–727. https://doi.org/10.1007/s10450-014-9615-y

10. Chen X, Zhang F, Wang Q et al (2015) The synthesis of ZnO/SnO_2 porous nanofibers for dye adsorption and degradation. Dalt Trans 44:3034–3042. https://doi.org/10.1039/c4dt03382e
11. Chronakis IS (2010) Micro-/nano-fibers by electrospinning technology: Processing, properties and applications, 1st edn. Elsevier Ltd.
12. Dai Y, Zhang N, Xing C et al (2019) The adsorption, regeneration and engineering applications of biochar for removal organic pollutants: a review. Chemosphere 223:12–27. https://doi.org/10.1016/j.chemosphere.2019.01.161
13. Dey S, Mohanta J, Dey B (2020) Sucrose-triggered, self-sustained combustive synthesis of magnetic nickel oxide nanoparticles and efficient removal of malachite green from water. ACS Omega 5:16510–16520. https://doi.org/10.1021/acsomega.0c00999
14. Du F, Sun L, Huang Z et al (2020) Electrospun reduced graphene oxide/TiO2/poly(acrylonitrile-co-maleic acid) composite nanofibers for efficient adsorption and photocatalytic removal of malachite green and leucomalachite green. Chemosphere 239:124764. https://doi.org/10.1016/j.chemosphere.2019.124764
15. Esfandiyari T, Nasirizadeh N, Ehrampoosh MH, Tabatabaee M (2017) Characterization and absorption studies of cationic dye on multi walled carbon nanotube–carbon ceramic composite. J Ind Eng Chem 46:35–43. https://doi.org/10.1016/j.jiec.2016.09.031
16. Gao Q, Luo J, Wang X et al (2015) Novel hollow α-Fe_2O_3 nanofibers via electrospinning for dye adsorption. Nanoscale Res Lett 10:1–8. https://doi.org/10.1186/s11671-015-0874-7
17. Ghasemi E, Ziyadi H, Afshar AM, Sillanpää M (2015) Iron oxide nanofibers: a new magnetic catalyst for azo dyes degradation in aqueous solution. Chem Eng J 264:146–151. https://doi.org/10.1016/j.cej.2014.11.021
18. Gouda M (2021) Removal of methylene blue dye using nickel oxide/carboxymethyl cellulose nanocomposite: kinetic, equilibrium and thermodynamic studies. J Text Sci Fash Technol 8:1–18. https://doi.org/10.33552/jtsft.2021.08.000685
19. Gusain R, Gupta K, Joshi P, Khatri OP (2019) Adsorptive removal and photocatalytic degradation of organic pollutants using metal oxides and their composites: a comprehensive review. Adv Colloid Interface Sci 272:102009. https://doi.org/10.1016/j.cis.2019.102009
20. Halder J, Islam N (2015) Water pollution and its impact on the human health. J Environ Hum 2:36–46. https://doi.org/10.15764/eh.2015.01005
21. Holkar CR, Jadhav AJ, Pinjari DV et al (2016) A critical review on textile wastewater treatments: possible approaches. J Environ Manage 182:351–366. https://doi.org/10.1016/j.jenvman.2016.07.090
22. Homaeigohar S, Zillohu AU, Abdelaziz R et al (2016) A novel nanohybrid nanofibrous adsorbent for water purification from dye pollutants. Materials (Basel) 9:1–16. https://doi.org/10.3390/ma9100848
23. Joo Kim H, Raj Pant H, Hee Kim J et al (2014) Fabrication of multifunctional TiO2-fly ash/polyurethane nanocomposite membrane via electrospinning. Ceram Int 40:3023–3029. https://doi.org/10.1016/j.ceramint.2013.10.005
24. Kapelewska J, Kotowska U, Karpińska J et al (2019) Water pollution indicators and chemometric expertise for the assessment of the impact of municipal solid waste landfills on groundwater located in their area. Chem Eng J 359:790–800. https://doi.org/10.1016/j.cej.2018.11.137
25. Kavithayeni V, Geetha K, Akash Prabhu S (2019) A review on dye reduction mechanism using nano adsorbents in waste water. Int J Recent Technol Eng 7:332–339
26. Khan S, Malik A (2018) Toxicity evaluation of textile effluents and role of native soil bacterium in biodegradation of a textile dye. Environ Sci Pollut Res 25:4446–4458. https://doi.org/10.1007/s11356-017-0783-7
27. Kim JH, Yoo SJ, Kwak DH et al (2014) Characterization and application of electrospun alumina nanofibers. Nanoscale Res Lett 9:1–6. https://doi.org/10.1186/1556-276X-9-44
28. Kulshreshtha SN (1998) A global outlook for water resources to the year 2025. Water Resour Manag 12:167–184. https://doi.org/10.1023/A:1007957229865
29. Kumar S, Meena HM, Verma K (2017) Water pollution in India: its impact on the human health: causes and remedies. Int J Appl Environ Sci 12:275–279

30. Lei D, Qu B, Lin HT, Wang T (2015) Facile approach to prepare porous GeO$_2$/SnO$_2$ nanofibers via a single spinneret electrospinning technique as anodes for Lithium-ion batteries. Ceram Int 41:10308–10313. https://doi.org/10.1016/j.ceramint.2015.04.085
31. Liu J, Yang H, Gosling SN et al (2017) Water scarcity assessments in the past, present, and future. Earth's Futur 5:545–559. https://doi.org/10.1002/2016EF000518
32. Mahapatra A, Mishra BG, Hota G (2013) Adsorptive removal of Congo red dye from wastewater by mixed iron oxide-alumina nanocomposites. Ceram Int 39:5443–5451. https://doi.org/10.1016/j.ceramint.2012.12.052
33. Malwal D, Gopinath P (2016) Fabrication and applications of ceramic nanofibers in water remediation: a review. Crit Rev Environ Sci Technol 46:500–534. https://doi.org/10.1080/10643389.2015.1109913
34. Malwal D, Gopinath P (2017) Efficient adsorption and antibacterial properties of electrospun CuO-ZnO composite nanofibers for water remediation. J Hazard Mater 321:611–621. https://doi.org/10.1016/j.jhazmat.2016.09.050
35. Marinho BA, de Souza SMAGU, de Souza AAU, Hotza D (2021) Electrospun TiO$_2$ nanofibers for water and wastewater treatment: a review. J Mater Sci 56:5428–5448. https://doi.org/10.1007/s10853-020-05610-6
36. Mondal K, Sharma A (2016) Recent advances in electrospun metal-oxide nanofiber based interfaces for electrochemical biosensing. RSC Adv 6:94595–94616. https://doi.org/10.1039/c6ra21477k
37. Pasini SM, Valério A, Yin G et al (2021) An overview on nanostructured TiO$_2$–containing fibers for photocatalytic degradation of organic pollutants in wastewater treatment. J Water Process Eng 40:101827. https://doi.org/10.1016/j.jwpe.2020.101827
38. Paździor K, Bilińska L, Ledakowicz S (2019) A review of the existing and emerging technologies in the combination of AOPs and biological processes in industrial textile wastewater treatment. Chem Eng J 376:120597. https://doi.org/10.1016/j.cej.2018.12.057
39. Peng C, Zhang J, Xiong Z et al (2015) Fabrication of porous hollow γ-Al$_2$O$_3$ nanofibers by facile electrospinning and its application for water remediation. Microporous Mesoporous Mater 215:133–142. https://doi.org/10.1016/j.micromeso.2015.05.026
40. Phan D-N, Kim I-S (2020) Composite nanofibers: recent progress in adsorptive removal and photocatalytic degradation of dyes. Compos Nanocomposite Mater From Knowl to Ind Appl. https://doi.org/10.5772/intechopen.91201
41. Saravanan R, Karthikeyan S, Gupta VK et al (2013) Enhanced photocatalytic activity of ZnO/CuO nanocomposite for the degradation of textile dye on visible light illumination. Mater Sci Eng C 33:91–98. https://doi.org/10.1016/j.msec.2012.08.011
42. Tabe S (2017) A review of electrospun nanofber membranes. J Membr Sci Res 3:228–239. https://doi.org/10.22079/jmsr.2017.56718.1124
43. Thenmozhi S, Dharmaraj N, Kadirvelu K, Kim HY (2017) Electrospun nanofibers: New generation materials for advanced applications. Mater Sci Eng B Solid-State Mater Adv Technol 217:36–48. https://doi.org/10.1016/j.mseb.2017.01.001
44. Tijing LD, Yao M, Ren J et al (2019) Nanofibers for water and wastewater treatment: recent advances and developments
45. Tripathy M, Hota G (2020) Maghemite and graphene oxide embedded polyacrylonitrile electrospun nanofiber matrix for remediation of arsenate ions. ACS Appl Polym Mater 2:604–617. https://doi.org/10.1021/acsapm.9b00982
46. Tripathy M, Padhiari S, Hota G (2020) L-Cysteine-functionalized mesoporous magnetite nanospheres: synthesis and adsorptive application toward arsenic remediation. J Chem Eng Data 65:3906–3919. https://doi.org/10.1021/acs.jced.0c00250
47. Walcarius A, Mercier L (2010) Mesoporous organosilica adsorbents: Nanoengineered materials for removal of organic and inorganic pollutants. J Mater Chem 20:4478–4511. https://doi.org/10.1039/b924316j
48. Xue J, Wu T, Dai Y, Xia Y (2019) Electrospinning and electrospun nanofibers: methods, materials, and applications. Chem Rev 119:5298–5415. https://doi.org/10.1021/acs.chemrev.8b00593

49. Xue J, Xie J, Liu W, Xia Y (2017) Electrospun nanofibers: new concepts, materials, and applications. Acc Chem Res 50:1976–1987. https://doi.org/10.1021/acs.accounts.7b00218
50. Yasin SA, Zeebaree SYS, Zeebaree AYS et al (2021) The efficient removal of methylene blue dye using CuO/PET nanocomposite in aqueous solutions. Catalysts 11:1–15. https://doi.org/10.3390/catal11020241
51. Zander NE (2013) Hierarchically structured electrospun fibers. Polymers (Basel) 5:19–44. https://doi.org/10.3390/polym5010019
52. Zhang J, Mensah A, Narh C et al (2020) Fabrication of flexible TiO_2-SiO_2 composite nanofibers with variable structure as efficient adsorbent. Ceram Int 46:3543–3549. https://doi.org/10.1016/j.ceramint.2019.10.071
53. Zhang Y, Shi R, Yang P et al (2016) Fabrication of electronspun porous CeO_2 nanofibers with large surface area for pollutants removal. Ceram Int 42:14028–14035. https://doi.org/10.1016/j.ceramint.2016.06.009

Natural Biodegradable Polymeric Bio-adsorbents for Textile Wastewater

Lopamudra Das, Papita Das, Avijit Bhowal, and Chiranjib Bhattacharjee

1 Introduction

Textiles factories account for one-fifth of the global trade water contamination and they consume huge amounts of water and use various toxic chemicals to produce tissue or fabrics. Among the numerous existing pollutants in textile discharges, dyes are reviewed as the worst contaminants [30]. Before the period of the industrial revolution with inventing technologies, textile materials were colored with natural biodegradable dyeing compounds obtained from plant/animal sources, which were not ever a danger toward the Mother Nature. However, acquirement of these natural dyes at enormous extents was constantly challenged a big difficulty. With developing technologies, effluents of the textile industry put up with worrying amounts of toxic dyes which impend the ecosystem through imposing harmful impacts on human beings and adjacent ecologies. The dye in the water instigated some damaging influences for example depletion in sunlight penetration and inhibitions to the growth of marine creatures by decreasing the amount of dissolved oxygen in deep water, additionally dye acts as anoxious, mutagenic, and hazardous agent in living cells [19, 79].

As a concern, there is a crucial necessity to eliminate the contaminated dyes from the untreated toxic effluents of the textile industries before discharging them into the open water to avert carcinogenic impacts on living bodies [69, 80]. Conventional physical and chemical techniques have been designed for the treatment of textile dye-contaminated water including adsorption, coagulation, ion exchange, etc. [37]. In spite of their instantaneous implication in wastewater treatment, application of

L. Das · P. Das (✉) · A. Bhowal · C. Bhattacharjee
School of Advanced Studies on Industrial Pollution Control Engineering, Jadavpur University, Kolkata, India
e-mail: papita.das@jadavpuruniversity.in

Department of Chemical Engineering, Jadavpur University, Kolkata, India

© The Author(s), under exclusive license to Springer Nature Singapore Pte Ltd. 2022 209
A. Khadir and S. S. Muthu (eds.), *Polymer Technology in Dye-containing Wastewater*,
Sustainable Textiles: Production, Processing, Manufacturing & Chemistry,
https://doi.org/10.1007/978-981-19-0886-6_9

these processes was limited because of the unavoidable sludge generation as end product or else by the requirement to repeatedly redevelop the adsorbent materials. Moreover, most of them showed poor adsorbate uptake capability by means of notable low-grade sensitivity and selectivity [11].

Last several years the researchers are evolving innovative advanced adsorbent materials that can efficiently encapsulate colored components from textile wastewater. Nowadays, natural polymer-based polymeric biosorbent had drawn great interest in textile wastewater treatment processes due to their impressive advantages such as biodegradability, economic viability, nontoxicity, feasibility, and renewability [20, 82]. Ecological biodegradable natural polymers, including starch, cellulose, and protein, are extensively applied for the preparation of the biodegradable polymeric biosorbens [82, 90]. Extensive contemplation is focused on the synthesis of the chemically modified multi-component hybrid polymer material derived by reinforcing or grafting the synthetic polymers, carbonaceous material, and metal oxide onto natural ones including starch, cellulose, chitosan/chitin, and alginate [18, 23, 90] [90]. Natural biopolymers contain several functional groups in their molecular chains, therefore they showed remarkable ability in the formation of the hybrid polymeric biosorbent by chemical modification with advantageous functional and commercial properties [58]. Polymeric biocomposites having different functional groups have been conferred in the following segment for this purpose.

Eco-friendly applications regarding biocomposites of biopolymers produced by grafting them with PVA (polyvinyl alcohol), poly (acrylamide), metal oxides, bentonite clay, graphene oxide nanoparticles, etc. have been reported earlier [75, 81, 85]. Last few years, numerous scientists reported the graft copolymerization of natural biopolymers to develop innovative tailored biodegradable polymer materials, having desirable improved properties for definite implementations such as Sharma et al., studied the synthesis of Gg-cl-poly(acrylic acid-aniline) polymer composite and their behavior in the dye removal process [73]. Sing et al. [75] described the crosslinked polymerization process to prepare the psyllium polysaccharide/acrylamide hydrogel [75]. Madhusudana Rao et al., developed PVC grafted sodium alginate microgels and analyzed characterization of those microgels [46]. Erizal et al., reported synthesis and characterization of the sodium alginate-based biopolymer (poly (acrylamide-co-acrylic acid)—NaAlg) [23].

Hence, the investigation intended for nontoxic, biocompatible, non-conventional, cost-effective, and recyclable remediation of the textile dye-containing wastewater instigated a rise in the importance of polymeric biosorbents. As stated by Ali et al. those alternative adsorbents should have certain general characteristics for example abundance and easy acquirement, commercial viability, and biodegradability [1].

To this purpose, a brief description of the textile dye classification, their toxic effects on ecosystem, ephemeral insight of the natural biodegradable biopolymers, and polymeric biosorbents adopted in treatment of the textile dye wastewater have been reviewed in the following sections.

2 Textile Dyes: Classification and Class

Commercial synthetic textile dyes are available in an excessive number, and so their identification and classification are difficult. Color Index is an important factor for identification of most the synthetic dyes. Dyes can be classified on the basis of chemical structure (azo dye, nitro dye, etc.), nuclear structure (anionic, cationic, or nonionic), or industrial application (reactive dye, acid dye, etc.) [56]. From a structural perspective, the dyes can be identified by two key groups: chromophores and auxochromes. The chromophore provides color to the dye molecule and auxochromes enrich the compatibility of the products [69].

Following paragraph describe some familiar type of dyes, classified on the subject of their industrial usage.

Basic Dye: Basic dyes are mainly water-soluble and known as cationic dye as they produce cations by ionization in solution. They are largely utilized in paper, medicine industry, and also used for dyeing polyacrylonitrile, silk, cotton, nylons, polyesters, etc. These dyes specifically belong to the chemical structures of diazahemicyanine, triarylmethane, cyanine, acridine, etc. Some examples of basic dye are Methylene blue, Malachite green, Basic Red 46, Brilliant Green, etc. [56].

Acid Dye: applied for coloring nylon, wool, silk, modified acrylics, paper printing, leather, food, and cosmetics. Most of these dyes are anionic type and have capability of dissolving in water. They mainly belong to the chemical structures of the azo, azine, xanthene, nitros, etc. Some examples of these dyes are Acid Red 18, Acid Yellow 23, Acid Black 48, etc. [7].

Disperse Dye: These water-insoluble are used mostly for dying polyester, cellulose, nylon, cellulose acetate, and acrylics. They are nonionic types of dyes and mainly contain chemical structures of benzodifuranones, nitro, azo, anthraquinones, etc. Some examples of these dyes are Disperse Red 1, Disperse Orange 3, Disperse Blue 35, etc. [90].

Direct Dye: applied for coloring of cotton, rayon, paper, nylon, leather, etc. These dyes have capability of dissolving in water and anionic type as they produce anions in solution. Mostly, these dyes are having chemical structure of the stilbene, phthalocyanine, oxazine, etc. [69]. Some examples of excessive used direct dyes are Direct Orange 34, Direct Red 23, Direct Black 38, etc.

Reactive Dye: these anionic dyes are largely used for coloring on cotton, wool, and other cellulosic fibers, etc. These dyes belong to the chemical groups including anthraquinone, azo, phthalocyanines, formazan, oxazines, etc. Some examples of these dyes are Reactive Black 5, Reactive Blue 4, Reactive Red 120, and Reactive Red 2 [10].

Solvent Dye: conventionally applied for fuels, resins, plastics, lubricants, woods, and waxes. They are specifically solvent-soluble but not water-soluble. They mainly contain chemical structural groups of azo and anthraquinones, phthalocyanines, etc.

[60]. Some examples of this type of dyes are Natural Black 1, Mordant Blue 14, and Mordant Red 3.

3 Toxicity of Textile Dye

Several dyes have visibility in water even at very less concentrations (1 mg/L^{-1}). The color concentrations in the textile effluents are varied in between 10 and 200 mg/L^{-1} [30]. Hence, dumping of untreated colored textile outflow in exposed water causes aesthetic problems to human beings and aquatic life. Additionally, discharged contaminated water is also responsible for environmental degradation and mostly affects the aquatic ecosystem as they hamper photosynthesis process by blocking the light penetration through water layer and decreasing the concentration of dissolved oxygen. Dyes having complex chemicals structure and photolytic stability are suspected to be toxic, carcinogenic, and non-biodegradable substances [69, 80]. For example, the dyes, which contain chemical structural class of azo group binds to an aromatic ring [7]. These dyes can form aromatic amines, arylamines by mineralization, which are considered to be hazardous [10]. Maximum numbers of the textile dyes are water-soluble and easily absorbed on living cell membrane by contact and by inhalation. This may cause a threat of cancer, splenic sarcomas, chromosomal abnormalities, nuclear anomalies, hypersensitivity, eye irritation, skin irritation, contact dermatitis, respiratory distress, lacrimations when inhaled or consumed [60, 80].

4 Biodegradable Natural Polymers

Natural biodegradable polymer derived from animal-origin for instance chitin, chitosan, collagens, and nucleic acids, or from vegetable-origin for example starch, cellulose, and alginates are nowadays being researched worldwide and applied as the effective biosorbent for decontamination of wastewater because they are renewable-resources based material, decomposable, inexpensive, non-hazardous, and generate less sludge at the post-treatment process, in addition, to avert the environmental concern [25, 63].

Most of the biopolymers have been fabricated and few of them are naturally produced in the natural environs throughout the microbial growth periods [45, 67]. At the primitive stage of the biopolymers generation most often they are derived from agro-feed stock like grain, root vegetables, and other carbohydrates through microbial fermentation. Though, on account of developments in technology, second-generation biopolymers progressed and deviate from food-based resources [38, 63].

Few types of biodegradable natural polymers are documented in the following section of the review literature.

Starch: This natural polymer is one of the utmost profuse and reachable biopolymers in the world. Characterization of starch molecules with 3D structural design are analyzed via crystallinity (in a range of 15%-45%) and by D-glucose units associated with macromolecules (amylopectin, branched and linear-linked α-D-glucan, and amylose) [98]. Crystalline starch is produced through hydrolysis of the crystal-like part of the starch [38, 55]. To attain precise biosorption efficiency, chemical modifications of starch are frequently required [97, 99] and is effectively suitable because of the presence of functional groups on the starch structure.

Cellulose: Cellulose molecule is a linear, crystal-like homo-polymer consisting of D-glucose linkages that are interconnected through β-1,4glycosidic bonds. They are mostly obtained from various plant sources like grasses, stalks of vegetables, wood pulps, cotton, jute, etc. is consist of repeating β-(1–4) D-glucose linkages [26]. Moreover, Animals, bacteria, algae, fungi, and cellular slime molds, are also huge sources for the fabrication of cellulosic fiber [5, 12]. Due to the orientation of the O–H groups in the cellulose chain, they have a tendency to stick out alongside the long-chain molecules and are readily accessible for H-bonding. Cellulose materials show appreciable strength and strong resistance to the solvents due to existing of the H-bonding in the crystalline regions.

Additionally, natural polymers, such as chitosan/chitin are also significant viable substitutes as precursor base material for preparing polymeric composites.

Chitin and Chitosan: Anther abundant natural biodegradable polysaccharide is chitin and is largely originate in the external skeletons of crustaceans and other invertebrates as well as in the specific fungi cell-wall [51]. Chitin molecules consist of linear β-1,4-linked polymer of N-acetyl-D-glucosamine. Whereas chitosan is fundamentally a derived form of chitin produced by enzymatic N-deacetylation and is possible to synthesize chemically from native chitin [91]. Chitosan is structurally categorized as the amino-polysaccharide of glucosamine and less crystalline compared to chitin because of the definite antiparallel chains in their structure and their crosslinking. Chitin is insoluble but chitosan is water-soluble. On account of having higher degree of hydration, Chitosan exhibited high biosorption efficiency, significant reactivity, considerable selectivity which assists the effective biosorption of pollutants in water purification process [13, 35].

Alginate: Alginate is one type of natural polysaccharide derived from cell-wall of brown algae and consists of two uronic acid units (1–4)-β-D-mannuronic acid and (1–4)-α-L-guluronic acid residues [66]. It is another exceptionally significant natural polymer that is paying attention to the scientists in the field of water purification technology, biomedical engineering, and biotechnology [41, 88]. Moreover, alginate was found to be cost-effective natural biopolymer for the synthesis of the polymeric biosorbent for biosorbtion process and it has considerable soluble capability in water [57]. In general, alginates have been chemically modified by applying graft copolymerization method to enhance their functional properties. Several environmental applications were recommended for alginate-based biosorbents.

5 Applications of Biodegradable Polymeric Biosorbent in Textile Dye Removal

Generally, grafted or impregnated natural polymers exhibited a higher biosorption efficiency compared to native natural polymers. Natural polymer-derived biodegradable polymeric biosorbents reduce the consumption of synthetic polymers and showed enormous significance in the current circumstance of lessening oil reserves and greenhouse gas emission associated with the production of synthetic polymers from petroleum and nonrenewable carbon sources [34]. Due to having nontoxic nature, flexibility, biodegradable capability, high uptake capacity, low density, and recyclability, natural biopolymer-derived biosorbents can be accessible and ecological alternatives in the field of remediation process of the textile dye-contaminated water [5, 25]. Polymeric biocomposites, in which at least one of the constituents either the polymer matrix or the reinforcement is prepared from biomaterial [86]. In this perspective, natural polymers have been drawn the attention of researchers worldwide to prepare biodegradable polymeric biosorbents by combining petroleum monomers or bio-based sources with natural polymers (for example cellulose [83, 90] starch [79] and polysaccharides [93]) by chemical modification or surface modification and search for integration the features of the individual constituents have also been paying interest. By crosslinking, biopolymers become highly stabilized and operational flexibility increased [16]. Some widely used crosslinking agents are Glutaraldehyde, formaldehyde, epichlorohydrin, and glyoxal.

The succeeding sections of this article assemble latest studies on the applications of some biodegradable natural polymer-derived composites for ecological biosorption of textile dye.

5.1 Starch-Based Polymeric Composites

A well-known operative method for preparing modified starch by chemical reformation is graft copolymerization with definite biological monomers, synthetic monomers, or carbonaceous material. Grafted starch showed remarkable improved functional properties such as increase in fractal dimension, high biosorption efficiency, and great bridging influences [22, 53]. Gong et al., characterized and evaluate the adsorption efficiency of magnetic carboxymethyl Starch (CMS)/PVA for removal of methylene blue dye and recycled for eight succeeding cycles [27]. Zhou et al. reported the dye adsorption behavior of the polyaluminum ferric chloride-starch graft polymeric composite for the treatment of textile wastewater [97]. Bhattacharyya et al. developed graphene oxide/potato starch composite and showed their dye adsorption potential [9]. Li et al., reported on the adsorption study of the multiple cationic dyes by applying Silica-sand/anionized-starch polymeric composite [39]. Sen et al. studied the preparation, characterization, and utilization of polyacrylamide grafted carboxymethyl starch for the treatment of colored textile industrial wastewater [72]. Preparation and characterization of Starch/Fe_3O_4 were studied by Stan et al. and they reported dye adsorption efficiency of more than 70% for encapsulation of Optilan Blue [79]. Li et al., reported the synthesis of magnetic starch-based composite and their removal potential for methylene blue dye adsorption [40]. Xia et al., reported adsorption potential of modified Starch/PVA biopolymer for the treatment of textile wastewater [89]. Azman et al., focused on the synthesis and operational efficiency evaluation of Starch-grafted polyacrylamide hybrid hydrogels [6]. Preparation and decolorization efficiency of chitosan-starch-based polymeric composite for adsorption of congo red dye were discussed by Sami et al. [71].

5.2   Cellulose-Based Polymeric Composites

Cellulose grafted polymeric composites for decolorization of textile wastewater, synthesized by reinforcing cellulosic fibers (nanofibers or macro-fibers) with a certain polymer matrix (such as PVA (polyvinyl alcohol), PAN (polyacrylonitrile), PEO (polyethylene oxide) [18, 74, 87] or biomaterials (such as carbonaceous material [44], bentonite clay or metal oxides [78] have been suggested and reported in so many literature. Apparently, the blending of those materials with cellulose extremely impacts the characteristics and performance of polymeric composite in terms of flexibility, absorptivity, resistance to biofouling [74, 92]. Karim et al. described efficient remediation of Rhodamine dye by nanocellulose crystals in accordance with H-bonding and electrostatic attraction. They suggested that negatively charged nanocelluloses formed a stable bond with cationic dyes which enhanced the adsorption efficiency [36]. Song et al., prepared magnetic amine/Fe3O4-resin by introducing amine groups onto cellulose structure and reinforcing Fe3O4 in the polymer matrix to improve adsorption efficiency for removal of anionic dyes [78]. Wang et al., prepared MnO2 coated cellulose biopolymer and utilized them as biosorbent in the decolorization process of methylene blue-containing wastewater and found removal efficiency of 99.8% [87]. De Castro Silva et al., developed the phosphate cellulose by integrating it with chloroethyl phosphate for the application in the Brilliant green dye removal study [21]. Zhao & Wang reported the grafting of carboxymethylcellulose with montmorillonite to prepare nanocomposites and their biosorption behavior in Congo red dye removal study [95]. Activated carbon-reinforced magnetic cellulose biosorbents were developed by applying external magnetic field and were employed for organic dye encapsulation from colored water [44]. Yang et al., prepared polyacrylamide/cellulose hydrogels and discussed their successive implication in decolorization process for removal of methylene blue dye [90]. [76] developed dimethylaminoethyl methacrylate was impregnated onto a cellulose material and explained adsorption performances of the composite for removal of direct violet 31dyes from wastewater [76].

5.3 Chitin and Chitosan-Based Polymeric Composites

Nowadays, chitosan biopolymer has become efficient possible source of base materials for synthesis of the chitosan-based polymeric composites for decolorization of textile contaminated water due to their nano-scale size, great specific surface area, and nonexistence of diffusion limitations [35, 65]. A great number of chitosan-grafted polymeric biosorbents had been experienced for their dye uptake proficiency under different operational conditions. Many research literatures described GO (graphene oxide) impregnated biopolymers could be utilized to reduce the practical limitations (such as problem in reactor cleaning, toxic effects of nanosize GO particle, high operational cost, etc.) associated with the applications of commercial GO alone in wastewater treatments [14]. Manna et al. studied multiple dye adsorptions by GO dots coated chitin and obtained adsorption efficiency of more than 75% [49]. Banerjee et al., showed adsorption of acid yellow and acid blue dye using chitosan-GO nanocomposite [7]. GO/Fe3O4/chitosan polymeric composite had been developed and applied as efficient biosorbent for decolorization of dyeing water by Tran et al., [84]. Chitosan reinforced with GO and PVA showed effective biosorption capability for the removal of congo red dye from colored wastewater [18]. Amir et al. focused on the synthesis process of polymer composite by combining chitosan polymer with TiO2 nanoparticles for encapsulation and photodegradation of Methyl Orange [2]. Modified chitosan-ethyl acrylate was synthesized by Sadeghi-Kiakhani et al., and they reported that biosorption ability for adsorption of basic dyes (Blue 41 and Basic Red 18) was remarkably enhanced after the introduction of a huge number of carboxyl(COOH) groups into the chitosan structure [68]. Zhou et al. investigated the removal study of acid dyes using modified magnetic chitosan composite and conveyed that the dye adsorption improved due to the strong electrostatic interaction among anionic dye and protonated amino groups present in chitosan structure [96]. Biosorption of reactive dyes by chemically crosslinked chitosan composite beads has been studied by Chiou and Li [15]. Jagaba et al., developed modified chitosan-based composites by combining with other materials, including polyacrylamide and bentonite clay and they obtained very high adsorption efficiency (99%) using those adsorbents for removal of duasyn direct dye under low pH conditions [29].

5.4 Alginate Based Polymeric Composites

Last few years, current studies have emphasized the widely applications of alginate-based polymeric composites in water purification process for removal of pollutants, because of their nontoxicity, mechanical stability, biocompatibility, and water absorbency [37, 41, 88]. Alginates can be chemically modified or grafted through several approaches, especially through polycondensation and radical polymerization [77]. Presence of divalent cations (for example calcium ions) helps to form alginates gel by crosslinking them. Crosslinking of alginates occurs because of the assembling of the glucuronic acid blocks in alginate chain [31]. Asadi et al., proposed suggested calcium alginate grafted polymeric beads for treatment of methyl violet dye-containing wastewater [4]. Parlayici synthesized alginate-based perlite composite beads by implementing sol–gel chemistry and use them for the effective biosorption of multiple dyes (methylene blue, malachite green, and methyl violet) from colored contaminated water [57]. Environment-friendly, biodegradable, cost-effective biosorbent named as Sodium Alginate/titania nanocomposites have been suggested by Mahmoodi et al., for the purification of textile dye water [47]. Pourjavadi et al. reported preparation and swelling behavior of chemically modified alginate-based composite (alginate-g-poly (sodium acrylate)/kaolin), which was grafted using sodium alginate and kaolin and polymeric composite was crosslinked by methylene biscrylamide [58]. Rezaei et al. focused on the preparation of Fe_3O_4 reinforced alginate composite beads for the decolorization of methylene blue dye from polluted water [64]. By applying solvent-casting method, Radoor et al. fabricated a unique, eco-friendly membrane adsorbent (PVA/SA/ZSM-5) with appreciable regeneration capability for the treatment of wastewater containing Congo red dye [61].

Different polymeric biosorbents and their adsorption capacity for textile dye removal are reviewed and presented in the following Table 1.

6 Conclusions and Future Perspectives

The literature survey reviewed the circumstances that there have been noticeable rises in manufacturing and consumption of color in last few years causing an immense warning of pollution. The present literature exhibited the literature review regarding the toxicity and the classification of textile dyes on the basis of their industrial usage. Additionally, this literature set the goal of assembling latest highlights in treatment of textile dye-containing wastewater by implementing different biodegradable polymeric biosorbents, mentioning their abundance, accessibility, feasibility, and characteristic features. An analysis of the natural polymers, with distinct focus on starch, cellulose, alginate, and chitin/chitosan, has confirmed that these biomaterials showed the significantly high potential for fabrication of biopolymer-based composite materials for wastewater purification. From our literature review, it was evident that the various documented research works using several polymeric biosorbents have demonstrated appreciable biosorption abilities for decorization of toxic dye-containing wastewater. It was surveyed that the dye biosorption capability of natural polymer-based polymeric biocomposites was generally dependent on biomaterial that used for grafting and on type of monomer induced and the characteristic features of the dye.

Though, the basic limitations of their applications are still a challenge for implementing the investigational methods on large scale and under inconstant operational conditions whereas maximum studies are principally limited to a lab-scale research.

Considerable determinations are currently being made in the investigation and advance in production of biodegradable biopolymer as the basic resources intended

Table1 Potential of polymeric composites in dye removal

Dye	Polymeric biosorbent	Adsorption capacity ($mg\ g^{-1}$)	References
Acid blue 25	Chitosan/cyclodextrin material	77.4	[51]
Malachite green	Poly(vinyl alcohol)/chitosan composite	380.65	[32]
Reactive orange 16	Chitosan/sepioliteclay	190.965	[50]
Malachite Green	NiO NPs/Chitin beads	370.37	[62]
Methylene Blue	Chitosan-based graphene oxide (GO) gels	1100	[14]
Acid Yellow 36	Chitosan-GO nanocomposite	68.86	[7]
Acid Red 2	Magnetic chitosan nanocomposites	90.06	[33]
Methylene Blue	Magnetic chitosan-GO composite	180.83	[24]
Methylene Blue	Chitosan/organic rectorite-Fe_3O_4	24.690	[93]
Malachite green	Native cellulosic polymers	2.422	[59]
Methyl orange	Natural carbohydrate polymeric biosorbents of rice flour	173.24	[52]
Rhodamine B	Carbohydrate polymeric adsorbent of wheat flour	142.26	[28]
Methylene Blue	GO reinforced PVA/Carboxymethyl cellulose	172.14	[17]
Methylene Blue	MnO_2 coated cellulose nanofibers	Not reported	[87]
Brillint green	Phosphate cellulose integrated with chloroethyl phosphate	114.2	[21]
Malachite green	Graphene oxide /cellulose bead	30.09	[94]
Reactive red dye	Aminoethanethiol modified cellulose surface	78.00	[74]
disperse dyes	Polyacrylamide Grafted Cellulose	25.84	[90]
Reactive brilliant red	Magnetic amine/Fe_3O_4-resin	101.0	[78]
Malachite green	Alginate/poly aspartate composite hydrogel beads	300–350	[77]
Malachite Green	Superparamagnetic sodium alginate-coated Fe_3O_4	47.84	[48]
Methylene blue	Alginate grafted poly-acrylonitrile beads	03.51	[70]
Methylene blue	Sodium alginate-coated perlite beads	74.6	[57]
Methylene Blue	GO/Calcium alginate	181.81	[42]
Crystal violet	hybrid sodium alginate-PVA -silica oxide	1756.2	[43]

(continued)

Table1 (continued)

Dye	Polymeric biosorbent	Adsorption capacity ($mg\ g^{-1}$)	References
Methylene Blue	Activated organo-bentonite /sodium alginate	414.1	[8]
Acid green 25	Sodium alginate/titania	151.5	[47]
Methylene Blue	Graphene oxide/potato starch composite	500	[9]
Reactive Orange 131	modified starch/polyvinyl alcohol composite	539	[89]
Crystal violet	Silica-sand/anionized-starch	1246.40	[39]
Crystal violet	MCNCs/starch-g-(AMPS-co-AA)	2500	[54]
Optilan Blue	Starch/Fe_3O_4	74.58	[79]
Malachite green	Polyacrylamide/bentonite composite	243.11	[3]

for innovative applications. Specifically, the increasing price of conventional adsorbents undeniably makes biopolymer-based biosorbents a matter of interest for employing them as efficient biosorbents in decolorization process of colored water.

Acknowledgements The authors acknowledge the Dalmia Holding Group for supporting us financially and also obliged to the Chemical Engg. Department and School of Advanced Studies on Industrial Pollution Control Engineering, Jadavpur University, Kolkata, India.

References

1. Ali I, Asim M, Khan TA (2012) Low cost adsorbents for the removal of organic pollutants from wastewater. J Environ Manage. https://doi.org/10.1016/j.jenvman.2012.08.028
2. Amir MNI, Muhd Julkapli N, Hamid SBA (2017) Effective adsorption and photodegradation of methyl orange by TiO_2-chitosan supported glass plate photocatalysis. Mater Technol. https://doi.org/10.1080/10667857.2016.1201635
3. Anirudhan TS, Suchithra PS (2009) Adsorption characteristics of humic acid-immobilized amine modified polyacrylamide/bentonite composite for cationic dyesin aqueous solutions. J Environ Sci. https://doi.org/10.1016/S1001-0742(08)62358-X
4. Asadi S, Eris S, Azizian S (2018) Alginate-based hydrogel beads as a biocompatible and efficient adsorbent for dye removal from aqueous solutions. ACS Omega. https://doi.org/10.1021/acsomega.8b02498
5. Azizi Samir MAS, Alloin F, Dufresne A (2005) Review of recent research into cellulosic whiskers, their properties and their application in nanocomposite field. Biomacromol 6:612–626. https://doi.org/10.1021/bm0493685
6. Azman I, Mutalib SA, Yusoff SFM, Fazry S, Noordin A, Kumaran M, Mat Lazim A (2016) Novel Dioscorea hispida starch-based hydrogels and their beneficial use as disinfectants. J Bioact Compat Polym 31:42–59. https://doi.org/10.1177/0883911515597704
7. Banerjee P, Barman SR, Mukhopadhayay A, Das P (2017) Ultrasound assisted mixed azo dye adsorption by chitosan–graphene oxide nanocomposite. Chem Eng Res Des. https://doi.org/10.1016/j.cherd.2016.10.009

8. Belhouchat N, Zaghouane-Boudiaf H, Viseras C (2017) Removal of anionic and cationic dyes from aqueous solution with activated organo-bentonite/sodium alginate encapsulated beads. Appl Clay Sci. https://doi.org/10.1016/j.clay.2016.08.031

9. Bhattacharyya A, Banerjee B, Ghorai S, Rana D, Roy I, Sarkar G, Saha NR, De S, Ghosh TK, Sadhukhan S, Chattopadhyay D (2018) Development of an auto-phase separable and reusable graphene oxide-potato starch based cross-linked bio-composite adsorbent for removal of methylene blue dye. Int J Biol Macromol. https://doi.org/10.1016/j.ijbiomac.2018.05.069

10. Bhupinderkaur, Chanchal (2016) Environmental and health concerns of the textile industry. Int J Civil Struct Environ Infrastruct Eng Res Dev

11. Bibi S, Farooqi A, Hussain K, Haider N (2015) Evaluation of industrial based adsorbents for simultaneous removal of arsenic and fluoride from drinking water. J Clean Prod. https://doi.org/10.1016/j.jclepro.2014.09.030

12. Brown RM (2004) Cellulose structure and biosynthesis: what is in store for the 21st century? J Polym Sci Part A Polym Chem. https://doi.org/10.1002/pola.10877

13. Chatterjee S, Chatterjee S, Chatterjee BP, Guha AK (2007) Adsorptive removal of congo red, a carcinogenic textile dye by chitosan hydrobeads: binding mechanism, equilibrium and kinetics. Colloids Surf A Physicochem Eng Asp. https://doi.org/10.1016/j.colsurfa.2006.11.036

14. Cheng CS, Deng J, Lei B, He A, Zhang X, Ma L, Li S, Zhao C (2013) Toward 3D graphene oxide gels based adsorbents for high-efficient water treatment via the promotion of biopolymers. J Hazard Mater. https://doi.org/10.1016/j.jhazmat.2013.09.065

15. Chiou MS, Li HY (2003) Adsorption behavior of reactive dye in aqueous solution on chemical cross-linked chitosan beads. Chemosphere. https://doi.org/10.1016/S0045-6535(02)00636-7

16. Crini G, Badot PM (2008) Application of chitosan, a natural aminopolysaccharide, for dye removal from aqueous solutions by adsorption processes using batch studies: a review of recent literature. Prog Polym Sci. https://doi.org/10.1016/j.progpolymsci.2007.11.001

17. Dai H, Huang Y, Huang H (2018) Eco-friendly polyvinyl alcohol/carboxymethyl cellulose hydrogels reinforced with graphene oxide and bentonite for enhanced adsorption of methylene blue. Carbohydr Polym. https://doi.org/10.1016/j.carbpol.2017.12.073

18. Das L, Das P, Bhowal A, Bhattachariee C (2020a) Synthesis of hybrid hydrogel nano-polymer composite using Graphene oxide, Chitosan and PVA and its application in waste water treatment. Environ Technol Innov. https://doi.org/10.1016/j.eti.2020.100664

19. Das L, Das P, Bhowal A, Bhattachariee C (2020) Treatment of malachite green dye containing solution using bio-degradable Sodium alginate/NaOH treated activated sugarcane baggsse charcoal beads: Batch, optimization using response surface methodology and continuous fixed bed column study. J Environ Manage. https://doi.org/10.1016/j.jenvman.2020.111272

20. Das L, Saha N, Saha PD, Bhowal A, Bhattacharya C (2020c) Application of synthesized nanocellulose material for removal of malachite green from wastewater. In: Recent trends in waste water treatment and water resource management. https://doi.org/10.1007/978-981-15-0706-9_2

21. De Castro Silva F, Da Silva MMF, Lima LCB, Osajima JA, Da Silva Filho EC (2016) Integrating chloroethyl phosphate with biopolymer cellulose and assessing their potential for absorbing brilliant green dye. J Environ Chem Eng. https://doi.org/10.1016/j.jece.2016.07.010

22. Du Q, Wang Y, Li A, Yang H (2018) Scale-inhibition and flocculation dual-functionality of poly(acrylic acid) grafted starch. J Environ Manage. https://doi.org/10.1016/j.jenvman.2018.01.016

23. Erizal S, Budianto E, Mahendra A, Yudianti R (2013) Radiation synthesis of superabsorbent poly(acrylamide-co-acrylic acid)-sodium alginate hydrogels. In: Advanced materials research. https://doi.org/10.4028/www.scientific.net/AMR.746.88

24. Fan L, Luo C, Sun M, Li X, Lu F, Qiu H (2012) Preparation of novel magnetic chitosan/graphene oxide composite as effective adsorbents toward methylene blue. Bioresour Technol. https://doi.org/10.1016/j.biortech.2012.02.067

25. Galiano F, Briceño K, Marino T, Molino A, Christensen KV, Figoli A (2018) Advances in biopolymer-based membrane preparation and applications. J Memb Sci. https://doi.org/10.1016/j.memsci.2018.07.059

26. Ganguly P, Sengupta S, Das P, Bhowal A (2020) Valorization of food waste: Extraction of cellulose, lignin and their application in energy use and water treatment. Fuel. https://doi.org/10.1016/j.fuel.2020.118581
27. Gong G, Zhang F, Cheng Z, Zhou L (2015) Facile fabrication of magnetic carboxymethyl starch/poly(vinyl alcohol) composite gel for methylene blue removal. Int J Biol Macromol. https://doi.org/10.1016/j.ijbiomac.2015.07.061
28. Hasan MM, Shenashen MA, Hasan MN, Znad H, Salman MS, Awual MR (2021) Natural biodegradable polymeric bioadsorbents for efficient cationic dye encapsulation from wastewater. J Mol Liq. https://doi.org/10.1016/j.molliq.2020.114587
29. Jagaba AH, Abubakar S, Lawal IM, Latiff AA, Umaru I (2018) Wastewater treatment using alum, the combinations of alum-ferric chloride, alum-chitosan, alum-zeolite and alum-moringa oleifera as adsorbent and coagulant. Int J Eng Manage. https://doi.org/10.11648/j.ijem.201802 03.13
30. Islam S (2020) A study on the solutions of environment pollutions and worker's health problems caused by textile manufacturing operations. Biomed J Sci Tech Res. https://doi.org/10.26717/bjstr.2020.28.004692
31. Jain A, Gupta Y, Jain SK (2007) Perspectives of biodegradable natural polysaccharides for site-specific drug delivery to the colon. J Pharm Pharm Sci
32. Jeon YS, Lei J, Kim JH (2008) Dye adsorption characteristics of alginate/polyaspartate hydrogels. J Ind Eng Chem. https://doi.org/10.1016/j.jiec.2008.07.007
33. Kadam AA, Lee DS (2015) Glutaraldehyde cross-linked magnetic chitosan nanocomposites: Reduction precipitation synthesis, characterization, and application for removal of hazardous textile dyes. Bioresour Technol. https://doi.org/10.1016/j.biortech.2015.06.148
34. Kalia S, Avérous L (2011) Biopolym Biomed Environ Appl. https://doi.org/10.1002/978111 8164792
35. Kardam A, Raj KR, Srivastava S, Srivastava MM (2014) Nanocellulose fibers for biosorption of cadmium, nickel, and lead ions from aqueous solution. Clean Technol Environ Policy 16:385–393. https://doi.org/10.1007/s10098-013-0634-2
36. Karim Z, Mathew AP, Grahn M, Mouzon J, Oksman K (2014) Nanoporous membranes with cellulose nanocrystals as functional entity in chitosan: Removal of dyes from water. Carbohydr Polym. https://doi.org/10.1016/j.carbpol.2014.06.048
37. Kuang Y, Du J, Zhou R, Chen Z, Megharaj M, Naidu R (2015) Calcium alginate encapsulated Ni/Fe nanoparticles beads for simultaneous removal of Cu (II) and monochlorobenzene. J Colloid Interface Sci. https://doi.org/10.1016/j.jcis.2015.01.080
38. Le Corre D, Bras J, Dufresne A (2010) Starch nanoparticles: a review. Biomacromol. https://doi.org/10.1021/bm901428y
39. Li P, Gao B, Li A, Yang H (2020) Evaluation of the selective adsorption of silica-sand/anionized-starch composite for removal of dyes and Cupper(II) from their aqueous mixtures. Int J Biol Macromol. https://doi.org/10.1016/j.ijbiomac.2020.02.047
40. Li W, Wei H, Liu Y, Li S, Wang G, Han H (2021) Fabrication of novel starch-based composite hydrogel microspheres combining Diels-Alder reaction with spray drying for MB adsorption. J Environ Chem Eng 9:105929. https://doi.org/10.1016/j.jece.2021.105929
41. Li X, Qi Y, Li Y, Zhang Y, He X, Wang Y (2013) Novel magnetic beads based on sodium alginate gel crosslinked by zirconium(IV) and their effective removal for Pb^{2+} in aqueous solutions by using a batch and continuous systems. Bioresour Technol. https://doi.org/10.1016/j.biortech.2013.05.081
42. Li Y, Du Q, Liu T, Sun J, Wang Y, Wu S, Wang Z, Xia Y, Xia L (2013) Methylene blue adsorption on graphene oxide/calcium alginate composites. Carbohydr Polym. https://doi.org/10.1016/j.carbpol.2013.01.094
43. Lin D, Shi M, Zhang Y, Wang D, Cao J, Yang J, Peng C (2019) 3D crateriform and honeycomb polymer capsule with nano re-entrant and screen mesh structures for the removal of Multi-component cationic dyes from water. Chem Eng J. https://doi.org/10.1016/j.cej.2019.121911
44. Luo X, Zhang L (2009) High effective adsorption of organic dyes on magnetic cellulose beads entrapping activated carbon. J Hazard Mater. https://doi.org/10.1016/j.jhazmat.2009.06.009

45. Ma Y, Qi P, Ju J, Wang Q, Hao L, Wang R, Sui K, Tan Y (2019) Gelatin/alginate composite nanofiber membranes for effective and even adsorption of cationic dyes. Compos Part B Eng. https://doi.org/10.1016/j.compositesb.2019.01.048

46. Madhusudana Rao K, Krishna Rao KSV, Sudhakar P, Chowdoji Rao K, Subha MCS (2013) Synthesis and characterization of biodegradable poly (vinyl caprolactam) grafted on to sodium alginate and its microgels for controlled release studies of an anticancer drug. J Appl Pharm Sci. https://doi.org/10.7324/JAPS.2013.3609

47. Mahmoodi NM, Hayati B, Arami M, Bahrami H (2011) Preparation, characterization and dye adsorption properties of biocompatible composite (alginate/titania nanoparticle). Desalination. https://doi.org/10.1016/j.desal.2011.02.034

48. Mohammadi A, Daemi H, Barikani M (2014) Fast removal of malachite green dye using novel superparamagnetic sodium alginate-coated Fe_3O_4 nanoparticles. Int J Biol Macromol. https://doi.org/10.1016/j.ijbiomac.2014.05.042

49. Manna S, Bhattacharya S, Sengupta S, Das P (2018) Synthesis of graphene oxide dots coated biomatrices and its application for the removal of multiple pollutants present in wastewater. J Clean Prod. https://doi.org/10.1016/j.jclepro.2018.08.261

50. Marrakchi F, Khanday WA, Asif M, Hameed BH (2016) Cross-linked chitosan/sepiolite composite for the adsorption of methylene blue and reactive orange 16. Int J Biol Macromol. https://doi.org/10.1016/j.ijbiomac.2016.09.069

51. Martel B, Devassine M, Crini G, Weltrowski M, Bourdonneau M, Morcellet M (2001) Preparation and sorption properties of a β-cyclodextrin-linked chitosan derivative. J Polym Sci Part A Polym Chem. https://doi.org/10.1002/1099-0518(20010101)39:1%3c169::AID-POL A190%3e3.0.CO;2-G

52. Munjur HM, Hasan MN, Awual MR, Islam MM, Shenashen MA, Iqbal J (2020) Biodegradable natural carbohydrate polymeric sustainable adsorbents for efficient toxic dye removal from wastewater. J Mol Liq https://doi.org/10.1016/j.molliq.2020.114356

53. Meimoun J, Wiatz V, Saint-Loup R, Parcq J, Favrelle A, Bonnet F, Zinck P (2018) Modification of starch by graft copolymerization. Starch/Staerke. https://doi.org/10.1002/star.201600351

54. Moharrami P, Motamedi E (2020) Application of cellulose nanocrystals prepared from agricultural wastes for synthesis of starch-based hydrogel nanocomposites: efficient and selective nanoadsorbent for removal of cationic dyes from water. Bioresour Technol. https://doi.org/10.1016/j.biortech.2020.123661

55. Nasrollahzadeh M, Sajjadi M, Iravani S, Varma RS (2021) Starch, cellulose, pectin, gum, alginate, chitin and chitosan derived (nano)materials for sustainable water treatment: a review. Carbohydr Polym. https://doi.org/10.1016/j.carbpol.2020.116986

56. Nikfar S, Jaberidoost M (2014) Dyes and colorants. In: Encyclopedia of toxicology, 3rd edn. https://doi.org/10.1016/B978-0-12-386454-3.00602-3

57. Parlayici Ş (2019) Alginate-coated perlite beads for the efficient removal of methylene blue, malachite green, and methyl violet from aqueous solutions: kinetic, thermodynamic, and equilibrium studies. J Anal Sci Technol. https://doi.org/10.1186/s40543-019-0165-5

58. Pourjavadi A, Ghasemzadeh H, Soleyman R (2007) Synthesis, characterization, and swelling behavior of alginate-g-poly(sodium acrylate)/kaolin superabsorbent hydrogel composites. J Appl Polym Sci 105:2631–2639. https://doi.org/10.1002/app.26345

59. Pradeep Sekhar C, Kalidhasan S, Rajesh V, Rajesh N (2009) Bio-polymer adsorbent for the removal of malachite green from aqueous solution. Chemosphere. https://doi.org/10.1016/j.chemosphere.2009.07.068

60. Puvaneswari N, Muthukrishnan J, Gunasekaran P (2006) Toxicity assessment and microbial degradation of azo dyes. Indian J Exp Biol

61. Radoor S, Karayil J, Parameswaranpillai J, Siengchin S (2020). Removal of anionic dye Congo red from aqueous environment using polyvinyl alcohol/sodium alginate/ZSM-5 zeolite membrane. Sci Rep. https://doi.org/10.1038/s41598-020-72398-5

62. Raval NP, Shah PU, Shah NK (2016) Nanoparticles loaded biopolymer as effective adsorbent for adsorptive removal of malachite green from aqueous solution. Water Conserv Sci Eng. https://doi.org/10.1007/s41101-016-0004-0

63. Reddy MM, Vivekanandhan S, Misra M, Bhatia SK, Mohanty AK (2013) Biobased plastics and bionanocomposites: current status and future opportunities. Prog Polym Sci. https://doi. org/10.1016/j.progpolymsci.2013.05.006

64. Rezaei H, Haghshenasfard M, Moheb A (2017) Optimization of dye adsorption using Fe_3O_4 nanoparticles encapsulated with alginate beads by Taguchi method. Adsorpt Sci Technol. https://doi.org/10.1177/0263617416667508

65. Rosa S, Laranjeira MCM, Riela HG, Fávere VT (2008) Cross-linked quaternary chitosan as an adsorbent for the removal of the reactive dye from aqueous solutions. J Hazard Mater. https:// doi.org/10.1016/j.jhazmat.2007.11.059

66. Russo T, Fucile P, Giacometti R, Sannino F (2021) Sustainable removal of contaminants by biopolymers: a novel approach for wastewater treatment. Current state and future perspectives. Processes 9:719. https://doi.org/10.3390/pr9040719

67. Sabarish R, Unnikrishnan G (2018) Polyvinyl alcohol/carboxymethyl cellulose/ZSM-5 zeolite biocomposite membranes for dye adsorption applications. Carbohydr Polym. https://doi.org/ 10.1016/j.carbpol.2018.06.123

68. Sadeghi-Kiakhani M, Arami M, Gharanjig K (2013) Preparation of chitosan-ethyl acrylate as a biopolymer adsorbent for basic dyes removal from colored solutions. J Environ Chem Eng. https://doi.org/10.1016/j.jece.2013.06.001

69. Salleh MAM, Mahmoud DK, Karim WAWA, Idris A (2011) Cationic and anionic dye adsorption by agricultural solid wastes: a comprehensive review. Desalination. https://doi.org/10. 1016/j.desal.2011.07.019

70. Salisu A, Sanagi MM, Naim AA, Karim KJ (2015) Removal of methylene blue dye from aqueous solution using alginate grafted polyacrylonitrile beads. Der Pharma Chem

71. Sami AJ, Khalid M, Iqbal S, Afzal M, Shakoori AR (2016) Synthesis and application of chitosan-starch based nanocomposite in wastewater treatment for the removal of anionic commercial dyes. Pak J Zool. https://doi.org/10.17582/journal.pjz/2017.49.1.21.26

72. Sen G, Ghosh S, Jha U, Pal S (2011) Hydrolyzed polyacrylamide grafted carboxymethylstarch (Hyd. CMS-g-PAM): an efficient flocculant for the treatment of textile industry wastewater. Chem Eng J 171:495–501. https://doi.org/10.1016/j.cej.2011.04.016

73. Sharma K, Kaith BS, Kumar V, Kalia S, Kumar V, Swart HC (2014) Water retention and dye adsorption behavior of Gg-cl-poly(acrylic acid-aniline) based conductive hydrogels. Geoderma. https://doi.org/10.1016/j.geoderma.2014.04.035

74. Silva LS, Lima LCB, Silva FC, Matos JME, Santos MRMC, Santos Júnior LS, Sousa KS, da Silva Filho EC (2013) Dye anionic sorption in aqueous solution onto a cellulose surface chemically modified with aminoethanethiol. Chem Eng J. https://doi.org/10.1016/j.cej.2012. 11.118

75. Singh B, Sharma V, Kumar S (2011) Synthesis of smart hydrogels by radiation polymerisation for use as slow drug delivery devices. Can J Chem Eng. https://doi.org/10.1002/cjce.20456

76. Sokker HH, Halim ESA, Aly AS, Hashem A (2004) Cellulosic fabric wastes grafted with DMAEMA for the removal of direct dyes. Adsorpt Sci Technol. https://doi.org/10.1260/026 3617043026497

77. Sonawane SH, Chaudhari PL, Ghodke SA, Parande MG, Bhandari VM, Mishra S, Kulkarni RD (2009) Ultrasound assisted synthesis of polyacrylic acid-nanoclay nanocomposite and its application in sonosorption studies of malachite green dye. Ultrason Sonochem. https://doi. org/10.1016/j.ultsonch.2008.10.008

78. Song W, Gao B, Xu X, Xing L, Han S, Duan P, Song W, Jia R (2016) Adsorption-desorption behavior of magnetic amine/Fe_3O_4 functionalized biopolymer resin toward anionic dyes from wastewater. Bioresour Technol. https://doi.org/10.1016/j.biortech.2016.01.078

79. Stan M, Lung I, Soran ML, Opris O, Leostean C, Popa A, Copaciu F, Lazar MD, Kacso I, Silipas TD, Porav AS (2019) Starch-coated green synthesized magnetite nanoparticles for removal of textile dye Optilan Blue from aqueous media. J Taiwan Inst Chem Eng. https://doi.org/10.1016/ j.jtice.2019.04.006

80. Sudha M, Saranya A (2014) Microbial degradation of azo dyes: a review. Int J Curr Microbiol 3:670–690. ISSN: 2319-7706

81. Swain SK, Patnaik T, Dey RK (2013) Efficient removal of fluoride using new composite material of biopolymer alginate entrapped mixed metal oxide nanomaterials. Desalin Water Treat. https://doi.org/10.1080/19443994.2012.749426

82. Thakur S, Sharma B, Verma A, Chaudhary J, Tamulevicius S, Thakur VK (2018) Recent progress in sodium alginate based sustainable hydrogels for environmental applications. J Clean Prod. https://doi.org/10.1016/j.jclepro.2018.06.259

83. Tian D, Zhang X, Lu C, Yuan G, Zhang W, Zhou Z (2014) Solvent-free synthesis of carboxylate-functionalized cellulose from waste cotton fabrics for the removal of cationic dyes from aqueous solutions. Cellulose. https://doi.org/10.1007/s10570-013-0112-3

84. Tran HV, Bui LT, Dinh TT, Le DH, Huynh CD, Trinh AX (2017) Graphene oxide/Fe$_3$O$_4$/chitosan nanocomposite: a recoverable and recyclable adsorbent for organic dyes removal. Application to methylene blue. Mater Res Express 4(3):035701

85. Varaprasad K, Vimala K, Ravindra S, Reddy NN, Raju KM (2011) Development of Sodium Carboxymethyl Cellulose-based Poly(acrylamide-co-2acrylamido-2-methyl-1-propane sulfonic acid) hydrogels for in vitro drug release studies of ranitidine hydrochloride an anti-ulcer drug. Polym Plast Technol Eng 50:1199–1207. https://doi.org/10.1080/03602559.2011.553872

86. Varghese AG, Paul SA, Latha MS (2019) Remediation of heavy metals and dyes from wastewater using cellulose-based adsorbents. Environ Chem Lett. https://doi.org/10.1007/s10311-018-00843-z

87. Wang Y, Zhang X, He X, Zhang W, Zhang X, Lu C (2014) In situ synthesis of MnO2 coated cellulose nanofibers hybrid for effective removal of methylene blue. Carbohydr Polym. https://doi.org/10.1016/j.carbpol.2014.04.008

88. Wu N, Wei H, Zhang L (2012) Efficient removal of heavy metal ions with biopolymer template synthesized mesoporous titania beads of hundreds of micrometers size. Environ Sci Technol 46:419–425. https://doi.org/10.1021/es202043u

89. Xia K, Liu X, Wang W, Yang X, Zhang X (2020) Synthesis of modified starch/polyvinyl alcohol composite for treating textile wastewater. Polymers (Basel). https://doi.org/10.3390/polym12020289

90. Yang X, Chen K, Zhang Y, Liu H, Chen W, Yao J (2017) Polyacrylamide grafted cellulose as an eco-friendly flocculant: efficient removal of organic dye from aqueous solution. Fibers Polym. https://doi.org/10.1007/s12221-017-1216-4

91. Yang X, Zhang Q, Wang Y, Chen H, Zhang H, Gao F, Liu L (2008) Self-aggregated nanoparticles from methoxy poly(ethylene glycol)-modified chitosan: Synthesis; characterization; aggregation and methotrexate release in vitro. Colloids Surfaces B Biointerfaces 61:125–131. https://doi.org/10.1016/j.colsurfb.2007.07.012

92. Yin J, Deng B (2015) Polymer-matrix nanocomposite membranes for water treatment. J Memb Sci. https://doi.org/10.1016/j.memsci.2014.11.019

93. Zeng L, Xie M, Zhang Q, Kang Y, Guo X, Xiao H, Peng Y, Luo J (2015) Chitosan/organic rectorite composite for the magnetic uptake of methylene blue and methyl orange. Carbohydr Polym. https://doi.org/10.1016/j.carbpol.2015.01.021

94. Zhang X, Yu H, Yang H, Wan Y, Hu H, Zhai Z, Qin J (2015) Graphene oxide caged in cellulose microbeads for removal of malachite green dye from aqueous solution. J Colloid Interface Sci. https://doi.org/10.1016/j.jcis.2014.09.048

95. Zhao Y, Wang L (2012) Adsorption characteristics of Congo red from aqueous solution on the carboxymethylcellulose/montmorillonite nanocomposite. In: Advanced materials research. https://doi.org/10.4028/www.scientific.net/AMR.450-451.769

96. Zhou L, Jin J, Liu Z, Liang X, Shang C (2011) Adsorption of acid dyes from aqueous solutions by the ethylenediamine-modified magnetic chitosan nanoparticles. J Hazard Mater. https://doi.org/10.1016/j.jhazmat.2010.10.012

97. Zhou L, Zhou H, Yang X (2019) Preparation and performance of a novel starch-based inorganic/organic composite coagulant for textile wastewater treatment. Sep Purif Technol. https://doi.org/10.1016/j.seppur.2018.07.089

98. Zobel HF (1988) Molecules to granules: a comprehensive starch review. Starch Stärke. https://doi.org/10.1002/star.19880400203
99. Zou J, Zhu H, Wang F, Sui H, Fan J (2011) Preparation of a new inorganic-organic composite flocculant used in solid-liquid separation for waste drilling fluid. Chem Eng J. https://doi.org/10.1016/j.cej.2011.03.100

Applications of Inorganic Polymers in Textile Wastewater Treatment

G. K. Prashanth, M. S. Dileep, P. A. Prashanth, Manoj Gadewar, B. M. Nagabhushana, and S. R. Boselin Prabhu

1 Introduction

Huge number of atoms are bound to each other to form a very long chain of macro-molecule in 1D (one dimensional) array is called a polymer. In Greek, the word poly means many and meros means parts/units, i.e., a part (monomer) is repeated several times in a single unit. Polyethylene is one of the basic examples of polymers where ethylene is the monomer. The molecular properties create the characteristics of solid materials, such as elasticity, strength, film-forming, or fiber-forming qualities, which are not found in small-molecule systems. Polymers are usually so high in molecular weights that they are non-volatile for all practical purposes. The extensive use of polymers in all aspects of modern technology underpins these characteristics. In this

G. K. Prashanth (✉)
Department of Chemistry, Sir M. Visvesvaraya Institute of Technology, Bengaluru 562 157, India
e-mail: prashaanthgk@gmail.com

G. K. Prashanth · M. S. Dileep · P. A. Prashanth · B. M. Nagabhushana
Visvesvaraya Technological University, Belagavi 590 018, India

M. S. Dileep
Department of Physics, Sir M. Visvesvaraya Institute of Technology, Bengaluru 562 157, India

P. A. Prashanth
Department of Chemistry, PES College of Engineering, Mandya 571 401, India

M. Gadewar
Department of Pharmacology, School of Pharmacy, Vishwakarma University, Pune 411048, India

B. M. Nagabhushana
Department of Chemistry, M.S. Ramaiah Institute of Technology, Bengaluru 560054, India

S. R. B. Prabhu
Department of ECE, Surya Engineering College, Mettukadai 638 107, India

© The Author(s), under exclusive license to Springer Nature Singapore Pte Ltd. 2022
A. Khadir and S. S. Muthu (eds.), *Polymer Technology in Dye-containing Wastewater*,
Sustainable Textiles: Production, Processing, Manufacturing & Chemistry,
https://doi.org/10.1007/978-981-19-0886-6_10

field, many of the basic sciences effort to characterize to understand the relation between the macromolecular structure and the unusual properties [31].

There are mainly two types of polymers: one derived from organic and another from inorganic sources. Organic polymers mainly derive from petroleum, plants, animals, and microorganism. Therefore, they are used in daily life in excess and without them, life is difficult to imagine too. Inorganic polymers have inorganic repeating units. The inorganic polymers have more advantages over the organic polymers. The backbone of the organic polymer may react with oxygen or ozone and lose its activity. Most of the organic polymers release the toxic smoke when burned. Many organic polymers degrade in the ultra-violet and gamma radiation. Most organic polymers dissolve in organic solvents and lose their activity. The second set of reasons for the rising interest in inorganic macromolecules is related to the differences they have known or anticipated from their completely organic counterparts. Inorganic elements produce various combinations of polymer properties compared to carbon atoms. For a thing, the bonds forming between inorganic elements are often longer, stronger, and more durable than carbon-formed bonds to free radical cleavage reactions [31].

In the last ten years, a substantial amount of work has been reported that involves the development and applications of partially inorganic and inorganic polymers, but certain products still remain as laboratory curiosities only [39]. The most economically significant polymer of inorganic, which has originated various uses is polysiloxane (silicones). In more recent years, polyphosphazene, carborane, and poly-silane polymers have been developed. Coordination polymers and organometallic polymers are other prominent polymers of inorganic section. There are some very interesting wholly inorganic polymers, as well as partial inorganic polymers, which include polymer sulfur, boron nitride, poly-sulfur nitride, asbestos, and glass [33]]. Polysiloxane, polyphosphazene, polysilane polymers have several applications like cosmetics, O-rings, gaskets, lubricating oils, hydraulic fluids, greases, biomedical devices, sealants, oil-resistant hoses, elastomers, high-temperature elastomers, and silicon-carbide [6, 13, 27, 30, 37, 43]. There is a large range of polymers in inorganic chemistry and these novel polymers are more attractive as they possess several properties such as biomedical application, electrical conductors, and engineering materials. Therefore it is an opportunity to expand the basic knowledge in the field of inorganic polymers to develop new materials which contribute to advancing technology.

For humans, plants, and animals, water is the fundamental commodity. Drinking water is a big problem in the recent years since population density is slowly increasing the mixing of wastewater into drainage from various industry sectors such as dyes, pigments, textiles, and photographic industries [17, 19, 55]. Majority of the wastewater from textile industries contains a variety of natural colors that are toxic and extremely harmful to nature, humans, and sea animals [53]. Diseases such as skin rashes, allergies, liver, and kidney damage may be caused by non-biodegradable dyes [51, 16].

Numerous scientists all over world are researching the area of wastewater treatment by using inorganic polymers. A chemist can develop infinite polymers by judiciously mixing various chemicals to react under desirable conditions, just as an architect chooses bricks, stones, and logs of wood in diverse forms, sizes, and patterns to create various designs. Researchers are synthesizing different types of inorganic polymers such as polymeric nanocompostion and polymeric metal complexes which are used for treatment of wastewater.

We cover the synthesis of several inorganic polymers and their applications in the field of wastewater treatment in this book chapter.

2 Synthetic Techniques of Inorganic Polymers

The inorganic polymer can be synthesized by different methods such as sol–gel method, solution casting, melt compounding, intercalation, in situ polymerization, spinning, template synthesis, phase separation, electrochemical synthesis, hydrothermal method, and slow evaporation method [5, 10, 12, 14, 23, 26, 36, 40, 44, 46, 47, 49].

Polymerization of zinc oxide has more advantages over zinc oxide nanoparticles. Polymeric ZnO was synthesized by Shukla et al. using the simple sol–gel method [46]. Nanoparticles have been used to improve the compatibility, ion exchange, and surface interaction of organic molecules with each other in a range of nanomaterials. Many times nanoparticles were also functionalized prior to dispersion in the matrix, enabling them to interact more effectively with organic molecules. Fe_3O_4@chitosan(CS)@graphene oxide(GS) was coated by the ionic liquid. Fe_3O_4@CS@GO (1 g) dissolved in ionic liquid and resulting mixture was sonicated for two hours [25].

3 Characterization of Inorganic Polymers

The characterization of inorganic polymer is a vital element of material study, and without it, we cannot be sure that polymer is developed. To understand the material scientifically, the basic procedure is essentially required. Methods for investigating material properties and microscopic structures, such as mechanical, thermal, and density investigations are included in characterization. Characterization aids in the definition of material structure and composition and allows us to assess whether the synthetic technique was successful. We utilize many characterization techniques such as UV–visible spectroscopy (Fig. 1a), FTIR (Fourier Transform Infrared) (Fig. 1b), TEM (Transmission Electron Microscopy) (Fig. 1c), SEM (Scanning Electron Microscopy) (Fig. 1d), XRD (X-ray Diffraction) (Fig. 2a), and Zeta potential (Fig. 2b) [8] in this part.

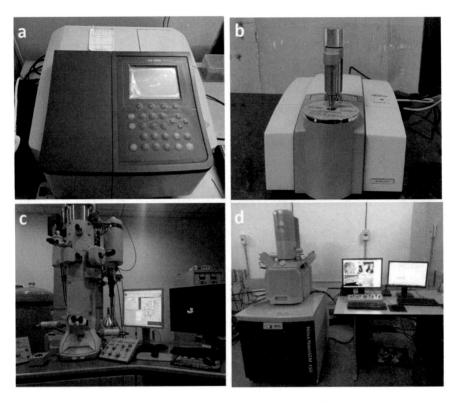

Fig. 1 **a** UV–visible spectroscope; **b** FTIR spectroscope; **c** TEM; **d** SEM

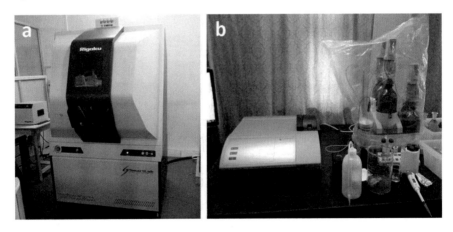

Fig. 2 **a** X-ray diffractometer; **b** Zeta particle size analysis

4 Some Examples with Applications

4.1 Ti³⁺ Base Polymer

Ghosh et al. synthesized two π-conjugation organic donor-aceptor1-donor aceptor2 co-polymers namely P1 and P2 [18]. The reaction is given in Fig. 3. The polymers were synthesized by using benzothia/selenadiazole and 1,5-dihydroxyanthraquinone by the conventional polycondensation reaction (Stille coupling). They also synthesized Ti^{3+}/TiO_2 nanorods (NRs) by using these organic semiconductors with Ti^{3+} self-drop TiO_2. The polymers were characterized by different techniques such as SEM, XRD; thermal stability of polymer; absorption and emission spectra.

The photocatalytic activity of Ti^{3+}/TiO_2 NRs, P1, and P2 was performed against 10 μM aqueous solution of organic pollutant dye rhodamine B (RhB) in presence of visible light. The reaction conditions were pH 7, 150 min, and in presence of visible light. The degradations of RhB after 50 min were 17.4, 32.46, and 25.33% by Ti^{3+}/TiO_2 NRs, P1, and P2, respectively. They concluded that these reactions occurred in the same first-order reaction kinetics. The rate constant values of Ti^{3+}/TiO_2 NRs, P1, and P2 were 11.97×10^{-4}, 24.65×10^{-4}, and 19.98×10^{-4} min^{-1}, respectively. Moreover, they continued their investigation on the same reaction in acidic medium (H_2SO_4). The degradation of RhB increased up to 48% and 45.70% for P1 and P2, respectively. The rate constants of P1 and P2 were 45.24×10^{-4} and 36.92×10^{-4} min^{-1}, respectively [18]. RhB degradation by $Ti^{3+}/TiO2$ NRs, P1, P2, P1 with 1 mM H_2SO_4 and P2 with 1 $_{mM}$ H_2SO_4 in pH 7, 150 min and visible light, kinetics of Ti^{3+}/TiO_2 NRs, P1 and P2. And kinetics of P1 with 1 mM H_2SO_4 and P2 with 1 $_{mM}$ H_2SO_4 are depicted in Fig. 4a–c, respectively.

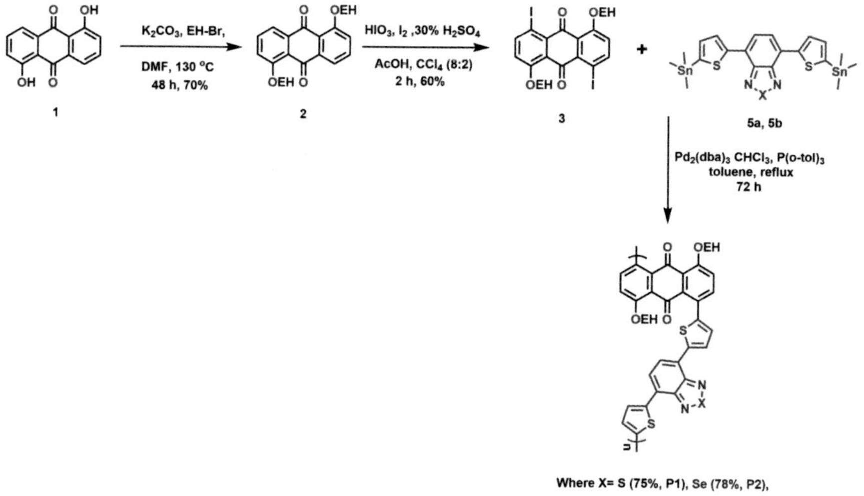

Where X= S (75%, P1), Se (78%, P2),

Fig. 3 Synthesis process of P1 and P2 Polymers. [Reprinted with permission from [18]]

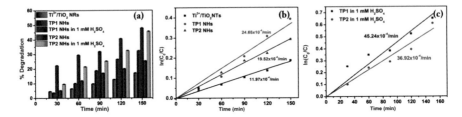

Fig. 4 **a** RhB degradation by Ti^{3+}/TiO_2 NRs, P1, P2, P1 with 1 $_{mM}$ H_2SO_4 and P2 with 1 $_{mM}$ H_2SO_4 in pH 7, 150 min and visible light. **b** Kinetics of Ti^{3+}/TiO_2 NRs, P1 and P2. **c** Kinetics of P1 with 1 mM H_2SO_4 and P2 with 1 $_{mM}$ H_2SO_4 [Reprinted with permission from [18]]

4.2 CNT/Cdot/FA/TiO₂ Polymer

Water purification involves the use of less poisonous and cost-significant catalysts like composites of organic–inorganic polymers. Mallakpour et al. have discussed the production of photocatalysts, including magnesium fluorohydroxyapatite (FA); carbon dots (Cdots); TiO_2 nanoparticles (CNT/Cdot/FA/TiO₂), and multi-walled carbon nanotubes (CNTs) for using them in dye like methylene blue (MB) degradation in the aqueous medium [29]. The polymers were characterized by EDX (Energy-dispersive X-rays analysis); Patterns of XRD; FE-SEM (Field emission scanning electron microscopy); TEM; and BET (Brunauer–Emmett–Teller). The optical bands of CAlg/CNT/Cdot/FA/TiO₂, CNT/Cdot/FA composite, pure TiO_2, CNT/Cdot/FA/TiO₂, and pure FA were 4.87 eV, 4.86 eV, 3.19 eV, 3.26 eV and 4.63 eV, respectively. They investigated the photocatalytic activity against methylene blue. They showed that the sanitization yield (40%) of this biocatalyst was higher with UV light (28%) than methylene blue sorption in the dark conditions. The presence of TiO_2 NPs ensured that CAlg/CNT/Cdot/FA/TiO₂ displayed a large efficient separation of dye in presence of UV than that of dark environments (sorption) (Fig. 5a and b). In order to assess the potential for degradation of biocatalyst, pH variation of the MB in aqueous medium (4, 6, and 8) was done and its results were described in Fig. 5c. Varying the pH values did not make any significant changes in MB degradation. Figure 5d shows the influence of biocatalyst polymer CAlg/CNT/Cdot/FA/TiO₂. The biocatalyst quantity was increased from 0.02 g to 0.2 g due to the availability of additional active surface sites on the catalyst surface (condition; MB, $t = 315$ min and 10 mg L^{-1}), photo-degradation increased accordingly. Among all polymers, CNT/Cdot/FA/TiO₂ was most effective, degrading MB in only 27 min as shown in Fig. 5e medium [29].

4.3 Bauxite Residues (BR) Polymer

The use of modified bauxite residues (BR) synthesized by porous inorganic polymers (IP) monoliths as the reactive fraction for the use as adsorbent for methylene

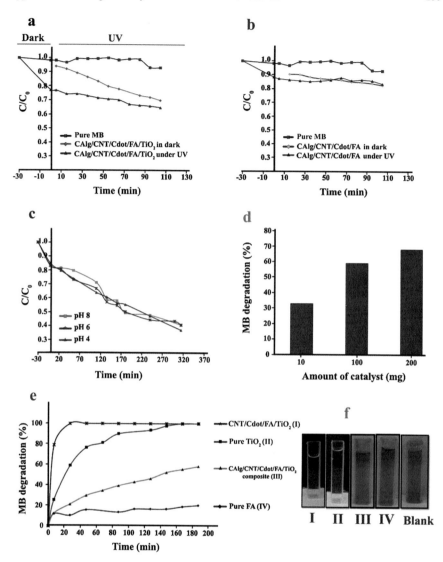

Fig. 5 Photo catalytic activity studies against MB in aqueous medium. **a** polymer CAlg/CNT/Cdot/FA/TiO$_2$; **b** polymer CAlg/CNT/Cdot/FA (MB-10 mgL^{-1}, Catalyst-20 mg Time-315 min); **c** pH; **d** catalyst amount (Methylene Blue-10mgL^{-1}, pH 6, Catalyst-20 mg Time-315 min); **e** comparative study; **f** some images of degradation. [Reprinted with permission from [29]]

blue (MB) in synthetic wastewater has been reported. The polymers were characterized by zeta potential; SEM; FTIR; and XRD. Firstly, un-reactive, as produced bauxite residues were turned into an appropriate glassy-precursor-material for inorganic polymers (IPs) through the blending of bauxite residues (BR) through small amounts of carbon (c), and CaSiO$_3$; after the thermal treatment with 1200 °C for 2 h, and quench H_2O. The activation of alkaline slags together through the foaming agent leads to an extremely porous-micro-structure with an overall porosity of up to 85%. The synthesized porous monoliths showed large methylene blue intakes (approximately 17 mg of MBg^{-1} IPs with an initial 75 mgL^{-1} concentration of methylene blue). Higher monolith porosity, higher pH, increased starting MB concentration along with an agitation of the test solution have a positive effect on the ability for adsorption; while optimal adsorption volume of the solution has been recognized. In addition, the adsorption tests up to the five cycles (adsorption and desorption) were repeated; with approximately a 30% drop in adsorption capability; but promising cumulative consumption of around 40 mg of methylene blue per gram IPs; demonstrated the recycle of these new monolithic adsorbents. The greatest benefit of the monoliths is that the adsorbents from wastewater tanks were not filtered because their integrity remained intact. The surface arrangement of the adsorbents; which can induce particulate collision by stirring, was not damaged compared to granular adsorbents. In the proposed process, no other reactive precursor needed to be added to the BR and made it independent of other industries. Furthermore, since previous waste materials are converted into a reactive precursor and a new functional binder for wastewater purification, the research could help to enhance social perception of BR [20]. Optimization of MB volume with sample (T, 23 ± 2 °C; C0, 50 mg/L; V, 200 mL) eft) and pH (0.3A/0.45; T, 23 ± 2 °C; C0, 35 mg/L; V, 200 mL) are given in Fig. 6.

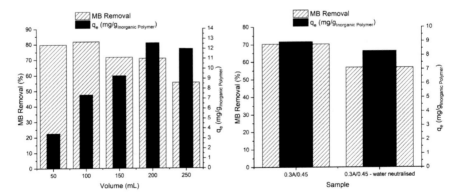

Fig. 6 Optimization of MB volume with sample (T, 23 ± 2 °C; C_0, 50 mg/L; V, 200 mL) (left) and pH (0.3A/0.45; T, 23 ± 2 °C; C_0, 35 mg/L; V, 200 mL) (right). [Reprinted with permission from [20]]

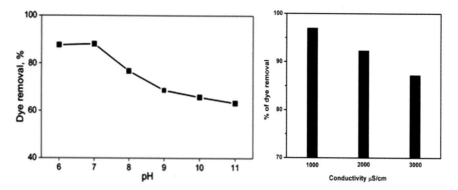

Fig. 7 Optimization of the pH on dye separation ability of AM-DMDAAC (PAFC-Starch-g-p). (Condition: 0.2 mLmg^{-1} dye; 100 mgL^{-1} reactive blue-KN R) (Left). Dye removal efficiency by conductivity of AM-DMDAAC (PAFC-starch-g-p) (Condition: 0.2 mLmg^{-1} dye, 100 mgL^{-1} reactive green-KE 4B) (right). [Reprinted with permission from [58]]

4.4 PAFC-Starch-G-Polymer

A new 'PAFC-Starch-g-p' (AM-DMDAAC) inorganic-natural composite of the polymer was prepared and applied to the treatment of textile wastewater. AM-DMDAAC (PAFC-Starch-g-p) was synthesized using starch, polyaluminum ferric chloride (PAFC), acrylamide (AM), and dimethyl diallyl ammonium chloride (DMDAAC). The polymer was characterized by inorganic species distribution, SEM, and Zeta potential tests.

It has been shown that the efficiency of AM-DMDAAC (PAFC-Starch-g-p) was larger and had huge stability than other polymeric-composite coagulants prepared. AM-DMDAAC (PAFC-Starch-g-p) can decrease the amount of the chemical by more than 50% in comparison to the conventional coagulation floction process for the treatment of synthetic textile wastewater. Textile wastewater often has high pH, as alkaline in pretreatment processes such as mercerizing and scouring was most needed [56]. AM-DMDAAC (PAFC-Starch-g-p) had an effect of pH on the efficiency of dye removal in Fig. 7. Accelerated high-temperature storage tests have demonstrated huge storage stability of AM-DMDAAC (PAFC-Starch-g-p) as shown in Table 1 [58].

4.5 Fe/Zr-Pillared-Clinoptilolite (Nano-Fe/Zr PC)

The surface water has excessive concentrsations of heavy materials ammonium and phosphate that could cause health and environmental problems. The novel Nano Fe/Zr PC (Fe/Zr pillared clinoptilolite) was synthesized through impregnating clinoptilolite with poly-hydroxy cations such as Fe^{3+}/Zr^{4+}. The XRD; BET analysis; FT-IR, and

Table 1 Storage stability of AM-DMDAAC (PAFC-starch-g-p)

Storage (d)	pH	Conductivity (mS/cm)	Classes distribution			Reactive blue-KN-R
			Ma %	Mb %	Mc %	(%)*
1	3.575	23.6	88.8	0.9	10.3	88.99
2	3.537	23.6	89.9	1.4	8.7	84.42
3	3.577	23.7	90.4	0.5	9.2	85.92
4	3.440	23.6	90.1	5.5	4.4	83.42
5	3.546	23.7	89.4	7.7	· 2.9	81.22

* 100 mgL^{-1} (initial concentration of dye); 0.2 mLmg^{-1} (dye coagulant)
[Reprinted with permission from [58]]

TEM, were used to characterize the material obtained, which indicated that the nano-scale Fe or Zr of the clinoptilolite was successfully formed at both surface and pores. The effects of various operational parameters on adsorption behavior were studied in batch experiments. Phosphate, cadmium (Cd), and ammonium were separated with Nano-Fe/Zr-PC from wastewater in the pH range between 3 and 10 and consequent results are shown in Fig. 8a. In these pH ranges, when pH of the solution was increased, the ammonium removal rate increased gradually from pH 4 to 8 and reached the maximum (97.13%) at pH 8 and then decreased significantly at pH 9–10. The efficiency of the phosphate recovery was maximized at 93% in acidic conditions (pH less than 7), while the recovery rate of phosphate decreased markedly when the pH increased to more than 7 as shown in Fig. 8a. The efficiencies of removal of cadmium (Cd^{+2}) were slightly increased with the pH range between 3 and 10. The capability of the cadmium (Cd^{+2}) recovery was maximized to 99.3% at pH 10. Figure 8b shows the test results for the measurement of the effect of dosage of an adsorbent on the removal of cadmium (Cd^{+2}), ammonium, and phosphate. The rate of deletion of cadmium (Cd^{+2}) and ammonium rapidly increased from 52.4% to 94.9% and 24.7% to 98.4%, respectively; with an increment the nano-Fe/Zr-PC dosages from 0.250 to 2.500 gL^{-1} as shown in Fig. 8b. The phosphate removal was

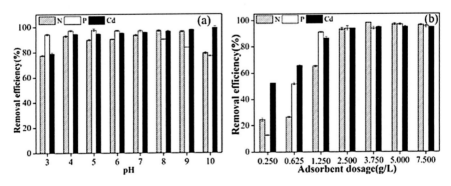

Fig. 8 Optimization **a** pH and **b** dosages of polymer. [Reprinted with permission from [57]]

Table 2 Kinetic of cadmium (Cd^{+2}), ammonium, and phosphate

Order	Kinetic-parameters	Ammonium	Phosphate	Cd^{+2}
Pseudo-first-order model	q_e (mg g^{-1})	26.57	12.72	3.42
	k_1 (mg(g·min)$^{-1}$	0.0029	0.1008	0.0210
	R^2	0.98	0.95	0.96
Pseudo-second-order model	q_e (mg g^{-1})	24.49	13.76	3.15
	k_2 (g (mg·min)$^{-1}$	0.0570	0.0091	0.0493
	R^2	0.99	0.99	0.98

[Reprinted with permission from [57]]

significantly increased from 12.9 to 94.0% with nano Fe/Zr-PC dosages rising from 0.250 to 1.250 gL^{-1} as shown in Fig. 8b. The kinetics of cadmium (Cd^{+2}), ammonium, and phosphate were followed by two models. One was a Pseudo-first-order model, another was a Pseudo-second-order model as shown in Table 2 [57].

4.6 Microspheres Inorganic Polymer

Tang et al. synthesized a new porous-microspheres inorganic polymer by using suspension dispersion solidification (SDS) process. The porous-microspheres showed an approximate diameter of 100 μm with good sphericity. The polymer was characterized by XRD; BET analysis; FT-IR; and SEM. The surface of the microsphere has been augmented to 100.99 m^{2}g^{-1} from 79.80 m^{2}g^{-1} with an additional foaming agent with a corresponding drop in their pores diameter of up to 7 nm. The experiment with adsorption batch was performed to assess the effect on the determined adsorption capacities of dosage, initial concentration, time, temperature, and pH. The adsorption rates for the unfoamed and foamed adsorbents in the solution of Pb (II) (20 mgL^{-1}) were compared, as exposed in Fig. 8, in order to investigate the effects of the forming-agent (H_2O_2, K_{12}) on adsorption performance. Additional adsorption amounts for non-foaming and foaming adsorbents were 305.43 mgg^{-1} and 362.84 mgg^{-1}, respectively; when the early concentration was augmented to 200 mg L^{-1}. Table 3 also examines the adsorption capabilities for various heavy-metal ions and self-governing adsorption abilities for Cd (II), Pb (II),

Table 3 Comparison of adsorption capacity, amount and competitive coefficient (a_i)

Ion	Cd (II)	Pb (II)	Cu (II)
Capacity of adsorption (mg g^{-1})	54.3	362.8	55.71
Adsorption amount	10	58.6	20
Coefficient of competitive (a_i)	0.08	0.65	0.28

[Reprinted with permission from [50]]

and Cu (II). The adsorption abilities of Cd (II), Pb (II), and Cu (II) were 54.26 mgg^{-1}, 362.84 mgg^{-1} and 55.71 mgg^{-1}, respectively. Adsorption experiments were carried out using a mixed solution that contains Cd (II), Pb (II), and Cu (II) to investigate adsorption-selectivity of the porous-microspheres inorganic polymer. Results showed that Cd (II), Pb (II), and Cu (II) were adsorbed by microspheres inorganic polymer at 10.00 mgg^{-1} 58.6 mgg^{-1}, and 20.00 mgg^{-1}, respectively [50].

The effect of time reaction on adsorption capability is shown in Fig. 8. The adsorption rate was very rapid during the first 2 h, with a removal rate of 74.43%. The adsorption reached equilibrium at five hours with entire adsorption capability of 374.80 mgg^{-1} and 99.95% which demonstrated the excellent performance of adsorption. The microsphere inorganic polymer adsorbents' kinetics of adsorption was studied by a pseudo-first-order kinetic model and pseudo-second-order kinetic model as presented in Table 4. This method for the synthesis of porous inorganic polymer adsorbents is simple to use, cost-effective, novel, and has high adsorption capability. Hence, these materials can efficiently be used as alternatives for the industrial elimination of Pb (II) ions from wastewater [50]. FT-IR spectra, before and after adsorption, and Optimization of time are given in Fig. 9.

Table 4 Kinetic model of the polymer

Kinetic model	Q_e ca (mg g^{-1})	k_1 or k_2	$(R)^2$	$(\chi)^2$
Pseudo-first-order	405.30	$k_1 = 18.72 \times 10^{-3}$	0.956	4.5
Pseudo-second-order	384.62	$k_2 = 1.63 \times 10^{-4}$	0.994	0.6

[Reprinted with permission from [50]]

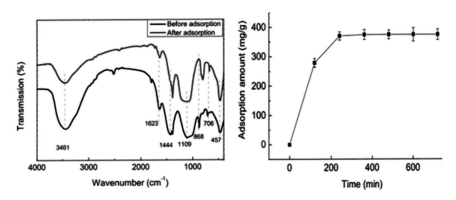

Fig. 9 FT-IR spectra: before and after adsorption (left); Optimization of time (right) [Reprinted with permission from [50]]

4.7 Polymeric Ferric Silicate Aluminum (PAFSi)

Polymeric ferric silicate aluminum (PAFSi), a composite coagulant, has been developed for the treatment of wastewater containing high oils. The structure of PAFSi coagulants was characterized by XRD, FT-IR, and SEM. The oil and chemical oxygen demand (COD) recovery rate was used for assessing PAFSi coagulation capacity in the treatment of wastewater containing high oil. The impact of the basicity on the oil and COD removal rates in wastewater is shown in Fig. 10(left). The basicity was increased, and then the removal capability of oil and COD from wastewater increased and then decreased. The maximum efficiencies of oil and COD removal were 93.0%, and 94.8%, respectively from the bases 0.6 and 0.5. Figure 10(right) shows the impacts of temperature on the oil and COD removal rates in oily wastes. Initially, removal capabilities of oil and COD were increased by increasing the preparation temperature. When the temperature of the preparation was over 70 °C, the oil and COD removal rate decreased slightly. The maximum removal efficiencies of oil and COD from wastewater were 94.21% and 95.64%, respectively at 70 °C. The zeta potential analysis showed that the neutralization of charge and the adsorption/bridging effect played important roles in the removal of COD and oil by coagulation. However, the main mechanism for reducing colloid particulate surface charges for oily wastewater treatment was charging neutralization. PAFSi showed superior performances of coagulation–flocculation in high-oil wastewater treatment. It may be used in the industrial treatment of wastewater as an alternative pretreatment for high-oil-containing wastes [48].

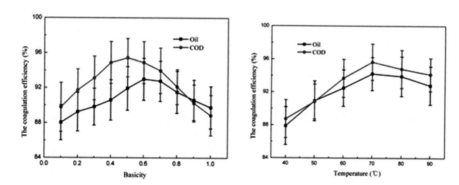

Fig. 10 (Left) Optimization effect of basicity for removal of oil and COD from wastewater (right) optimization effect of temperature for removal of oil and COD from wastewater. (condition: pH: 7; G value, 400 s^{-1}; dosages 60 mgL^{-1}; meanwhile; Al/Fe molar ratio, 1:1; Si/Fe molar ratio, 1:4; 60 °C) [Reprinted with permission from [48]]

5 Some Other Examples

The ground and surface water are polluted by organic dyes, heavy metals, and radioactive elements. Some other examples of inorganic polymers and their use in wastewater to remove dyes, heavy metals, and radioactive elements are shown in Table 5. The inorganic polymers have been used for purification of wastewater due to their tunable surface, small size, catalytic activity, magnetic properties, and reusable nature. Several methods have been employed for the purification of water by inorganic polymers like adsorption, chemical, coagulation, precipitation, filtration, flocculation, photo, biochemical degradation, catalytic, and ion exchange. For dye removal from waste, $\gamma Fe_2O_3/CSCs$ [60], $\gamma Fe_2O_3/SiO_2$/chitosan composite [59], Chitosan/Fe_2O_3/MWCNTs [61] Biopolymer/clay/conducting polymer nanocomposite [34] and Montmorillonite/$CoFe_2O_4$ composite [2] were used. Bistriazinylpyridine Polyvinyl alcohol/ titanium oxide [1] and TiO_2/Poly (acrylamide–styrene sodium sulfonate) detect the radioactive elements from the wastewater. Heavy metals in waste are identified using Chitosan gold nanocomposite [32], Polypyrrole decorated graphene/β-cyclodextrin composite [35], Mg-Fe layered double hydroxide [22], and other inorganic polymers. 2,4-dichlorophenol and paranitrophenol are detected by graphene oxide/molecularly imprinted polymer and Polyaniline-polyvinyl sulphonic acid composite, respectively [28, 41].

6 Conclusion

In the last 30 years, inorganic polymer science has progressively expanded through a series of synthesis advances, creating new materials with interesting features. The organic moiety and inorganic compounds have brought the evolution of inorganic polymers which have effective applications in the field of material science. To develop this area, it is necessary to combine the conventional, organometallic, organic, polymer chemistry and the manufacturing techniques for low-dimensional inorganic nanostructures with an interdisciplinary combination of conventional synthetic processes to create new materials. In this chapter, we have summarized the development of few new inorganic polymers with their uses in wastewater. We assume that this chapter can provide basic information regarding the latest inorganic polymers with their application in wastewater treatment and encourages active researchers in this field to use nanoparticles for the synthesis of inorganic/organic compounds.

Table 5 Name of some inorganic polymers with their applications in wastewater treatment

Entry	Inorganic polymer	Application	References
1	Bistriazinylpyridine Polyvinyl alcohol/titanium oxide	Thorium (IV) separation	[1]
2	TiO$_2$/Poly (acrylamide–styrene sodium sulfonate)	Cesium, Cobalt, and Europium ions removal	[7]
3	γFe$_2$O$_3$/CSCs	Methyl orange removal	[60]
4	Nanoparticles composites Chitosan gold nanocomposite	Lead nitrate removal	[32]
5	polyaniline/graphene composite	Hydrazine chemical removal	[4]
6	LCht/γFe$_2$O$_3$	Remazol red 198 removal	[11]
7	Polypyrrole decorated graphene/β-cyclodextrin composite	Hg(II) removal	[35]
8	Mg-Fe layered double hydroxide	Pb^{2+} removal	[22]
9	Geopolymer/alginate hybrid sphere	Pb^{2+} removal	[15]
10	γFe$_2$O$_3$/SiO$_2$/chitosan composite	Methyl orange removal	[59]
11	Graphene oxide/molecularly imprinted polymer	2,4-dichlorophenol removal	[28]
12	Fly ash-based geopolymer	Pb^{2+} removal	[3]
13	Porous biocomposites	Pb^{2+} removal	[21]
14	Metakaolin-based geopolymer	Pb^{2+} removal	[9]
15	CNTs/activated carbon fiber	Basic violet 10 removal	[54]
16	Silver imprinted polyvinyl alcohol nanocomposite	Hg(II) removal	[42]
17	Graphene oxide-based polymer nanocomposites	Pollutant removal	[52]
18	Hexacyanoferrate (II) polymeric nanocomposite	Cesium removal	[45]
19	Chitosan/Fe$_2$O$_3$/MWCNTs	Methyl orange removal	[61]
20	Biopolymer/clay/conducting polymer nanocomposite	Dyes removal	[34]
21	Piezoresistive polymer nanocomposite	Explosive vapor detection	[38]
22	Montmorillonite/CoFe$_2$O$_4$ composite	Removal of methylene blue	[2]
23	Polyaniline-polyvinyl sulphonic acid composite	Para-nitrophenol detection	[41]
24	Silver nanoparticle-coated polyurethane	Antibacterial water filter	[24]

References

1. Abbasizadeh S, Keshtkar AR, Mousavian MA (2013) Preparation of a novel electrospun polyvinyl alcohol/titanium oxide nanofiber adsorbent modified with mercapto groups for uranium (VI) and thorium (IV) removal from aqueous solution. Chem Eng J 220:161–171. https://doi.org/10.1016/j.cej.2013.01.029

2. Ai L, Zhou Y, Jiang J (2011) Removal of methylene blue from aqueous solution by montmorillonite/CoFe$_2$O$_4$ composite with magnetic separation performance. Desalination 266(1–3):72–77. https://doi.org/10.1016/j.desal.2010.08.004

3. Al-Zboon K, Al-Harahsheh MS, Hani FB (2011) Fly ash-based geopolymer for Pb removal from aqueous solution. J Hazard Mater 188(1–3):414–421. https://doi.org/10.1016/j.jhazmat.2011.01.133

4. Ameen S, Akhtar MS, Shin HS (2012) Hydrazine chemical sensing by modified electrode based on in situ electrochemically synthesized polyaniline/graphene composite thin film. Sens Actuators, B Chem 173:177–183. https://doi.org/10.1016/j.snb.2012.06.065

5. Arash B, Park HS, Rabczuk T (2015) Mechanical properties of carbon nanotube reinforced polymer nanocomposites: a coarse-grained model. Compos B Eng 80:92–100. https://doi.org/10.1016/j.compositesb.2015.05.038

6. Bakunin VN, Suslov AY, Kuzmina GN, Parenago OP, Topchiev AV (2004) Synthesis and application of inorganic nanoparticles as lubricant components—a review. J Nanopart Res 6(2):273–284. https://doi.org/10.1023/B:NANO.0000034720.79452.e3

7. Borai EH, Breky MME, Sayed MS, Abo-Aly MM (2015) Synthesis, characterization and application of titanium oxide nanocomposites for removal of radioactive cesium, cobalt and europium ions. J Colloid Interface Sci 450:17–25. https://doi.org/10.1016/j.jcis.2015.02.062

8. Chandraker SK, Ghosh MK, Lal M, Shukla R (2021) A review on plant-mediated synthesis of silver nanoparticles, their characterization and applications. Nano Express. https://doi.org/10.1088/2632-959X/ac0355

9. Cheng TW, Lee ML, Ko MS, Ueng TH, Yang SF (2012) The heavy metal adsorption characteristics on metakaolin-based geopolymer. Appl Clay Sci 56:90–96. https://doi.org/10.1016/j.clay.2011.11.027

10. David G, Drobota M, Simionescu BC (2015) Preparation approach effect on polyurethane/montmorillonite nanocomposites characteristics. High Perform Polym 27(5):555–562. https://doi.org/10.1177/0954008315584184

11. Debrassi A, Baccarin T, Demarchi CA, Nedelko N, Ślawska-Waniewska A, Dłużewski P, Bilska M, Rodrigues CA (2012) Adsorption of Remazol red 198 onto magnetic N-lauryl chitosan particles: equilibrium, kinetics, reuse and factorial design. Environ Sci Pollut Res 19(5):1594–1604. https://doi.org/10.1007/s11356-011-0662-6

12. Dong H, Sliozberg YR, Snyder JF, Steele J, Chantawansri TL, Orlicki JA, Walck SD, Reiner RS, Rudie AW (2015) Highly transparent and toughened poly (methyl methacrylate) nanocomposite films containing networks of cellulose nanofibrils. ACS Appl Mater Interfaces 7(45):25464–25472. https://doi.org/10.1021/acsami.5b08317

13. Esmaeilirad N, White S, Terry C, Prior A, Carlson K (2016) Influence of inorganic ions in recycled produced water on gel-based hydraulic fracturing fluid viscosity. J Petrol Sci Eng 139:104–111. https://doi.org/10.1016/j.petrol.2015.12.021

14. Fu Z, He C, Li H, Yan C, Chen L, Huang J, Liu YN (2015) A novel hydrophilic–hydrophobic magnetic interpenetrating polymer networks (IPNs) and its adsorption towards salicylic acid from aqueous solution. Chem Eng J 279:250–257. https://doi.org/10.1016/j.cej.2015.04.146

15. Ge Y, Cui X, Liao C, Li Z (2017) Facile fabrication of green geopolymer/alginate hybrid spheres for efficient removal of Cu (II) in water: batch and column studies. Chem Eng J 311:126–134. https://doi.org/10.1016/j.cej.2016.11.079

16. Ghosh MK, Jain K, Khan S, Das K, Ghorai TK (2020) New dual-functional and reusable bimetallic Y2ZnO4 nanocatalyst for organic transformation under microwave/green conditions. ACS Omega 5(10):4973–4981. https://doi.org/10.1021/acsomega.9b03875

17. Ghosh MK, Sahu S, Gupta I, Ghorai TK (2020) Green synthesis of copper nanoparticles from an extract of Jatropha curcas leaves: characterization, optical properties, CT-DNA binding and photocatalytic activity. RSC Adv 10(37):22027–22035. https://doi.org/10.1039/D0RA03186K
18. Ghosh NG, Sarkar A, Zade SS (2021) The type-II n-n inorganic/organic nano-heterojunction of Ti^{3+} self-doped TiO_2 nanorods and conjugated co-polymers for photoelectrochemical water splitting and photocatalytic dye degradation. Chem Eng J 407:127227. https://doi.org/10.1016/j.cej.2020.127227
19. Herrera G, Montoya N, Domenech-Carbo A, Alarcón J (2013) Synthesis, characterization and electrochemical properties of iron-zirconia solid solution nanoparticles prepared using a sol–gel technique. Phys Chem Chem Phys 15(44):19312–19321. https://doi.org/10.1039/C3CP53216J
20. Hertel T, Novais RM, Alarcón RM, Labrincha JA, Pontikes Y (2019) Use of modified bauxite residue-based porous inorganic polymer monoliths as adsorbents of methylene blue. J Clean Prod 227:877–889. https://doi.org/10.1016/j.jclepro.2019.04.084
21. Huang H, Liang W, Li R, Ali A, Zhang X, Xiao R, Zhang Z, Awasthi MK, Du D, Dang P, Huang D (2018) Converting spent battery anode waste into a porous biocomposite with high Pb (II) ion capture capacity from solution. J Clean Prod 20(184):622–631. https://doi.org/10.1016/j.jclepro.2018.03.017
22. Hudcová B, Veselská V, Filip J, Číhalová S, Komárek M (2018) Highly effective Zn (II) and Pb (II) removal from aqueous solutions using Mg–Fe layered double hydroxides: comprehensive adsorption modeling coupled with solid state analyses. J Clean Prod 171:944–953. https://doi.org/10.1016/j.jclepro.2017.10.104
23. Hussain F, Hojjati M, Okamoto M, Gorga RE (2006) Polymer-matrix nanocomposites, processing, manufacturing, and application: an overview. J Compos Mater 40(17):1511–1575. https://doi.org/10.1177/0021998306067321
24. Jain P, Pradeep T (2005) Potential of silver nanoparticle coated polyurethane foam as an antibacterial water filter. Biotechnol Bioeng 90(1):59–63. https://doi.org/10.1002/bit.20368
25. Li L, Duan H, Wang X, Wang C (2015) Fabrication of novel magnetic nanocomposite with a number of adsorption sites for the removal of dye. Int J Biol Macromol 78:17–22
26. Li C, Bai H, Shi G (2009) Conducting polymer nanomaterials: electrosynthesis and applications. Chem Soc Rev 38(8):2397–2409. https://doi.org/10.1039/B816681C
27. Li HY, Huang DN, Ren KF, Ji J (2020) Inorganic-polymer composite coatings for biomedical devices. Smart Mater Med. https://doi.org/10.1016/j.smaim.2020.10.002
28. Liang Y, Yu L, Yang R, Li X, Qu L, Li J (2017) High sensitive and selective graphene oxide/molecularly imprinted polymer electrochemical sensor for 2, 4-dichlorophenol in water. Sens Actuators, B Chem 240:1330–1335. https://doi.org/10.1016/j.snb.2016.08.137
29. Mallakpour S, Behranvand V, Mallakpour F (2019) Synthesis of alginate/carbon nanotube/carbon dot/fluoroapatite/TiO_2 beads for dye photocatalytic degradation under ultraviolet light. Carbohyd Polym 224:115138. https://doi.org/10.1016/j.carbpol.2019.115138
30. Mark JE (1991) Some novel polysiloxane elastomers and inorganic-organic composites. J Inorg Organomet Polym 1(4):431–448. https://doi.org/10.1007/BF00683510
31. Mark JE, Allcock HR, West R (2005) Inorganic polymers, 2nd edn. Oxford University Press
32. Mathew M, Sureshkumar S, Sandhyarani N (2012) Synthesis and characterization of gold–chitosan nanocomposite and application of resultant nanocomposite in sensors. Colloids Surf, B 93:143–147. https://doi.org/10.1016/j.colsurfb.2011.12.028
33. Mishra UN, Das S, Kandali R (2020) Bioremediation of synthetic polymers: present and future prospects of plastic biodegradation. Int J Curr Microbiol App Sci 9(12):1234–1247. https://doi.org/10.20546/ijcmas.2020.912.152
34. Olad A, Azhar FF (2014) Eco-friendly biopolymer/clay/conducting polymer nanocomposite: characterization and its application in reactive dye removal. Fibers Polym 15(6):1321–1329. https://doi.org/10.1007/s12221-014-1321-6
35. Palanisamy S, Thangavelu K, Chen SM, Velusamy V, Chang MH, Chen TW, Al-Hemaid FM, Ali MA, Ramaraj SK (2017) Synthesis and characterization of polypyrrole decorated graphene/β-cyclodextrin composite for low level electrochemical detection of mercury (II) in water. Sens Actuators, B Chem 243:888–894. https://doi.org/10.1016/j.snb.2016.12.068

36. Pathak S, Jana B, Ghosh MK, Ghorai TK (2017) ($C_7H_7NO_4Mo$) n: synthesis, characterization and thermal stability of a new oxo-bridged helical-1D-polymer cluster. J Mol Struct 1149:662–668. https://doi.org/10.1016/j.molstruc.2017.08.013

37. Patil A, Ferritto MS (2013) Polymers for personal care and cosmetics: overview. Polym Pers Care Cosmet 3–11. https://doi.org/10.1021/bk-2013-1148.ch001

38. Patil SJ, Duragkar N, Rao VR (2014) An ultra-sensitive piezoresistive polymer nano-composite microcantilever sensor electronic nose platform for explosive vapor detection. Sens Actuators, B Chem 192:444–451. https://doi.org/10.1016/j.snb.2013.10.111

39. Rahimi A, Shokrolahi P (2001) Application of inorganic polymeric materials: I. Polysiloxanes. Int J Inorg Mater 3(7):843–847. https://doi.org/10.1016/S1466-6049(01)00162-3

40. Rahmat M, Hubert P (2011) Carbon nanotube–polymer interactions in nanocomposites: a review. Compos Sci Technol 72(1):72–84. https://doi.org/10.1016/j.compscitech.2011.10.002

41. Roy AC, Nisha VS, Dhand C, Ali MA, Malhotra BD (2013) Molecularly imprinted polyaniline-polyvinyl sulphonic acid composite based sensor for para-nitrophenol detection. Anal Chim Acta 777:63–71. https://doi.org/10.1016/j.aca.2013.03.014

42. Sahu D, Sarkar N, Sahoo G, Mohapatra P, Swain SK (2017) Nano silver imprinted polyvinyl alcohol nanocomposite thin films for Hg^{2+} sensor. Sens Actuators, B Chem 246:96–107. https://doi.org/10.1016/j.snb.2017.01.038

43. Schall JD, Edo-Hernandez E (2012) Structure-property investigation of functional resins for UV-curable gaskets. RadTech Rep 26(2):34. https://www.radtech.org/proceedings/2012/papers/Session%209%20-%20Chemistry/JSchall_Henkel.pdf

44. Shawky HA, Chae SR, Lin S, Wiesner MR (2011) Synthesis and characterization of a carbon nanotube/polymer nanocomposite membrane for water treatment. Desalination 272(1–3):46–50. https://doi.org/10.1016/j.desal.2010.12.051

45. Sheha RR (2012) Synthesis and characterization of magnetic hexacyanoferrate (II) polymeric nanocomposite for separation of cesium from radioactive waste solutions. J Colloid Interface Sci 388(1):21–30. https://doi.org/10.1016/j.jcis.2012.08.042

46. Shukla SK, Govender PP, Agorku ES (2016) A resistive type humidity sensor based on crystalline tin oxide nanoparticles encapsulated in polyaniline matrix. Microchim Acta 183(2):573–580. https://doi.org/10.1007/s00604-015-1678-2

47. Sun F, Lin M, Dong Z, Zhang J, Wang C, Wang S, Song F (2015) Nanosilica-induced high mechanical strength of nanocomposite hydrogel for killing fluids. J Colloid Interface Sci 458:45–52. https://doi.org/10.1016/j.jcis.2015.07.006

48. Sun Y, Zhu C, Zheng H, Sun W, Xu Y, Xiao X, You Z, Liu C (2017) Characterization and coagulation behavior of polymeric aluminum ferric silicate for high-concentration oily wastewater treatment. Chem Eng Res Des 119:23–32. https://doi.org/10.1016/j.cherd.2017.01.009

49. Surudžić R, Janković A, Bibić N, Vukašinović-Sekulić M, Perić-Grujić A, Mišković-Stanković V, Park SJ, Rhee KY (2016) Physico–chemical and mechanical properties and antibacterial activity of silver/poly (vinyl alcohol)/graphene nanocomposites obtained by electrochemical method. Compos B Eng 85:102–112. https://doi.org/10.1016/j.compositesb.2015.09.029

50. Tang Q, Wang K, Yaseen M, Tong Z, Cui X (2018) Synthesis of highly efficient porous inorganic polymer microspheres for the adsorptive removal of Pb2+ from wastewater. J Clean Prod 193:351–362. https://doi.org/10.1016/j.jclepro.2018.05.094

51. Umamaheswari C, Lakshmanan A, Nagarajan NS (2018) Green synthesis, characterization and catalytic degradation studies of gold nanoparticles against congo red and methyl orange. J Photochem Photobiol, B 178:33–39. https://doi.org/10.1016/j.jphotobiol.2017.10.017

52. Wan Q, Liu M, Xie Y, Tian J, Huang Q, Deng F, Mao L, Zhang Q, Zhang X, Wei Y (2017) Facile and highly efficient fabrication of graphene oxide-based polymer nanocomposites through mussel-inspired chemistry and their environmental pollutant removal application. J Mater Sci 52(1):504–518. https://doi.org/10.1007/s10853-016-0349-y

53. Wang D, Duan Y, Luo Q, Li X, An J, Bao L, Shi L (2012) Novel preparation method for a new visible light photocatalyst: mesoporous TiO_2 supported Ag/AgBr. J Mater Chem 22(11):4847–4854. https://doi.org/10.1039/C2JM14628B

54. Wang JP, Yang HC, Hsieh CT (2010) Adsorption of phenol and basic dye on carbon nanotubes/carbon fabric composites from aqueous solution. Sep Sci Technol 46(2):340–348. https://doi.org/10.1080/01496395.2010.508066

55. World Health Organization (1993) Guidelines for drinking-water quality. World Health Organization

56. Yang DF, Kong XJ, Gao D, Cui HS, Huang TT, Lin JX (2017) Dyeing of cotton fabric with reactive disperse dye contain acyl fluoride group in supercritical carbon dioxide. Dyes Pigm 139:566–574. https://doi.org/10.1016/j.dyepig.2016.12.050

57. Zhou K, Wu B, Dai X, Chai X (2018) Development of polymeric iron/zirconium-pillared clinoptilolite for simultaneous removal of multiple inorganic contaminants from wastewater. Chem Eng J 347:819–827. https://doi.org/10.1016/j.cej.2018.04.104

58. Zhou L, Zhou H, Yang X (2019) Preparation and performance of a novel starch-based inorganic/organic composite coagulant for textile wastewater treatment. Sep Purif Technol 210:93–99. https://doi.org/10.1016/j.seppur.2018.07.089

59. Zhu HY, Jiang R, Fu YQ, Jiang JH, Xiao L, Zeng GM (2011) Preparation, characterization and dye adsorption properties of γ-Fe_2O_3/SiO_2/chitosan composite. Appl Surf Sci 258(4):1337–1344. https://doi.org/10.1016/j.apsusc.2011.09.045

60. Zhu HY, Jiang R, Xiao L, Li W (2010) A novel magnetically separable γ-Fe_2O_3/crosslinked chitosan adsorbent: preparation, characterization and adsorption application for removal of hazardous azo dye. J Hazard Mater 179(1–3):251–257. https://doi.org/10.1016/j.jhazmat.2010.02.087

61. Zhu HY, Jiang R, Xiao L, Zeng GM (2010) Preparation, characterization, adsorption kinetics and thermodynamics of novel magnetic chitosan enwrapping nanosized γ-Fe_2O_3 and multi-walled carbon nanotubes with enhanced adsorption properties for methyl orange. Biores Technol 101(14):5063–5069. https://doi.org/10.1016/j.biortech.2010.01.107

Printed in the United States
by Baker & Taylor Publisher Services